European Whales, Dolphins, and Porpoises

European Whales, Dolphins, and Porpoises

Marine Mammal Conservation in Practice

Peter G.H. Evans

Sea Watch Foundation & School of Ocean Sciences,
Bangor University, United Kingdom

ELSEVIER

ACADEMIC PRESS
An imprint of Elsevier

ASCOBANS

Academic Press is an imprint of Elsevier
125 London Wall, London EC2Y 5AS, United Kingdom
525 B Street, Suite 1650, San Diego, CA 92101, United States
50 Hampshire Street, 5th Floor, Cambridge, MA 02139, United States
The Boulevard, Langford Lane, Kidlington, Oxford OX5 1GB, United Kingdom

British Library Cataloguing-in-Publication Data
A catalogue record for this book is available from the British Library

Library of Congress Cataloging-in-Publication Data
A catalog record for this book is available from the Library of Congress

ISBN: 978-0-12-819053-1

For Information on all Academic Press publications
visit our website at https://www.elsevier.com/books-and-journals

Publisher: Charlotte Cockle
Acquisition Editor: Anna Valutkevich
Editorial Project Manager: Lindsay Lawrence
Production Project Manager: Poulouse Joseph
Cover Designer: Miles Hitchen

Typeset by MPS Limited, Chennai, India

Working together
to grow libraries in
developing countries

www.elsevier.com • www.bookaid.org

Contents

List of contributors   ix
Foreword   xi
Preface   xiii
Acknowledgements   xv

1. History of the ASCOBANS agreement   1

How does an international Conservation
  Agreement function?   9
References   23

2. Conservation agreements for the protection of whales, dolphins and porpoises   25

European international agreements applicable
  to cetaceans   25
  Habitats Directive   25
  Marine Strategy Framework Directive   25
  OSPAR (Oslo and Paris Convention)   27
  Helsinki Convention   27
  ACCOBAMS   27
  Barcelona Convention   30
Other international agreements applicable
  to cetaceans   32
  International Whaling Commission   32
  The US Marine Mammal Protection Act   35
  Conservation legislation in Canada   37
  Conservation legislation in Australia and
    New Zealand   39
The need for evidence-based consensus
  at the broadest level   42
Overview   43
References   43

3. Regional review of cetaceans in Northwest Europe   45

The Baltic   45
The Belt Seas   47
Skagerrak and Kattegat   50
Norwegian Atlantic   50
North Sea   50
English Channel   55

Irish Sea   55
Atlantic Scotland and Ireland   60
Bay of Biscay   64
Iberian Peninsula   65
References   68

4. Systematic list of European cetacean species   73

References   73

Family Phocoenidae (Porpoises)
Harbour porpoise *Phocoena phocoena*   73
References   77

Family Delphinidae
Rough-toothed dolphin *Steno bredanensis*   78
References   79

Common bottlenose dolphin *Tursiops truncatus*
References   83

Atlantic spotted dolphin *Stenella frontalis*
Reference   86

Striped dolphin *Stenella coeruleoalba*
References   88

Common dolphin *Delphinus delphis*
References   92

Fraser's dolphin *Lagenodelphis hosei*
References   94

White-beaked dolphin *Lagenorhynchus albirostris*
References   96

Atlantic white-sided dolphin *Lagenorhynchus acutus*
References   99

Risso's dolphin *Grampus griseus*
References   102

Melon-headed whale *Peponocephala electra*
Reference   103

Pygmy killer whale *Feresa attenuata*
References   104

False killer whale *Pseudorca crassidens*
References   105

Killer whale or orca *Orcinus orca*
References   107

Long-finned pilot whale *Globicephala melas*
References   110

Short-finned pilot whale *Globicephala
macrorhynchus*
References   112

Family Monodontidae

Narwhal *Monodon monoceros*
References   113

Beluga or white whale *Delphinapterus leucas*
References   114

Family Ziphiidae

Cuvier's beaked whale *Ziphius cavirostris*
References   117

Northern bottlenose whale *Hyperoodon
ampullatus*
References   121

True's beaked whale *Mesoplodon mirus*
References   123

Gervais' beaked whale *Mesoplodon
europaeus*
References   124

Sowerby's beaked whale *Mesoplodon bidens*
References   126

Gray's beaked whale *Mesoplodon grayi*
Reference   126

Blainville's beaked whale *Mesoplodon
densirostris*
References   128

Family Kogiidae

Pygmy sperm whale *Kogia breviceps*
References   129

Dwarf sperm whale *Kogia sima*
References   131

Family Physeteridae

Sperm whale *Physeter macrocephalus*
References   135

INFRAORDER Mysticeti, the baleen whales

Family Balaenidae (right whales)

North Atlantic right whale *Eubalaena
glacialis*
References   137

Bowhead whale *Balaena mysticetus*
References   139

Family Balaenopteridae (rorquals)

Humpback whale *Megaptera novaeangliae*
References   142

Common minke whale *Balaenoptera
acutorostrata*
References   146

Sei whale *Balaenoptera borealis*
References   149

Bryde's whale *Balaenoptera brydei*
Reference   151

Fin whale *Balaenoptera physalus*
References   154

Blue whale *Balaenoptera musculus*
References   157

5. **Conservation threats**   159

Hunting   159
Reduction of prey from fishing   162
Incidental capture in fishing gear   163
Chemical pollutants and other hazardous
    substances   170
Marine litter   172
Noise disturbance   174
    Shipping   176
    Seismic   177
    Marine construction: explosions, dredging
        and drilling   180
    Active sonar   184
    Risk assessment   186
Vessel strikes   187
Recreational disturbance   189
Climate change   191
References   192

6. **Conservation research**   203

Population structure   203
Counting animals   208
Photoidentification   219
Acoustics   220

Postmortem examinations and other
    approaches to investigating health status   228
References   231

7. Conservation actions   237

Hunting   237
Reduction of prey from fishing   237
Incidental capture in fishing gear   239
Chemical pollutants   248
Marine litter   248
Noise disturbance   249
Vessel strikes   254
Recreational activities   255
Climate change   260
Marine protected areas   260

Conservation plans   267
References   274

8. Focus on the future   277

Signing up to a conservation agreement   280
The mechanics of a conservation agreement   285
Focus on species and issues   286
Converting conservation legislation into action   292
Success or failure?   294
References   294

**Appendix: Agreement on the conservation of small
Cetaceans of the Baltic, North East Atlantic, Irish
and North Seas**   295
Index   301

Supported by:

Federal Ministry
for the Environment, Nature Conservation
and Nuclear Safety

based on a decision of the German Bundestag

List of contributors

Gerhard Adams (Retired) Species Protection Division, Nature Conservation and Nuclear Safety, German Federal Ministry for Environment, Bonn, Germany (Chapter 1)

Mats Amundin (Retired) Destination Kolmården, Kolmården, Sweden (Chapter 6)

Penina Blankett Ministry of the Environment, Helsinki, Finland (Chapter 8)

Stefan Bräger German Oceanographic Museum (DMM), Stralsund, Germany (Chapter 7)

Ida Carlén Coalition Clean Baltic, Uppsala, Sweden (Chapter 6)

Julia Carlström Swedish Museum of Natural History, Stockholm, Sweden (Chapter 6)

Petra Deimer-Schütte Society for the Conservation of Marine Mammals (Gesellschaft zum Schutz der Meeressäugetiere - GSM), Hasloh, Germany (Chapter 7)

Florence Descroix-Comanducci ACCOBAMS Permanent Secretariat, Jardin de L'UNESCO, Monaco (Chapter 2)

Geneviève Desportes North Atlantic Marine Mammal Commission, Tromsø, Norway (Chapter 7)

Folchert R. Van Dijken (Retired) Ministry of Agriculture, Nature and Food Quality, The Hague, The Netherlands (Chapter 7)

Greg Donovan International Whaling Commission, Cambridge, United Kingdom (Chapter 2)

Heidrun Frisch-Nwakanma CMS, Bonn, Germany (Chapter 1)

Marie-Christine Grillo (Retired) ACCOBAMS Permanent Secretariat, Jardin de L'UNESCO, Monaco (Chapter 2)

Jan Haelters Royal Belgian Institute of Natural Sciences, Ostend, Belgium (Chapter 2)

Philip Hammond Sea Mammal Research Unit, University of St Andrews, St Andrews, United Kingdom (Chapter 6)

Sami Hassani Oceanopolis, Brest, France (Chapter 8)

Jette Donovan Jensen[†] Department of Environmental Science Secretariat, Aarhus University, Roskilde, Denmark (Chapter 1)

Christina Lockyer Age Dynamics, Kongens Lyngby, Denmark (Chapter 1)

Stephan Lutter WWF Germany, Berlin, Germany (Chapter 8)

Yvon Morizur (Retired) IFREMER, Plouzané, France (Chapter 5)

Maj Munk Ministry of Environment and Food of Denmark, Odense, Denmark (Chapter 8)

Peter Reijnders (Retired) Chairgroup Aquatic Ecology & Waterquality, Wageningen University, Texel, The Netherlands and Wageningen Marine Research Institute, Den Helder, The Netherlands (Chapter 1)

Meike Scheidat Wageningen Marine Research, Wageningen University & Research, IJmuiden, The Netherlands (Chapter 5)

Mark Simmonds Humane Society International, London, United Kingdom and School of Veterinary Sciences, University of Bristol, Bristol, United Kingdom (Chapter 1)

Krzysztof Skóra[†] Hel Marine Station, Department of Oceanography and Geography, University of Gdańsk, Hel, Poland (Chapter 7)

---

† deceased

Ralf Sonntag International Fund for Animal Welfare, Hamburg, Germany (Chapter 8)

Patricia Stadié (Retired) ASCOBANS Secretariat, Bonn, Germany (Chapter 1)

Rüdiger Strempel HELCOM Secretariat, Helsinki, Finland (Chapter 1)

Mark Tasker (Retired) Joint Nature Conservation Committee, Aberdeen, United Kingdom and ICES, Copenhagen, Denmark (Chapter 8)

Robert Vagg CMS, Bonn, Germany (Chapter 1)

Foreword

Whales, dolphins and porpoises—all part of the order known as cetaceans—spend their lives in the Earth's oceans and seas. These marine mammals, which evolved from land animals, are highly intelligent, with strong family structures, complex communications and demonstrable emotions, such as grief, empathy and joy.

The cetaceans of north-west Europe, as with marine species worldwide, face numerous threats, and these pressures are becoming more severe. They are injured or killed in fisheries as bycatch and they suffer from ship strikes in busy transport lanes. Noise—from seismic testing, offshore development, boat engines and active sonar—can impair their hearing and ability to navigate, leading to mortality. Examinations of some stranded whales have found large amounts of plastic litter in their stomachs.

In light of these threats, in 1992, the countries of north-west Europe negotiated what became the UN Agreement on the Conservation of Small Cetaceans of the Baltic, North East Atlantic, Irish and North Seas, known as ASCOBANS. The Agreement entered into force in 1994, creating a platform for regional cooperation and action. It covers over 20 species of cetaceans, including the Harbour Porpoise, Common Dolphin, and Long-finned Pilot Whale.

The commitments under ASCOBANS have led to valuable conservation and research measures. For example, ASCOBANS parties have enacted legislation to minimise cetacean bycatch, and they have introduced regulations on acceptable noise levels, designated protected areas, and conducted abundance surveys. Throughout its existence, the Agreement has enjoyed enthusiastic support from many civil society organisations, research centres, intergovernmental organisations, and dedicated individuals such as Peter Evans, who has compiled this book.

This publication is a celebration of the 25th anniversary of ASCOBANS. It explores the history of the agreement, provides valuable information on the biology of these marine mammals, and discusses threats, research and conservation actions.

While we mark the past successes of ASCOBANS, we must look to the future, in which regional cooperation will be ever more needed to address threats to cetaceans. With the strong commitment of Member States and the many organisations and individuals engaged in this work, ASCOBANS can continue to play its unique role in ensuring the conservation of the remarkable small cetaceans of the Baltic, North East Atlantic, Irish and North Seas.

Amy Fraenkel
Acting Executive Secretary,
Convention on the Conservation of Migratory Species of Wild Animals and ASCOBANS

Preface

Arising from the Caribbean, warm current flows northwards up the east coast of the United States and then eastwards across the Atlantic, bathing the waters surrounding the continent of Europe and its outlying islands. It is called the North Atlantic Current and is an extension of the Gulf Stream. Those warm waters are the reason why about 90% of the Earth's human population north of 50°N lives not in Asia nor in North America but in Europe. Temperatures here are 5°C−10°C warmer than the global average for these latitudes and this has had profound consequences upon human history. As far back as Neolithic times, more than 5000 years ago, communities were living along the coast and exploiting nearby marine resources — fish and whales. This long history of exploitation led to the extinction of the North Atlantic grey whale and the near extermination of the eastern North Atlantic population of the right whale. In more recent times it may have also contributed to the present fragile conservation status of the harbour porpoise in the Baltic Sea.

Some of the capitals of Europe were built upon the exploitation of marine resources, and they have led to several centres of high human density. The result has been that pressures upon the marine environment particularly in the coastal zone have long been intense. With the industrial revolution in the late 18th and 19th century came further pressures, and during the course of the 20th century, new technologies led to ever more efficient means of exploiting fish and invertebrates as well as increased pollution from a steady stream of newly produced chemicals. By the latter half of the last century, major fish stocks had been severely reduced, entanglements of nontarget species in fishing gear were commonplace, and both pollution and disturbance from shipping and industrial activities were widespread. Focus was placed upon the North Sea and, since the mid-1980s, a series of Ministerial Conferences have been held to debate the state of the marine environment in the region. At the second North Sea Ministerial Conference, held in London in 1987, representations were made concerning the unfavourable status in Europe of two small cetacean species, the harbour porpoise and the bottlenose dolphin. These were repeated at the 1990 North Sea Ministerial Conference in The Hague. With the general climate of concern for the conservation status of small cetaceans,[1] particularly the harbour porpoise, raised by these meetings and pressure for action from NGOs, a regional Agreement for the Conservation of Small Cetaceans of the Baltic and North Sea (shortened to the acronym ASCOBANS) was born under UNEP's Convention on Migratory Species.

The year 2017 marked 25 years of the existence of ASCOBANS, and this seemed a good time to take a long hard look at the Agreement, examine its strengths and its weaknesses, and hopefully provide some insight as to how it could do better in the future.

A book about a conservation agreement may sound a trifle dull. But to make such an agreement work requires balancing a whole range of perspectives — government, industry, recreation and the conservation lobby, each with its own needs and objectives. Obtaining scientific information on the status and trends of a particular dolphin or whale population is only the start of the process towards effective conservation. The relative importance of different human activities has to be assessed — no easy task. And then the implementation of practical conservation measures requires addressing the often conflicting, demands imposed by other users of the sea. In order to give a flavour of these different perspectives and approaches, contributions have been sought from more than two dozen persons representing the United Nations, intergovernmental bodies, national governments, marine mammal and fisheries scientists, and conservation campaigners, all of whom have been directly involved in the Agreement at one time or another. The reader will find that different individuals often have different ideas as to where problems lie, as well as having different solutions. Those closest to the Agreement may be inclined to be more optimistic about what it has achieved, whereas independent campaigners are likely to be more critical of progress. Members of government delegations will have a variety of other issues to address outside marine mammal conservation, some of which may be considered higher priority or which may cause a conflict. And, inevitably, those government representatives will be ruled by their political masters and they

---

1. Cetaceans are whales, dolphins and porpoises that comprise the Order Cetartiodactyla, and SubOrder Cetacea; small cetaceans are members of the InfraOrder Odontoceti, the Toothed Whales (which includes dolphins and porpoises) but for the purpose of the Agreement excludes the sperm whale.

have even wider demands to juggle as they try to keep their electorates happy. It is a fact of life that when it comes to the crunch, voting for a political party will more likely be dictated by what they can deliver economically, or towards health and education, rather than for the environment because these are what affect the public most obviously on a daily basis. The successive failures of summits on climate change are testament to that. It is therefore no surprise that those of us seeking speedy action to aid the environment are often frustrated.

In this book I hope to recount the history of an international conservation agreement, show the varying perspectives on priorities and indicate where progress has been made as well as where in my view there remain important shortcomings. But most of all, I want to demonstrate the rich diversity of species of whale and dolphin that we have in our European seas, and the fragile marine ecosystem in which they live. Whales and dolphins comprise an iconic group of animals that capture the public eye, but they form only part of a wider web of life that we should just as much want to protect. If they are doing well, it is likely that so are many of the species of marine organisms with which they interact.

Acknowledgements

I would like to thank the following who contributed their thoughts on European cetacean conservation and the ASCOBANS Agreement for this book: Gerhard Adams, Mats Amundin, Penina Blankett, Stefan Brager, Ida Carlén, Julia Carlström, Petra Deimer, Florence Descroix-Comanducci, Genevieve Desportes, Greg Donovan, Jette Donovan-Jensen, Folchert van Dijken, Heidrun Frisch-Nwakanma, Marie-Christine Grillo, Jan Haelters, Philip Hammond, Sami Hassani, Christina Lockyer, Stephan Lutter, Yvon Morizur, Maj Munk, Peter Reijnders, Meike Scheidat, Mark Simmonds, Krzysztof Skóra, Ralf Sonntag, Pat Stadie, Rüdiger Strempel, Mark Tasker and Robert Vagg. Sadly, Jette Donovan-Jensen and Krzysztof Skóra both died before this book could be published.

Thanks also go to the following who commented on sections of text, and/or provided photographs or graphics: ACCOBAMS Secretariat, Natacha Aguilar, Sabina Airoldi, Mats Amundin, Pia Anderwald, ASCOBANS Secretariat, Mick Baines, Aydin Bahramlouian/ASCOBANS, Bio Consult SH, Anna Bird, Ian Birks, Penina Blankett, Marijke de Boer/WDC, Philippe Bouchet, Miriam Brandt, Bernd Brederlau, Jacopo Bridda, Ida Carlén, Inés Carvalho, Keenan Chadwick, UK Cetacean Strandings Investigation Programme (CSIP-ZSL), Molly Czachur, Michael Dähne, Nick Davies, Rob Deaville, Geneviève Desportes, Ansgar Diederichs, Paul Ensor, European Topic Centre (Laura-Patricia Gavilan Iglesias), Antonio Fernández, Fjord&Bælt Center, Luis Freitas, Anders Galatius, Steve Geelhoed, Lucy Gilbert, Anita Gilles, Florian Graner, GREMM, Marta Guerra/ULL, Philip Hammond, Sergio Hanquet, Christian Harboe-Hansen, Justin Hart, Mads Peter Heide-Jørgensen, Prof. Krzysztof Skóra Hel Marine Station (Iwona Pawliczka), HELCOM Secretariat, Helena Herr, Kevin Hepworth, Hydrotechnik Lübeck GmbH (Jürgen Hepper), Iroise Natural Marine Park (Philippe Le Niliot), John Irvine, Saana Isojunno, Kathy James, Paul Jepson/IoZ, Ari Laine/Parks & Wildlife Finland, Luca Lamoni, Caterina Lanfredi, Jennifer Learmonth, Katrin Lohrengel, Klaus Lucke, Marine Discovery Penzance, Marine Traffic, Patrick Miller, New England Aquarium (Scott Kraus), OCEANOPOLIS-Brest (Sami Hassani), David Ord, OSPAR Secretariat, Bjarne Mikkelsen/Faroese Museum of Natural History, Aevar Petersen, PGS (Anders Otnes), Nino Pierantonio, Graham Pierce, Enrico Pirotta, Antonio Portales, Marianne Rasmussen, Maren Reichelt, SAMBAH Project, Begoña Santos, Meike Scheidat, Sea Shepherd (Rob Read), Ursula Siebert/ITAW-TiHo, Marije Siemensma, Mark Simmonds, the late Krzysztof Skóra, Mark Somerville, Souffleurs d'Ecume (Morgane Ratel), Brandon Southall, Wouter Jan Strietmann, Signe Sveegaard, Christopher Swann, Petra Szlama, Jonas Teilmann, Tethys Research Institute (Simone Panigada), Mike Tetley, Kirsten Thompson, Nick Tregenza, Turku University, Fernando Ugarte, Gemma Veneruso, Heike Vester/Ocean Sounds, James Waggitt, and Caroline Weir.

Finally, I am very grateful to the ASCOBANS Secretariat, particularly Heidrun Frisch-Nwakamma, Aline Kühl-Stenzel, Mrinalini Shindi, Bettina Reinartz and Melanie Virtue, for giving me the opportunity to write this book and for all their help along the route to its publication.

Chapter 1

History of the ASCOBANS agreement

During the 1960s around the world there was an environmental awakening that our planet had finite resources; the burgeoning global human population and technological developments taking place were applying pressure that simply could not be sustained. It was the decade of Rachel Carson's book 'Silent Spring', drawing attention to the effects that pesticides were having around the world, and following it 10 years later in 1972 was 'A Blueprint for Survival', a special edition of the Ecologist magazine, later published in book form, which presented the case for societal action to help protect our environment. In northern Europe, the collapse of fish stocks such as herring and mackerel drew attention to the effects of overexploitation of marine resources, and the impact of modern whaling during the first half of the 20th century resulted in campaigns to ban commercial hunts. In 1976 the Food and Agriculture Organisation of the United Nations invited marine mammal specialists from around the world to a meeting in Bergen, Norway, called 'Mammals in the Seas' to review and debate the global status of marine mammals. With concerted pressure also from a number of environmental NGOs, this heralded moves towards a moratorium on commercial whaling, which came into effect in 1986.

Whilst attention during the 1970s was focused upon the effects of human exploitation of whales, concerns were being expressed in many parts of Europe for the decline of its smallest cetacean, the harbour porpoise (Fig. 1.1), for other reasons. The species was becoming scarce in the Baltic Sea, the southern North Sea, the Channel and around the Biscay coasts of France and Spain south to Portugal. It was extremely rare in the Mediterranean, and seriously threatened in the Black Sea. Several possible reasons were put forward: pollution from a range of new chemicals introduced on a large scale during the 1950s and 1960s; entanglement in fishing gear with the widespread introduction of monofilament nets over the same period; excessive exploitation of fish resources with ever-larger and more technologically advanced vessels in the 1960s; seismic exploration of the North Sea for oil and gas during the 1960s; and increased vessel traffic particularly in areas such as the Channel, Bay of Biscay, and Straits of Dover and Gibraltar, throughout the second half of the 20th century.

During the 1970s and 1980s a number of publications highlighted porpoise declines around Europe (Belgium — De Smet 1974, 1981; Denmark — Andersen 1982; Kinze 1987; France — Collet and Duguy 1987; Germany — Benke, Siebert, Lick, Bandomir, & Weiss, 1998; Kremer 1987; Kroger 1986; The Netherlands — Addink and Smeenk 1999; Smeenk 1987; Portugal — Teixeira 1979; Spain — Casinos and Vericad 1976; Sweden — Berggren and Arrhenius 1995; Lindstedt and Lindstedt 1988, 1989; United Kingdom — Evans 1980, 1990, 1992; Evans, Harding, Tyler, & Hall, 1986; Kayes 1985). The result was pressure from scientists and NGOs for political initiatives in Europe to help protect the species. A special meeting was arranged by WWF Germany at the Alfred Wegener Institute in Bremerhaven (Kroger, 1986) in order to coordinate research and conservation efforts, and what was initially going to become a European Harbour Porpoise Working Group led to the founding of the European Cetacean Society (ECS) in Hirtshals, Denmark in January 1987. One of the society's first initiatives was a statement of concern presented to the second North Sea Ministerial Conference, held in London in 1987 (Evans, Kinze, Kroger, & Smeenk, 1987), with support from NGOs such as WWF Germany, WWF Sweden and Greenpeace, and with continued pressure, a Memorandum of Understanding on Small Cetaceans was agreed at the third North Sea Ministerial Conference, held in The Hague in March 1990.

In this climate of environmental concern, moves were afoot to establish a number of international conservation agreements and directives. One of the first regional agreements under Article IV of the Convention on the Conservation of Migratory Species of Wild Animals (UNEP/CMS or Bonn Convention as it was termed), which had come into force in 1983, was ASCOBANS, the Agreement on the Conservation of Small Cetaceans of the Baltic and North Seas

European Whales, Dolphins, and Porpoises. DOI: https://doi.org/10.1016/B978-0-12-819053-1.00001-6

FIGURE 1.1 Concern for the conservation status of the harbour porpoise led to the establishment of ASCOBANS. *Photo: Florian Graner.*

(ASCOBANS). The Final Act was signed on 13 September 1991, and it was then opened for signature by Range States, at the UN Headquarters in New York on 17 March 1992.

The British government offered to host an interim Secretariat based at the Sea Mammal Research Unit, which was occupying part of the British Antarctic Survey building in Cambridge. The role of Executive Secretary was taken on by Dr Christina Lockyer in 1992, supported subsequently by Sara Heimlich. The Agreement would come into force once at least six Parties had signed and ratified. This occurred 2 years later on 29 March 1994, the early Parties being Belgium, Denmark, Germany, The Netherlands, Sweden and the United Kingdom. Since then, four other countries have signed and ratified as Parties to the Agreement. These are Poland (January 1996), Finland (September 1999), Lithuania (June 2005) and France (October 2005). The Secretary General of the United Nations has assumed the functions of Depository of the Agreement.

Initially the Agreement Area was confined to the Baltic, North Sea and Channel, but on 3 February 2008, an Atlantic extension of the area came into force westwards and southwards to the southwest tip of Portugal, which changed the name to the 'Agreement on the Conservation of Small Cetaceans of the Baltic, Northeast Atlantic, Irish and North Seas'. The original boundaries and those of the Extension Area are depicted in Fig. 1.2.

ASCOBANS is open for accession by all Range States. These have been defined as States exercising jurisdiction over any part of the range of a species covered by the Agreement, or States whose flag vessels are active in the Agreement area and cause adverse effects for the species. In addition, they include Riparian States not yet Party to ASCOBANS: Estonia, Ireland, Latvia, Norway, Portugal, Russian Federation and Spain, and regional economic integration organisations such as the European Union. Although the European Union has the possibility to accede to the Agreement, it has not done so as yet.

All non-Party Range States are encouraged to join the ASCOBANS Parties in their efforts to conserve the small cetacean species they share with other countries in the ASCOBANS Area, conscious that the management of threats to their existence, such as bycatch, habitat deterioration and other anthropogenic disturbance, requires concerted and coordinated responses. And in fact all of these non-Party states as well as the European Commission have sent representatives to ASCOBANS meetings at one time or another.

The full text of the revised Agreement is given at the end of the book in an Appendix. Every 3 to 4 years there is a formal Meeting of the Parties (MOPs) at which Resolutions are passed, addressing priority conservation actions as well as internal administrative matters. The locations of each of these meetings are given in Tables 1.1 and 1.2 respectively. Every year, an Advisory Committee (AC) meets comprising representatives of Range States including advisors, along with intergovernmental organisations and NGOs as observers. The main role of this body is to provide scientific advice

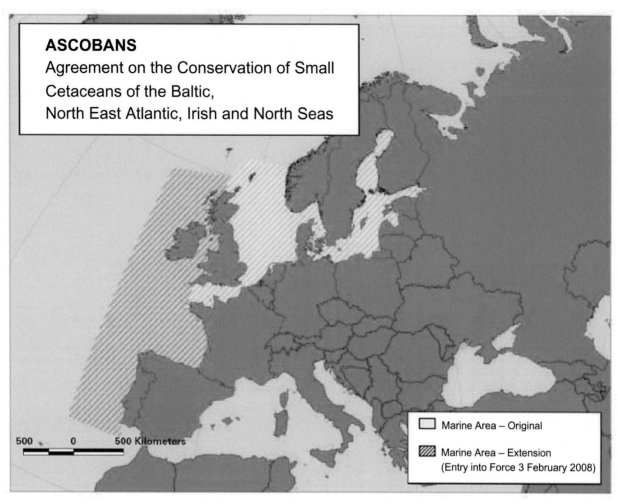

FIGURE 1.2   Map of the agreement area (along with the Atlantic area extension).

TABLE 1.1 Locations of Meeting of the Parties.

Meeting No.	Date	Location
MOP1	26−28 September 1994	Stockholm, Sweden
MOP2	17−19 November 1997	Bonn, Germany
MOP3	26−28 July 2000	Bristol, United Kingdom
MOP4	19−22 August 2003	Esbjerg, Denmark
MOP5	18−22 September 200612 December 2006	Egmond aan Zee, The NetherlandsThe Hague, The Netherlands
MOP6	16−18 September 2009	UN Campus, Bonn, Germany
MOP7	22−24 October 2012	Brighton, United Kingdom
MOP8	30 August−1 September 2016	Helsinki, Finland

TABLE 1.2 Locations of Advisory Committee meetings.

Meeting No.	Date	Location
AC1	8–10 March 1995	Cambridge, United Kingdom
AC2	29 November–1 December 1995	Cambridge, United Kingdom
AC3	13–15 November 1996	Copenhagen, Denmark
AC4	30 June–2 July 1997	Texel, The Netherlands
AC5	22–24 April 1998	Hel, Poland
AC6	12–14 April 1999	Aberdeen, United Kingdom
AC7	13–16 March 2000	Bruges, Belgium
AC8	2–5 April 2001	Nymindegab, Denmark
AC9	10–12 June 2002	Hindås, Sweden
AC10	9–11 April 2003	Bonn, Germany
AC11	27–29 April 2004	Jastrzebia Góra, Poland
AC12	12–14 April 2005	Brest, France
AC13	25–27 April 2006	Tampere, Finland
AC14	19–21 April 2007	San Sebastian, Spain
AC15	31 March–3 April 2008	UN Campus, Bonn, Germany
AC16	20–24 April 2009	Bruges, Belgium
AC17	4–6 October 2010	UN Campus, Bonn, Germany
AC18	4–6 May 2011	UN Campus, Bonn, Germany
AC19	20–22 March 2012	Galway, Ireland
AC20	27–29 August 2013	Warsaw, Poland
AC21	29 September–1 October 2014	Gothenburg, Sweden
AC22	29 September–1 October 2015	The Hague, Netherlands
AC23	5–7 September 2017	Le Conquet, France
AC24	25–27 September 2018	Vilnius, Lithuania

and information to the Secretariat and the Parties. The first meeting was held in Stockholm in September 1994 (Fig. 1.3), and Fig. 1.4 illustrates the venues of the ASCOBANS Secretariat over the years.

The ASCOBANS Secretariat remained in Cambridge, United Kingdom until June 1998 when it relocated to the premises of the UN in Bonn, Germany. Then, on 1 January 2001 it was integrated into the Convention on Migratory Species (CMS) Agreements Unit of the CMS Secretariat, and from 1 January 2007, it was decided that the UNEP/CMS Secretariat would serve as the Secretariat for the ASCOBANS agreement; and the Executive Secretary of UNEP/CMS would be the acting Executive Secretary of ASCOBANS. Originally foreseen for a provisional 3-year period, this arrangement was since confirmed further. Chairs and Vice-Chairs of the AC in Table 1.3, and the Secretariat staff from the beginning are listed in Table 1.4. Several of these have provided their own personal insights on the Agreement in this and subsequent chapters.

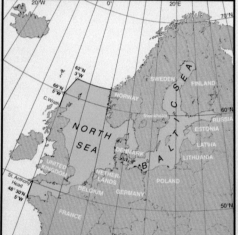

FIGURE 1.3   Poster advertising the first meeting of the parties to ASCOBANS in Stockholm, Sweden, September 1994. *Photo: Peter GH Evans.*

FIGURE 1.4 ASCOBANS Secretariat venues: (A) British Antarctic Survey, Cambridge, 1992−98; (B) Haus Carstanjen, Bonn-Bad Godesberg, 1998−2001; (C) Langer Eugen, Bonn, 2006 to present. Between 2001 and 2006, the Secretariat was in a smaller building in Kennedyallee, Bonn. *Photos: ASCOBANS Secretariat.*

TABLE 1.3 ASCOBANS AC chairs and vice-chairs.

AC Chairs	Country	Dates
Peter Reijnders	The Netherlands	1995−2001
Mark Tasker	United Kingdom	2001−07
Stefan Bräger	Germany	2007−10
Sami Hassani	France	2010 to present
AC Vice-Chairs	**Country**	**Dates**
Mark Tasker	United Kingdom	1994−96
Peter Reijnders	The Netherlands	2001−07
Jan Haelters	Belgium	2007−10
Penina Blankett	Finland	2010 to present

TABLE 1.4 ASCOBANS secretariat staff.

Professional staff	Function	Dates
Christina Lockyer	Executive Secretary	1992−96
Sara Heimlich	Executive Secretary	1996
Jette Jensen	Executive Secretary	1996−98
Holger Auel	Executive Secretary	1998−99
Rüdiger Strempel	Executive Secretary	1999−2006
Robert Hepworth	Acting Exec. Secretary (3%)	2007−09
Marco Barbieri	Senior Advisor (15%)	2007−09
Heidrun Frisch-Nwakanma	Coordinator (75%)	2007−16
Elizabeth Maruma Mrema	Acting Exec. Secretary (3%)	2009−12
Borja Heredia	Senior Advisor (15%)	2010−13
Bradnee Chambers	Executive Secretary (3%)	2012−19
Melanie Virtue	Senior Advisor (15%)	2013 to present
Aline Kühl-Stenzel	Coordinator (75%)	2017−18
Jenny Renell	Coordinator (75%)	2019 to present
Assistants	**Time**	**Dates**
Sara Heimlich	50%	1994−96
Clare Last	50%	1997−98
Patricia Stadié	100%	1998−2007
Tine Lindberg-Roncari	50%	2007−10
Bettina Reinartz	50%	2010 to present

The early years

The birth of ASCOBANS started formally with the Memorandum of Understanding on Small Cetaceans in the North Sea, signed by Ministers present at the third International Conference on the Protection of the North Sea. At the time I was enjoying a 2-year sabbatical leave from the Sea Mammal Research Unit (Cambridge, United Kingdom), in San Diego, California, USA, investigating age techniques in pilot whales, and completing some long-term life history studies on fin whales off Iceland. However, all good things end eventually, and I returned to Cambridge in 1990 to learn that my new assignment was soon to set up an Interim Secretariat for a new organisation, ASCOBANS. I worked part-time on getting everything in place for when the Agreement would come into force. A lot of the work was contacting national representatives of Range States and reminding them about signing and eventually ratifying the Agreement. I also created a letterhead and small logo with a cheeky dolphin leaping from the sea — since replaced with a sleek futuristic design. However, the Agreement signed on 17 March 1992 came into force once six Parties had signed and ratified. The early member Parties were Belgium, Denmark, Germany, The Netherlands, Sweden and the United Kingdom. By now, the Interim Secretariat was being run from a small office with a separate computer, phone line and fax and copy machine, and most importantly, with an automatic answer machine. As I was also working part-time in a laboratory elsewhere in the building, I could not be there to take calls but would contact everyone with a call back. My working life was therefore somewhat schizophrenic.

Around this time the planning of the SCANS survey was ongoing, and it would take place in July 1994. This survey would provide significant information on harbour porpoise distribution and status. An initial meeting — a precursor to the AC — was

(Continued)

(Continued)

held in Cambridge among national representatives to discuss progress on different issues, and also to draft texts for possible resolutions that might be presented at the first Meeting of Parties. I worked closely with Arnulf Müller-Helmbrecht of CMS on developing Rules of Procedure and finalising texts for Resolutions, also developing a budget, and other advisory documents. In advance of the first meeting, he and I travelled to Sweden to view the facilities and talk to the representative there about preparations. The first Meeting of Parties was held in Solna, Stockholm, Sweden, 26–28 September 1994, under the chairmanship of Lars Björkbom from Sweden. Six Resolutions were passed, including one that confirmed the location of the Permanent Secretariat in Cambridge. Another defined the elusive 'St Anthony's Head' — not found on any chart, and a key location in defining the area of jurisdiction, by pinpointing the location with geographical coordinates. Many Range States attended, including the European Commission. However, the EC had not shown much interest in ASCOBANS, and it was much later at the second Meeting of the Parties (MOP2, Bonn 1997), when discussing matters relating to bycatch and fisheries regulations, that a potential statement (MOP2: Resolution on Incidental Take of Small Cetaceans] from ASCOBANS inflamed the EC sufficiently to send a delegate immediately to attend the meeting and explain that ASCOBANS did not have relevant competence in relation to the Common Fisheries Policy (CEP). This incident reminds me of the inconsistencies in international agreements.

In the early years Norway was always in attendance of ASCOBANS meetings, and contributed actively to its work. However, it was long known that Norway would not sign the Agreement because of the clause referring to not allowing sustainable use. This was, and still is, a matter that Norway cannot accept. When I held the position of General Secretary of NAMMCO (2005–15), which does allow for sustainable use, I could see that there are instances of apparent paradoxes among statements in international agreements, which often contradict or conflict with each other and lead to ambiguity in interpretation for Parties. I made this point clearly as an invited speaker at the ECS conference in Stralsund, Germany, 20–24 March 2010, in my talk *Conservation and Management of Marine Mammals — are they compatible issues at regional and global levels?* Often international law is left unclear when organisations and treaties have conflicting statements. Furthermore, such disagreements do not foster cooperation among organisations dealing with similar matters. However, even recognising basic differences, there are still many matters that could benefit in efficiency and effectiveness through more cooperation and collaboration among organisations in research and action planning, and exchange of information. We do not all have to agree on everything, but we can still be open to cooperation.

In order to begin some dialogue and a plan of action after the MOP1, the first AC meeting was held in Cambridge on 8–10 March 1995. The first Chair was Peter Reijnders from The Netherlands. We adopted Rules of Procedure for the AC, and also decided that the work of the group would be shared among members by specifying tasks to each. This would hopefully ensure that matters would be acted upon and not forgotten.

In 1996 the Sea Mammal Research Unit officially started its relocation to the Gatty Marine Lab, St Andrews University in Scotland, and I decided to relocate to Denmark to work for the Danish Institute for Fisheries Research, leading a project on mitigation of porpoise bycatch in cod and halibut fisheries. I was pleased to continue participating in and contributing to ASCOBANS activities, but as a Danish delegate. I was also very pleased when Poland and Finland joined ASCOBANS, after my canvassing from the early days of the Agreement. Initially after I left England, my colleague Sara Heimlich took over as interim Secretary. However, later Jette Jensen from Denmark stepped in until the Sea Mammal Research Unit shut down in Cambridge, and the decision was made to transfer the ASCOBANS Secretariat to Bonn in partnership with CMS, and a new Secretariat structure was set up.

Dr Christina Lockyer
Director, Age Dynamics, Kongens Lyngby, Denmark, and Tromsø, Norway,
Executive Secretary, ASCOBANS, 1992–96

My life as an executive secretary to ASCOBANS

I have the great pleasure and honour of having been Executive Secretary to ASCOBANS during the years 1996–98, the 'good old' early days of its existence. I say pleasure and honour because I feel thankful for the opportunity to assist with the important issue of the conservation of the precious small cetaceans inhabiting our waters.

I moved from Denmark to Cambridge in England, where the Secretariat was located in those days. Moving to another country is a big decision, but, for a job where I believed that I could contribute to small cetacean conservation, then it was an easy decision. It is important to try to contribute where we can to the conservation of our planet's resources, whether it is by dedicating our professional or private lives directly or by being responsible users of its many resources.

My biggest moment was when we succeeded in making the first resolution that actually committed all member countries to take specific measures to reduce bycatch of harbour porpoises, rather than just talking about it at yet another meeting. Being Danish, it was also a personal pleasure to be part of this important first conservation step, because at that time research showed that the bycatch of harbour porpoises was highest in the Danish fishery for turbot.

(Continued)

(Continued)

Things only move when you work with very dedicated people who really care. Politicians rarely take actions with economic costs if they are not provided with firm scientific advice, and shown by lobbyists and the general public that it will not affect their vote. Fortunately, there were some very dedicated people in those days not only within ASCOBANS itself, but also outside from nonmember countries, other agreements and NGOs. I would specially like to thank ASCOBANS Chairman Peter Reijnders, nonmember country representative Arne Bjørge, and the International Whaling Commission representative Greg Donovan for their enthusiasm and commitment; they were a joy to work with and inspired me to keep going when I felt progress was too slow.

I moved on to work with fish and seafood research, and I am a strong advocate of the health benefits of eating fish … my grandchildren perhaps feel that they suffer from my efforts to persuade them, saying I am too strong a campaigner ☺ … However, given my time in ASCOBANS, and knowing what the turbot fishery did to harbour porpoises in those days, I stress to all my colleagues the importance of sustainable fisheries that do not adversely affect the marine environment and especially cetaceans.

I wish ASCOBANS congratulations on its 25th anniversary, and I trust that future generations will be able to enjoy the pleasure of small cetaceans. Let us all take care.

Jette Donovan Jensen
Executive Secretary, ASCOBANS, 1996—98
Deceased on April 2019

How does an international Conservation Agreement function?

Activities centre around the annual meetings of the AC and the more formal triennial Meetings of the Parties. When I first started attending these, I was struck by all the formal procedures[1]. Everyone is usually seated in a large rectangular arrangement, which may number as many as 50 persons (Fig. 1.5). There is some hierarchy in where the attendees are seated. The Chairpersons and Secretariat are situated at the front; the official representatives of the Parties nearest to them; and the IGO and NGO Observers furthest away. Similarly, the convention is for Parties to be able to speak first, followed by Observers, and voting only involves Parties. Each person has a banner with their name and affiliation on the table in front of them, which they raise if they wish to say something. They then wait for the Chair to call them, upon which they switch on a nearby microphone. Unlike many international meetings, proceedings are always conducted in English, which can be tiring for nonnative English speakers.

The general procedure has been that each meeting starts with various introductory remarks and opening statements, and then, after rules of procedure and the provisional agenda are agreed, the main business starts. Parties are first invited to present the key points from their annual reports including information on any changes in national legislation. Some discussion may ensue amongst Party members and Observers. Any Accession and Agreement amendments are reported upon by the Secretariat. One of the main outputs from the AC meetings is the development of a Work Plan for the next Triennium. This is refined each year leading up to the Meeting of the Parties where agreement is reached for where priorities should lie for the coming 3 years. Related to this is a review of each of the major conservation initiatives — currently these are the Baltic Harbour Porpoise Recovery Plan (known as the Jastarnia Plan), the Conservation Plan for harbour porpoises in the North Sea, bycatch issues and negative effects of sound. The Committee recognises the importance of publicity and outreach activities, and progress by the Secretariat and then each of the Parties is reviewed, and from these discussions new initiatives are proposed. Other aspects of the Triennium Work Plan are then addressed, such as reviews of new information on population size and distribution, population structure and causes of any changes; on pollution and its effects; on negative effects of vessels and other forms of disturbance; management of marine protected areas; aspects relating to the recent extension area of the Agreement; and a report from the informal working group on large cetaceans.

A varying amount of funding is available each year for projects, and the meeting considers first progress on projects already funded, and then reviews new proposals, deciding which should have priority. An important element of the AC meeting is the supporting information presented both in the form of documents (which frequently may number more than 50) and as short PowerPoint presentations. These serve as excellent, up-to-date papers and topic reviews that anyone interested in cetacean research and conservation in Europe can readily access through the ASCOBANS website (go to the following link: http://www.ascobans.org/en/meetings/advisory-committee). It is a resource that I think is greatly

1. These are not specific to ASCOBANS but to any large international intergovernmental meeting. Indeed, I have since found that the atmosphere was more informal than most.

FIGURE 1.5   An advisory committee meeting in progress. *Photo: ASCOBANS Secretariat.*

underexploited and yet not only provides important literature sources but often presents results that have not yet reached the public domain through scientific journals.

The final sessions of the scientific aspects of AC meetings cover relations with other bodies, drafting a list of dates of meetings likely to be of interest, and in the lead-up to the MOPs, consideration and preparation of draft resolutions. The Secretariat then circulates a list of action points and decisions from this Science and Conservation session for comment and approval.

Nowadays the AC meetings are divided into two main sessions, first Science and Conservation and then an Administrative session. This latter session is primarily for the Parties themselves, although Observers are usually allowed to attend. The Secretariat presents a report on financial and administrative matters, and Parties reach agreement on expenditures. They decide how funds should be apportioned to projects; draft resolutions on budgetary matters for any upcoming MOPs; the date and venue for the next meeting; and adoption of the list of action points considered at the session.

The meetings involve long days with work sessions often going late into the evening, particularly for any working groups that have been established (Figs. 1.6A and 1.6B). But those that have to work hardest are members of staff of the Secretariat. An insight into the challenges they face at each meeting is provided by the contributions to this chapter from Pat Stadié who was assistant to the Executive Secretary from 1998 to 2007, and Robert Vagg who has been report writer for many meetings from 2001 to the present. The principal roles of the Secretariat are, amongst others, the preparation and compilation of content, reporting of progress, support to the Chair in guiding the proceedings, and ensuring the meeting's objectives are met.

Not all the time at these meetings is spent in work. Besides the opportunity to socialise at coffee breaks and meal times, the host country usually offers a reception to meeting participants one evening, and there might be a half-day outing to one or more sites of interest.

(A)

(B)

FIGURE 1.6 (A) Bruges (Belgium): the venue of the 16th advisory committee meeting, (B) a working dinner in Bruges. *(A) Photo: Peter GH Evans, (B) Photo: ASCOBANS Secretariat.*

TABLE 1.5 List of scientific workshops supported by ASCOBANS.

Date and location	Topic	Outputs
21 April 2007, Spain	Offshore wind farms and their impacts	Evans (2008a)
22 April 2007, Spain	Selection criteria for marine protected areas	Evans (2008b)
8–9 October 2007, Germany	Small cetacean population structure	Evans and Teilmann, (2009)
20 March 2010, Germany	Cetacean bycatch mitigation	Evans, Siemensma and Bräger (2010)
20 March 2011, Spain	Chemical pollution	Evans (2014), Evans and Simmonds (2011)
24 March 2012, Ireland	EU Habitats Directive & its implementation	
6 April 2013, Portugal	Cetacean population structure	Gaspari, Evans and Montiglio (2013)
6 April 2013, Portugal	Towards a conservation strategy for white-beaked dolphins in the Northeast Atlantic	Tetley and Dolman (2013)
7 April 2013, Portugal	The challenge of spatially managing cetaceans – a highly mobile animal group	Evans et al. (2013)
6 April 2014, Belgium	Introducing noise into the marine environment – what are the requirements for an impact assessment for marine mammals?	Evans (2015)
12 March 2016, Madeira	Conserving Europe's cetaceans through synergy-building between the relevant legislative frameworks	
13 March 2016, Madeira	Cetacean pathology: necropsy technique and tissue sampling	IJsseldijk and Brownlow (2018)
29 April 2017, Denmark	Fostering interregional cooperation in underwater noise monitoring and impact assessment in waters around Europe, within the context of the European Marine Strategy Framework Directive	ASCOBANS-ACCOBAMS Joint Working Group on Noise (2017)

Every 3 to 4 years there is additionally a Meeting of the Parties. Inevitably, these are more political and more formal, although a high proportion of the attendees are the same persons involved in the AC meetings. The purpose of these meetings is to review progress in the various areas considered by the AC, refine where necessary and then adopt Resolutions, agree upon the Triennial Work Plan, set the budget and finalise various administrative matters.

Reports and papers associated with these can be found at the following link: http://www.ascobans.org/en/meetings/meetings-of-parties.

Although the meetings are the focus for activity of the Agreement, a lot happens in between meetings, mainly stemming from the Secretariat and AC Chairs but frequently also involving intersessional work conducted by NGOs and representatives of the Parties. There has also been close collaboration, particularly in recent times, with the marine mammal science community, for example, through the ECS, since these are generally the ones collecting the necessary information to advise on trends in cetacean status and conservation threats. One such fertile relationship has been the sponsorship of specialist workshops usually in partnership with the ECS and frequently also with its sister agreement, ACCOBAMS. Table 1.5 gives details of those workshops and the outputs from them.

From Aberdeen to San Sebastián — the inside story

The routine, if one may call it that, of the ASCOBANS Secretariat (Fig. 1.7) was routinely interrupted by the *Meetings*. The word is so innocuous, suggestive of cups of tea, laughter and chatter in a sunny room. Alas, 'Meetings' are the monsters that punctuate the years in many UN offices, and my life certainly revolved around those at ASCOBANS. There were the regulars with their code names, so mystifying to outsiders: the annual AC, the triennial MOP. These were later joined by yearly meetings of the Jastarnia Group, not to mention the ad hoc get-togethers. Delegates and observers gathered, usually for 3 days, often at a venue offered by the hosting Party. The dates were fixed at least a year in advance, and from then on, they loomed on the horizon like distant peaks we knew we must scale, drawing closer by the week.

My first experience of a UN meeting had been the sixth of the ASCOBANS AC, in Aberdeen. It was April 1999, I was still new to the job, and somehow Holger Auel, the then Executive Secretary, and Mark Tasker in Scotland had managed to organise the entire event without enlisting my help, or letting on what was about to happen. But all went reasonably well and, having had no hand in the arrangements, I didn't feel overresponsible. I remember Aberdeen, however, because it was there that I first met many of the 'ASCOBANS people' who would later become such familiar faces and good friends; and because it cured my fear of flying. Tickets had been booked: from Cologne to Amsterdam and thence to Aberdeen, returning by the same route. The thought of travelling on four planes within as many days petrified me, but as the meeting approached I decided there was no escape, and managed somehow to adopt a nonchalant façade. As it happened, I was cured of my phobia by the time we left Amsterdam, and I never looked back. Thank you, ASCOBANS, for that too.

FIGURE 1.7 Pat Stadié and other members of staff of the ASCOBANS secretariat. *Photo: ASCOBANS Secretariat.*

It was in Aberdeen that I first experienced the worry sometimes generated by meeting delegates. There was an elderly Russian cetologist, attending at the invitation of the German government, I believe. When he arrived, it seemed apparent that he spoke not a word of English. Holger worried because he couldn't join in the proceedings, and we both worried when he started disappearing for long periods. We wondered where he went and how he managed to get around, but he always reappeared eventually, and we didn't have to call the police and admit we had somehow lost a Russian scientist...

(Continued)

(Continued)

Rüdiger Strempel took over as Executive Secretary in October 1999. The next AC was approaching and I was no longer 'new'; that was when meetings became an important part of my life. Year after year I'd spend the months and weeks of preparation in a state of increasing frenzy. Rüdiger drafted the Secretariat documents, others were to come from outside and of course only trickled in, deadlines being ignored or just stretched. All had to be numbered and formatted, and in those days photocopied and sent to each participant by post. I also looked after registrations, sorting and filing them, making endless lists, checking and rechecking the numbers: hotel rooms, vegetarians, late arrivals, early departures, taxis needed, numbers for the excursion, seating, badges, pigeon-holes, computers, coffee breaks, lunches and the reception. As the time approached I would count and recount, check and recheck. And yet, still, one dreadful evening in Sweden, we discovered there was one more person waiting in the lobby than the number of rooms in the hotel. I still don't know how it happened, and the delegate was found a bed elsewhere in the town, but I still cringe with remorse when I recall it. (I'm still very sorry, Jonas. And thank you for forgiving me eventually.)

One of my jobs during meetings was keeping track of the documents, which as if by magic began to blossom and multiply as soon as the first participants arrived. 'Hi, Pat, lovely to see you again. This is the revised version of the document I sent you a month ago, could you make sure everyone gets it?' And I would modify our list of documents, make a new cover page with the new number (Rev. 1), recopy the maybe 50 pages, staple and distribute them, plus of course 'Rev. 1' of the List of Documents. Half an hour later, another kindly smiling person would hand me 'Rev. 2'.

Did I mention photocopying? The meeting venues were often adult education centres or government offices, and even sometimes hotels with real conference facilities. There was always a copier of sorts ('But of course, you can use ours'), and in our very early days Rüdiger and I would shrug and agree that we'd manage somehow. But experience soon taught us that these machines were never ever prepared for the copying avalanche that a UN meeting, even a small one like ours, will unleash. Sometimes they were located on the floor below (or in a different building), inevitably the staff of the establishment would also need them constantly, and they were just too slow. I dreaded the final hours of meetings: working groups were producing ever-new drafts for the plenary's approval, final versions of documents needed circulating, tempers were fraying and the machine in the copy room laboured on and on, pulsing its slow rhythm and spewing scalding sheets from its red-hot insides, while the document's progenitors hovered over me wailing, 'Why is it taking so long?'

Invariably the copiers broke down, sometimes even on the first day. Soon we began to insist that we needed at least two, to be on the safe side. In Brest our generous hosts hired two identical, superfast new machines for our sole use. Both broke down early on the first morning, within 30 minutes of each other. This confirmed my suspicion that these devices are malevolent aliens. When the first broke down, the backup machine, although located in the next room, heard us congratulating ourselves on our foresight, or was jealous of the attention the other was getting, and packed up too. The technician was called and spent the morning fixing them, and to be on the safe side stayed with us for the rest of the day, himself scarcely believing what had happened.

We had some close shaves with disaster. In Bruges Rüdiger and I were working our customary night shift, writing the report of the day's proceedings for review by the Chairman early next morning, when suddenly everything went dark; both the lights and the computers went off. The electricity supply in the conference wing was programmed to switch off automatically at 1 a.m., but no-one from the hotel had thought to tell us. We groped our way to the still illuminated hotel lobby, and persuaded the night porter that this was a matter of life and death. We did survive, and worse things happened.

One of our meetings was in an adult education establishment using computers that were programmed to delete any newly added data as they were switched off. This was probably a sensible way to prevent students clogging up the hard drives with useless files and virus-laden downloads. It was less helpful for a 3-day meeting where I needed a machine to keep track of dozens of documents that were being revised by the hour. I had brought all the originals on a CD-Rom, and when we arrived the evening before the meeting began, I carefully copied our documents to the hard drive before going for dinner. Luckily I went back later to add a new item I'd been given, to discover that the files I had saved had vanished, were nowhere to be found. I repeated the process, carefully saving them again and switching off the machine once more. Again they disappeared. Was I losing my mind already? When the school's computer expert was called, he said, 'Well, that's how it is here. Just don't switch off'. Great! Glad I had discovered this idiosyncrasy in time to prevent a catastrophe, I spent the next 3 days seriously fearing a power cut, or just someone tripping over the lead.

Hindås was the scene of the ultimate near-cataclysm, when our computer was abducted during dinner. It was the final evening. The meeting was over, the documents were safe, the report drafted, everyone happy. I was hungry and joined the rest of the group in the dining room, just yards away, secure in the knowledge that my final task, copying the final versions of everything from the hired PC to a CD-Rom for the journey home to Bonn, could easily be left until I'd eaten. I didn't stay long, wanting to finish and pack up my things before enjoying a relaxing evening. When I went back into my temporary office, I was greeted by a gaping empty space on the desk. The computer, the screen, the printer, had disappeared. Panic! Sabotage! This was serious!

It took some time to find someone, anyone, who could reconstruct events. Yes, a man from the computer hire company had arrived and collected their equipment. Urgent phone calls went out into the quiet Swedish hills. We were miles from anywhere, and it was getting late. Who had hired the computer, and, more important, from which company? How could they be contacted at this time of day? Where was the man with the van? Hours went by, I huddled in a corner, wishing, hoping, praying even, while Rüdiger conferred with kindly Swedes, who made more calls and did their best to help. Eventually the man with the van brought everything back, apologising profusely. He'd been passing, and thought he'd pop in now to save him coming back the

(Continued)

next morning. And, he beamed, we were especially lucky, because he'd driven home instead of going via the company premises. His first job in the workshop would have been to make sure the hard drive was wiped absolutely clean.

Yes, we had some adventures, plenty of laughs and sometimes a few tears. Rüdiger's unerring calm, even in the face of such calamities, was remarkable, and his refusal to be ruffled, though sometimes maddening, had a soothing influence even on those of us toiling behind the scenes. ASCOBANS meetings were certainly very special, but can I claim to have enjoyed them? Yes! And what was my favourite agenda item? *'Close of Meeting'*!

Patricia Stadié
ASCOBANS Secretariat Assistant, 1998—2007

A Report Writer's View

It is strange how seemingly insignificant incidents in one's life turn out to have a long-lasting impact. For the past 12 years, I have to a large extent made a secondary career of report writing (Fig. 1.8) at meetings of various Committees and Councils of the CMS Family, and all because the report writer of choice of the ASCOBANS Secretariat was already booked and was unable to attend the AC in Nymindegab, Denmark in the spring of 2001. At the time, I was approaching the end of my two-and-a-half-year secondment from the British Ministry of the Environment (what is now DEFRA), and as the resident native English speaker, I was called upon to step in and fill the breach.

To digress a while: in my experience, I have to say, that native English speakers are quite often not the best people to take notes or write meeting reports — the British, Americans et al. have a reputation for not being terribly good at learning other languages and as an indirect consequence fail to understand the mechanics of their own mother tongue — how

FIGURE 1.8 Robert Vagg, ASCOBANS report writer. *Photo: ASCOBANS Secretariat.*

often do you hear even educated English speakers utter grammatical catastrophes such as 'between you and I'[2] (chosen, incidentally, as the title of a book about common mistakes in English); and they tend to write for other native speakers rather than for an international audience for whom English is the second or maybe even third or fourth language. My

(Continued)

---

2. It should of course be 'between you and me'.

(Continued)

secret is that I have studied foreign languages, lived in other countries and have qualified as a teacher of English as a foreign language.

So it was that amidst the wind-swept dunes of the coast of West Jutland (where the late, lamented Krzysztof Skóra found the leather diving boot now on display at the visitor centre of the Marine Station/Seal Sanctuary in Hel), my career took a fateful turn. Since my return to Bonn in September 2006, initially to compile, write and edit the CMS Encyclopaedia (the CMS Family Guide), I have myself become the report writer of choice for ASCOBANS and have attended MOPs in The Hague, Bonn, Brighton and Helsinki, AC meetings in Pasaia (San Sebastian), Bruges, Bonn and Gothenburg, and Jastarnia Group meetings in Kolmården, a deeply snowy Turku, Hel when it had frozen over, a bitterly cold Copenhagen and picturesque Stralsund. My knowledge of North Sea and Baltic geography has increased immeasurably over the years, and I know the difference between my Darß and my Elbe and on one occasion almost pronounced Wladyslawowo correctly. Should my association with ASCOBANS continue, I hope to witness a definitive conclusion being reached on the rights and wrongs of excluder devices and a decision on whether or not ASCOBANS should align its zoning policy with HELCOM rather than ICES.

Since my return to CMS, I have overseen the editing of three editions of the Family Guide and witnessed the publication evolve into a purely electronic format. I also collaborated with CMS Ambassador Stanley Johnson in transforming the text to accompany the wonderful photographs selected by publishers Stacey International for the book 'Survival: Saving Endangered Migratory Species'. To my great surprise, 'Survival' won a prize from the American Independent Book Publishers' Association. I wonder how much of all that would have transpired, if by chance the other report writer had been available to go to Nymindegab.

Robert Vagg
ASCOBANS Report Writer, 2001 to present

ASCOBANS – Agreement on the Move

In our contribution to the ASCOBANS 10th Anniversary volume, Peter Reijnders, the long-standing Chair and Vice-Chair of the ASCOBANS AC and winner of the 2009 ASCOBANS Lifetime Award, and the author of these lines noted that 'On its 10th anniversary, the ASCOBANS can look back on a decade of continuous change and development'.[3] Much the same can be said today, as the Agreement celebrates its 25th anniversary.

The Agreement's first years were marked by an intensive process of 'institution building'. The first four Executive Secretaries, Christina Lockyer, Sara Heimlich-Boran, Jette Jensen and Holger Auel, successfully guided the fledgling Agreement through its infancy and tackled challenges such as setting up and managing the Secretariat, preparing and servicing the first MOP and of the AC – which were also still in the process of finding their bearings – and organising the Secretariat's relocation from the United Kingdom to Germany.[4] When I assumed the function of Executive Secretary of the Agreement in 1999, I was privileged to find an operational Secretariat colocated with the UNEP/CMS Secretariat at the UN premises in Bonn and well-functioning advisory and decision-making bodies. A further, far-reaching institutional realignment took place at the beginning of 2001, when the Secretariat was integrated into the UNEP system, as decided by the third Meeting of the Parties (MOP3, Bristol, United Kingdom, July 2000).[5] It was not until 2006, when the ASCOBANS Secretariat was merged with the CMS Secretariat, that organisational issues once again figured prominently on the ASCOBANS agenda.

By contrast, the first years of the new millennium were a period of extensive substantive activities for ASCOBANS. One of the major challenges to be tackled was the development of the Baltic dimension of the Agreement. MOP2 (Bonn, Germany, November 1997) adopted a resolution inviting Parties and Range States 'to develop (by 2000) a recovery plan for harbour porpoises in the Baltic Sea'.[6] Initial steps in this direction included the establishment of an ASCOBANS Baltic Discussion Group (ABDG) in 1998. Chaired by Finn Larsen, the group consisted of scientific experts. Following MOP3, which reiterated the call for a Baltic Sea Recovery Plan,[7] the process gathered new momentum, with governments and institutions from several Baltic Sea Parties providing funding and support. In January 2001 a voluntary contribution from the Swedish government enabled the ABDG to hold a final meeting, hosted by the Danish Institute for Fisheries Research in Charlottenlund, Denmark. The

(Continued)

---

3. Peter J.H. Reijnders & Rüdiger Strempel, Looking back, looking ahead, in: R. Strempel (Ed.) From Idea to Implementation, ASCOBANS, Ten Years On, Bonn, 2001, p. 25.

4. Christina Lockyer offers a vivid account of the trials and tribulations faced by a part-time Executive Secretary in a one-person Secretariat in Starting Out – Memories of the Early Years of the ASCOBANS Secretariat, in: From Idea to Implementation, ASCOBANS, 10 Years On, Bonn, 2001, p. 7.

5. MOP3 Resolution No. 1.

6. MOP2 Resolution on Incidental Take in Small Cetaceans.

7. MOP3 Resolution No. 3.

(Continued)

deliberations of this group and the report of its final meeting significantly influenced the development of the Baltic Harbour Porpoise Recovery Plan. In October 2001 a preparatory meeting of environment and fishery agencies and fishermen's organisations from the various Nordic Parties to ASCOBANS, funded by Sweden and the Nordic Council, was organised in Sweden. In January 2002 a workshop aimed at drafting a recovery plan was held in the Polish coastal town of Jastarnia. Hosted by the Foundation for the Development of the University of Gdańsk and the University of Gdańsk's Hel Marine Station, and funded by the Danish government, the workshop was attended by representatives of ministries, NGOs, fishermen's organisations, and public and private institutions from six Baltic Sea countries, as well as regional international organisations. Based on the outcome of this workshop and in cooperation with the Secretariat, Dr Randall R. Reeves, the facilitator of the workshop, produced the draft Baltic Harbour Porpoise Recovery Plan that was presented to MOP4 (Esbjerg, Denmark, August 2003), and has since become known as the Jastarnia Plan. Due to concerns about competency issues raised by the European Commission at MOP4, Parties expressed their support for but did not formally adopt the plan in Esbjerg.[8] A revised version of the Plan, produced by the ASCOBANS Baltic Sea Steering Group (Jastarnia Group), was, however, adopted by MOP6 (Bonn, Germany, October 2009).[9] A further revision was adopted by the eighth Meeting of the Parties (Helsinki, Finland, August/September 2016).[10] Since 2005 the ASCOBANS steering group for the Baltic Sea region, known as the Jastarnia Group, has met annually. True to its terms of reference, it has developed into an efficient forum for promoting and evaluating progress in the implementation of the plan and establishing further implementation priorities and has contributed significantly to the two periodic revisions of the Jastarnia Plan.

Once the Jastarnia Plan was in place, ASCOBANS focused on developing a Harbour Porpoise Conservation Plan for the North Sea, as called for by the fifth International Conference for the Protection of the North Sea (Bergen, Norway, March 2002).[11] In 2003, Germany took the lead on this project, funding an expert tasked with drafting the plan and hosting a first stakeholder workshop in Hamburg, Germany, in December 2004. Norway also played a very active role in the elaboration of the plan. In the course of its development, the plan evolved from the Recovery Plan envisioned by the Bergen meeting into a Conservation Plan, which was adopted by MOP6 (Bonn, Germany, October 2009). A North Sea Steering Group was established and for several years the North Sea Parties funded a North Sea Coordinator. Regrettably, this position had to be suspended, while plans for the appointment of a Baltic Sea Coordinator only materialised in 2018.

While the two plans cover the majority of the original ASCOBANS agreement area, they leave a stretch of sea from the Western Baltic through the Belt Sea to the Kattegat unprotected. The problem of this area, colloquially referred to as the 'gap area', has, however, also been addressed by ASCOBANS. At its seventh meeting (Copenhagen, Denmark, February 2011), the Jastarnia Group recommended that a draft paper containing background information and proposed objectives and measures for this region be commissioned 'and refined by the eighth meeting of the Jastarnia Group with a view to enabling formal adoption of such objectives and measures by the seventh Meeting of the Parties'.[12] The 18th meeting of the AC (AC18, Bonn, Germany, May 2011) endorsed this recommendation and made funds available. A draft plan authored by a team of scientists from Aarhus University[13] was reviewed by the eighth meeting of the Jastarnia Group (Bonn, Germany, January/February 2012) and again following the 19th meeting of the AC (Galway, Ireland, March 2012). The seventh Meeting of the Parties (Brighton, United Kingdom, October 2012) adopted the plan, which is formally entitled Conservation Plan for the Harbour Porpoise Population in the Western Baltic, the Belt Sea and the Kattegat.[14] The Jastarnia Group (without its Finnish, Lithuanian and Polish members) acts as the steering group for this plan.

Another gap had been closed some years previously. In 2003 ASCOBANS took the decisive step towards realising the Agreement's long-standing aim of extending the Agreement area to the southwest, thereby making it contiguous with that of its southern sister Agreement, ACCOBAMS. MOP4 Resolution No. 4, adopted following a process of coordination involving Ireland, Spain, Portugal and ACCOBAMS, extended the Agreement area accordingly. This development was expressly welcomed by ACCOBAMS at the time.[15] The extension entered into force in February 2008, following the deposit of the fifth instrument of ratification.

However, not only the agreement area has grown. The number of Parties has also increased, although growth in this respect has been somewhat sluggish. In 1996 Poland was the first country to join the six 'founding members',[16] followed by Finland in 1999, Lithuania and France (both in 2005). More than half of the 17 ASCOBANS Range States have therefore joined the Agreement.

(Continued)

8. MOP4 Resolution No. 8.

9. MOP6 Resolution No. 1.

10. MOP8 Resolution No. 3.

11. Ministerial Declaration of the fifth International Conference for the Protection of the North Sea (Bergen Declaration), Paragraph 30.

12. Report of the seventh Meeting of the ASCOBANS Jastarnia Group, p. 18.

13. Signe Sveegaard, Jonas Teilmann, Line A. Kyhn, Anders Galatius and Jakob Tougaard

14. MOP7 Resolution No. 1.

15. Cf. Report of the 10th Meeting of the Advisory Committee to ASCOBANS, p. 16. The extension has meanwhile given rise to some controversy and ACCOBAMS has extended its Agreement area northwards to overlap with the ASCOBANS area.

16. Belgium, Denmark, Germany, The Netherlands, Sweden, United Kingdom. The European Community was also among the original signatories but has never become a Party to the Agreement.

(Continued)

Education and Outreach is another area in which ASCOBANS has continuously developed and evolved, in particular since the beginning of its second decade. MOP3[17] and AC7 approved the Secretariat's suggestion to step up activities in this sector and, in the absence of a dedicated budget line, the annual voluntary contribution generously provided by Germany initially served to fund many of these measures. 2001 was a particularly active year in this respect. Activities included the development of a new ASCOBANS brochure which was later translated into the languages of all Range States (and an updated version of which is still available from the Secretariat), a 10th anniversary publication, a 10th anniversary exhibition shown in Bonn and in Gdańsk, Poland and a relaunch of the ASCOBANS website. The logo designed by the Secretariat on the occasion of the 10th anniversary was subsequently adopted as the new Agreement logo. In the following years, ASCOBANS continued its information and outreach campaign for a variety of target groups. Fact sheets and the ASCOBANS Guide to Accession, published in 2003, were aimed at policy makers in national authorities. The International Day of the Baltic Harbour Porpoise (IDBHP), observed annually since 2003 by the Secretariat in cooperation with government authorities and scientific institutions from around the Baltic Sea, targets the general public in Baltic Sea countries. The IDBHP also inspired the 'Year of the Dolphin' campaign, successfully carried out jointly with the CMS Secretariat in 2006/2007 throughout the agreement area. Other activities include a further relaunch of the website in 2011 (and again in 2014), and the development of a comprehensive Communication, Education and Public Awareness Plan (endorsed by AC17, Bonn, Germany, October 2011).

The above list of achievements shows that in those first two decades, ASCOBANS made a continuous contribution to the cause of small cetacean conservation in Europe's northern regional seas. The Agreement has established itself as a unique forum of expertise, bringing together scientists, civil society representatives and policy makers from across the region, who pinpoint the threats to the species concerned and clearly formulate measures required to address them. It can also safely be assumed that ASCOBANS has contributed to enhancing awareness of the situation of small cetaceans throughout the region. Nevertheless, it is equally evident that the aim of the Agreement laid down in Article 2.1 ('to achieve and maintain a favourable conservation status for small cetaceans') has yet to be achieved. The overall conservation status of these species in the ASCOBANS area is uncertain, and some populations are known to be in decline. This applies in particular to the harbour porpoise populations in the Baltic Sea.[18] The highly successful SAMBAH (Static Acoustic Monitoring of the Baltic Sea Harbour Porpoise) project (carried out from 2011 to 2013) has confirmed beyond doubt that the harbour porpoise population in the so-called Baltic Proper is dramatically small and comprises approximately 500 individuals.[19] The listing of this population by IUCN as critically endangered is therefore fully justified.

It is unrealistic to expect that ASCOBANS could singlehandedly ameliorate this situation. The political and legal environment in which the Agreement operates is not conducive to swift and decisive conservation measures. The Agreement area comprises almost exclusively EU Member States and all current Parties are EU countries and therefore subject to the restrictions imposed on them by European law. Notably, Article 3 (d) The Treaty on the Functioning of the European Union (TFEU) assigns the exclusive competence for the conservation of marine biological resources under the CFP to the EU. However, whereas this limits the leeway of Member States, it does not condemn them to inactivity. Parties should therefore resist the temptation to hide behind the EU and, instead, continue to seek to leverage the full potential of ASCOBANS. The above examples demonstrate that ASCOBANS operates most successfully when Parties take an active interest in its work and assume ownership of the Agreement. In addition to ensuring viable Secretariat arrangements and securing adequate funding and support for the Secretariat and the agreement's activities, ownership translates into action at various levels. At the ASCOBANS level, Parties should agree on the most far-reaching and forward-looking measures possible (within the framework delineated by European law). At the EU level, they should actively and consistently promote the positions adopted by the Agreement and seek to influence relevant EU policies with a view to achieving optimum conservation. At the national level, they should strive to overcome 'implementation gaps' between the measures propagated by the Agreement and actual action.

In summary, it can be stated that ASCOBANS is a dynamic multilateral environmental agreement that has continuously evolved and developed since its inception in 1991 and has influenced cetacean conservation in its Agreement area. But the Agreement's mission is not accomplished. More could and needs to be achieved. Halfway through its third decade, ASCOBANS and its Parties will need to make continued efforts to bring this project to fruition. Like the species that it is intended to protect, the Agreement will need to stay on the move.

Rüdiger Strempel
Executive Secretary of the Common Wadden Sea Secretariat (CWSS)[20],
Chair of the Jastarnia Group, 2009-17,
Executive Secretary of ASCOBANS, 1999-2007

---

17. MOP3 Resolution No. 6.

18. Listed as critically endangered by the International Union for the Conservation of Nature (IUCN) since 2008.

19. More information and the final report of the project can be found at www.sambah.org

20. The views and opinions expressed in this article are those of the author and do not necessarily reflect the official policy or position of CWSS.

Running the 'New' Secretariat – or: the attempt to do more with less

At the fifth Meeting of the Parties in 2006, governments had to make a tough decision: either pay higher annual fees in order to keep an independent Secretariat, or go for a radically different solution. Pressing financial considerations aside, there were hopes that new arrangements might not only be more cost-effective but could also increase the international profile of the Agreement and enhance synergies with related work under other treaties. How secretariat services to an intergovernmental treaty are delivered is by no means a trivial matter – much of the success of the meetings and of the intersessional follow-up of decisions depends on the level of service that can be provided.

It was therefore with mixed feelings that ASCOBANS Parties decided in December 2006 to entrust the Secretariat of the mother convention, the CMS, with fulfilling the secretariat function for the Agreement (Fig. 1.9). This new arrangement took effect only a few weeks later, on 1 January 2007. Thus, ASCOBANS very suddenly went from a Secretariat with its own Executive Secretary and full-time assistant, both very experienced and dedicated solely to the Agreement, to being a subunit within a much larger team, with four staff members contributing different percentages of their time to ASCOBANS, but totalling the equivalent of less than 1.5 full-time posts. None of the new team had previously been directly involved with running the Agreement. The new junior position of Coordinator was created, with 75% of time dedicated to ASCOBANS and the remainder to CMS. Recruitment was completed in early 2007 and I took office in April of that year.

As was to be expected, the transition into this new arrangement did not go entirely smoothly, despite the best efforts of those directly involved. Parties were severely criticised by civil society for not being willing to spend more for a functioning cetacean agreement. Tensions also arose because the reduced staff complement was not reflected in a less demanding work programme for the Secretariat, rather to the contrary. It took time and patience on all sides to come to realistic views and expectations, and a great deal of dedicated effort to deliver what is necessary for a well-functioning treaty.

ASCOBANS is now in the 12th year of this new arrangement, and Parties have made it permanent, expressing their contentment with the current set-up. But does this mean that the change was a success?

FIGURE 1.9   (A) UN Campus in Bonn. *Photo: A. Bahramiouian* (B) View from UN building across the river Rhine. *Photo: K. Chadwick.*

(Continued)

(Continued)

FIGURE 1.9 *(Continued)*

From a Secretariat perspective, the answer is a clear 'yes and no'. There definitely would be a need for more staff in order to address adequately the needs of the Agreement, and of the species it is concerned with. However, in the current economic climate, this problem is not likely to be addressed very soon. The same is of course true of virtually every environmental organisation, whether it is intergovernmental like ASCOBANS, governmental, or nongovernmental.

That we have been able to satisfy our Parties is of course a nice reward for the hard work put in over the last years. And, indeed, I believe that overall the arrangement has been a success and has contributed significantly to making the most of synergies within the CMS family of marine mammal treaties. There is frequent and intense interaction with the CMS staff and a free flow of information, which allows ASCOBANS to draw from the wide range of expertise available in the joint Secretariat. For its part, CMS benefits from direct access to additional cetacean specialists, both in the Secretariat and through the ASCOBANS AC. There is now close cooperation on a number of issues and projects, ranging from the development of joint policy on underwater noise to common approaches to information management and outreach. All this is a result of having the same management and staff members involved in cetacean-related activities of both CMS and ASCOBANS.

Outside criticism has shifted away from the Secretariat arrangements towards actual implementation of the conservation decisions in the waters covered by the Agreement. We take this as an indication that the Agreement's internal troubles resulting from the change in Secretariat arrangements are over. Now the focus is fully on helping Parties achieve the good conservation status for small whales, dolphins and porpoises that is the aim of ASCOBANS — a goal well worth all our efforts!

Heidrun Frisch-Nwakanma
ASCOBANS Coordinator/CMS Marine Mammals Officer, 2007—16

Has ASCOBANS had a nice future behind, or one ahead of it?

Ever since its inception, I was confident that ASCOBANS was a timely answer and would be a useful instrument to improve the level of small cetacean conservation and protection. The first initiative leading to what later became ASCOBANS was taken in 1984 in Geneva, at a planning meeting for the upcoming first Conference of Parties (COP) to CMS, held in 1985. At that meeting, a keynote was presented on the relevance for nature conservation of the relatively recently published Directive on the

(Continued)

(Continued)

Conservation of Wild Birds (later amended into the Birds Directive). The general opinion after that presentation was that agreements for other species, for example mammals, were needed and that CMS might play a role in that. The meeting decided to put forward to the COP two proposals: one on bats and one on small cetaceans. My Dutch colleague at that time, Kees Lankester, was one of the initiators so this information is first hand. It was clear that the focus should in the first instance be on the North Sea and Baltic Sea area. This was primarily because of the deplorable conservation status of the harbour porpoise in these regions. Via many meetings, the Agreement on the Conservation of Small Cetaceans of the Baltic and North Sea was finally concluded under UNEP/CMS in 1992.

My involvement in ASCOBANS started shortly after the agreement was concluded, notably at a late bar gathering at the annual meeting of the International Whaling Commission. We agreed that it indeed was a good thing to establish a special conservation body for small cetaceans, to focus for the time being on harbour porpoises and that it would be beneficial for ASCOBANS to cooperate with cetacean scientists also involved in the IWC. I got involved in developing that idea and transforming the agreement text into a real conservation plan. One of the first meetings of countries intending to join ASCOBANS was held in Sweden shortly after the Agreement was signed. At that meeting, a rather vague document was put on the table, which became ASCOBANS' Objectives. The classification 'vague' refers to the rather administrative nature of it and inherent prominence of organisational and financial matters. I joined that meeting as an advisor to the head of the Dutch delegation, who happened to be one of the 'founding fathers' of the ASCOBANS idea, and I criticised the proposed plan and made a plea to improve and rewrite that document. 'All right' he said, 'but then you have to have a different text by tomorrow morning when this issue is on the agenda'. So it was, and after a full night's intense work, a draft of alternative objectives was available. To our surprise the meeting quickly chose these alternative objectives and adopted them with minor amendments. That was my first real input in ASCOBANS, instigated by a strong wish to really do something for the protection of small cetaceans, and to put the animals themselves in the core business of ASCOBANS, not the organisation itself. This attitude has been my motivation and mission throughout my activities within ASCOBANS, and in other international conservation management organisations and bodies. These activities became more intense when I was elected Chair of the AC to ASCOBANS in 1995, which lasted for two terms of 3 years each, and then became Vice-Chair for another two terms, stepping down in 2007 (Fig. 1.10).

With respect to the content of the Agreement, I am very positive of the progress we have made over the last two decades. Though still not ideal, the conservation status of porpoises would have been much worse if ASCOBANS had not existed. The number of Contracting Parties and interested Range States has increased, and, equally important, the range of coverage of the Agreement has also increased. Harbour porpoises have continued to be the core species of ASCOBANS, without neglecting the other small cetaceans occurring in the Agreement area. Considerable progress has been made in identifying and addressing the most imminent threats to small cetaceans. Addressing means, in this context, investigating possible solutions and mitigation measures. In particular, I would like to mention the issue of bycatch. At the start of the Agreement, the problem was recognised, but strangely enough when addressing the problem nearly every responsible authority would reply, 'it may occur

FIGURE 1.10 Peter Reijnders (on the right of the picture) with his co-Chair, Mark Tasker (on the left), and Rob Hepworth, ASCOBANS Executive Secretary in the centre. *Photo: ASCOBANS Secretariat.*

(Continued)

(Continued)

but on a very low scale in our waters'. So the importance of this issue was denied and ignored, primarily fed by the presumption that changing fishing practices would not be without economic consequences. However, when more and better statistics became available, with our Danish colleagues as pioneers, it gradually became clear that bycatch was a serious problem. A valuable contribution to the work in this field came from many scientists and organisations in and outside ASCOBANS. The involvement of AC scientists occurred through their respective Parties, and directly in, for example, ICES and Scientific Committees of the European Commission. It was a long process but finally the unsustainable bycatch of harbour porpoises in certain fisheries was addressed in an important European Commission Council Regulation (EC 812/2004). This is actually a good example of how I see ASCOBANS should operate. ASCOBANS should identify and signal existing and emerging threats to small cetaceans, but in addressing and providing solutions it needs to seek close cooperation with other organisations, including NGOs and responsible state management authorities (e.g. Nature Management, Fisheries, Environment) at the regional scale and the European level. To achieve this, I endeavoured from the beginning to obtain an optimal composition of the AC: a mix of scientists and administrators, each with a specific role, but making the combination more relevant and efficient, than the sum of each of them operating singly. Moreover, I consider cooperation and active participation of Intergovernmental Organisations and NGOs essential. It has proven to be of vital importance for the achievements of ASCOBANS.

Are there only successes to be noted? That question is really a rhetorical one. If everything we aimed at had been achieved, ASCOBANS would have become redundant by now. That is far from being the case. It is my observation that the initial progress has slowed down somewhat. Gradually over the last decade, the organisation of ASCOBANS has changed, and inherently that has affected its effectiveness. One essential aspect in my view is that the change from a stand-alone Secretariat to a CMS-based Secretariat did not really bring the benefits expected. The relations among and between the Parties became complicated and the decision-making processes become treacly and distant. Another critical point is the fact that the Parties have limited their efforts to enable relevant scientists to participate in AC activities. Also the relatively recent developments that more and more countries are using the EU Habitats Directive to set their level of conservation efforts (including towards marine mammals and their habitats) at a minimum level required by the European Commission is regrettable. This is a serious drawback and contrasts with the spirit of why ASCOBANS was created. The Parties have to realise that they own ASCOBANS. It is not someone else's idea or invention, and neither does it exist to serve, for example, CMS/UNEP.

The present challenge is to reverse this attitude of allocating a low priority to nature conservation activities. It is not a phenomenon unique to ASCOBANS; it is known in other conservation areas and referred to as the 'Backlash of Nature'.

I continue to believe in ASCOBANS and that its existence can easily be justified. It is encouraging to see that progress has been made with the implementation of the harbour porpoise plans: the 'Conservation Plan for Harbour Porpoises in the North Sea', the 'Recovery Plan for Harbour Porpoises in the Baltic Sea' (Jastarnia Plan) and the 'Conservation Plan for the Harbour Porpoise Population in the western Baltic, the Belt Sea and the Kattegat'. These offer ample background documentation and viable solutions for an adequate conservation management and recovery of at least harbour porpoises and bottlenose dolphins in the Agreement area. The bycatch issue is still one of the pressing problems to be solved. The ASCOBANS AC's involvement in the European Commission's efforts of revising EU-Regulation 812/2004 and the Data Collection Framework under the Common Fishery Policy, and the implementation of the Marine Strategy Framework Directive, should definitely be acknowledged.

Irrespective of these matters relating to how ASCOBANS functions, I have warm memories of working with many good colleagues and friends in the AC, and I especially appreciate the hard work and involvement of the IGOs and NGOs, whose input has been crucial to the successes obtained. I also have enjoyed working with the several persons running the Secretariat, and last but not least, running the AC and keeping in check the Meeting of Parties would not have been so easy to stand and so pleasant as well without my co-Chair, Mark Tasker.

Prof. Dr Ir. Peter Reijnders
Chair, Advisory Committee, 1995–2001,
Vice Chair, Advisory Committee, 2001–07

The 25th anniversary of ASCOBANS

On 17 March 1992, the ASCOBANS was opened for signing by Parties.

Germany was keen to cooperate with neighbouring countries and the CMS Secretariat to develop the Agreement and later on to sign and ratify it. Germany was and is committed to working on a healthier situation for small cetaceans, especially the harbour porpoise. This Agreement pinpoints the difficulties which these marine mammals face particularly in the North Sea and the Baltic, and has been welcomed to address their conservation needs.

Since entering into force, ASCOBANS has taken important decisions and recommendations for the protection of the small cetaceans, in particular the harbour porpoise (Fig. 1.11).

(Continued)

(Continued)

FIGURE 1.11 Harbour porpoises suffer mortality particularly from entanglement in bottom set gillnets. *Photo: Prof. Krzysztof Skóra Hel Marine Station, Department of Oceanography and Geography, University of Gdańsk.*

For many years, bycatch has been identified as the major factor to affect small cetaceans. Some steps have been taken to overcome this problem. Today we recognise that in many regions of the agreement, bycatch rates are still unacceptable. We need to continue to work on this.

Another critical factor is man-made underwater noise. The use of offshore wind energy has been growing and was a new challenge for Germany as the related pile-driving to establish wind energy plants causes significant underwater noise of major concern. In this context, ASCOBANS was helpful to ensure that due recognition was given to the damage to porpoises that loud sounds or general noise pollution can cause. The use of air bubble curtains was tested to diminish the noise peaks and this technique turned out to be helpful in mitigating those problems.

Important successes of ASCOBANS were the Jastarnia Plan for the Baltic Sea and a corresponding Plan for the North Sea. Now it is important, urgent and our duty to implement those plans nationally and within the EU. And despite all the obstacles and problems to achieve a better status for the cetaceans concerned, we should not lose our vision, hope and engagement.

Contracting Parties to ASCOBANS accepted the invitation of the German Government to host the ASCOBANS Secretariat in Bonn. The German Government is proud that this Secretariat is run in Bonn together with other Secretariats of the CMS family.

I seize the chance to thank all governmental and nongovernmental bodies, NGOs, scientific institutions, persons and all the others involved and engaged during the last quarter of a century for their cooperation with the aim of caring for whales and dolphins in the ASCOBANS area. The tasks of ASCOBANS have brought together friends, working enthusiastically in a good spirit for endangered small cetaceans.

Gerhard Adams
Member of German Delegation to ASCOBANS Head of the Division for Species Protection,
Federal Ministry for the Environment, Nature Conservation and Nuclear Safety

FIGURE 1.12 The future for the conservation of marine mammals lies with the future generation: young seawatchers observing porpoises in NW Wales. *Photo: Peter GH Evans.*

In these first few decades of the 21st century, international conservation agreements such as ASCOBANS assume ever greater importance as human-linked pressures upon the environment steadily grow. Inaction now will have consequences for future generations (Fig. 1.12). The human species is slow to respond to environmental threats, as witnessed by the continued destruction of rain forests, plunder of the seas' resources, and inexorable increase in global threats from climate change.

References

Addink, M., & Smeenk, C. S. (1999). The harbour porpoise *Phocoena phocoena* in Dutch coastal waters: Analysis of stranding records for the period 1920-1994. *Lutra, 41*, 55—80.

Andersen, S. H. (1982). Changes of occurrence of the harbour porpoise, *Phocoena phocena*, in Danish waters, as illustrated by catch statistics from 1834-1970. *FAO Fisheries Series (5) [Mammals in the Seas], 4*, 131—133.

ASCOBANS-ACCOBAMS Joint Working Group on Noise (2017). Report of the Joint ECS-ASCOBANS-ACCOBAMS Workshop on "Fostering inter-regional cooperation in underwater noise monitoring and impact assessment in waters around Europe, within the context of the European Marine Strategy Framework Directive", *the European Cetacean Society's 31st ECS Conference, Middelfart, Denmark, 29 April 2017*, ASCOBANS AC23/Inf. 5.1.1.b, 26 pp.

Benke, H., Siebert, U., Lick, R., Bandomir, B., & Weiss, R. (1998). The current status of harbour porpoises (*Phocoena phocoena*) in German waters. *Archive of Fishery and Marine Research, 46*(2), 97—123.

Berggren, P., & Arrhenius, F. (1995). Sightings of harbour porpoises (*Phocoena phocoena*) in Swedish waters before 1990. *Reports of the International Whaling Commission* (Special Issue 16), 99–108.

Casinos, A., & Vericad, J.-R. (1976). The cetaceans of the Spanish coasts: A survey. *Mammalia, 40,* 267–289.

Collet, A. & Duguy, R. (1987). French research on cetaceans. In J.W. Broekema & C. Smeenk (Eds.) *The European Cetacean Society Report of the 1987 meeting* (41–44), Hirtshals, Denmark, ECS Newsletter, 1.

De Smet, W. M. A. (1974). Inventaris van de walvisachtigen (Cetacea) van de Vlammse kust en de Schelde. *Bulletin van het Kominklijk Belgisch Instituut voor Natuurwetenschappen, 50*(1), 1–156.

De Smet, W. M. A. (1981). Gegevens over dewalvisachtigen (Cetacea) van de Vlaamse kust en de Schelde uit de periode 1969-1975. *Bulletin van het Kominklijk Belgisch Instituut voor Natuurwetenschappen, 54*(4), 1–34.

Evans, P. G. H. (1980). Cetaceans in British Waters. *Mammal Review, 10,* 1–52.

Evans, P. G. H. (1990). European cetaceans and seabirds in an oceanographic context. *Lutra, 33,* 95–125.

Evans, P. G. H. (1992). *Status Review of Cetaceans in British and Irish Waters.* London: UK Department of the Environment, 98 pp.

Evans, P.G.H. (Ed.) (2008a). Offshore wind farms and marine mammals: impacts and methodologies for assessing impacts. *Proceedings of the ASCOBANS/ECS Workshop, San Sebastián, Spain, 22 April 2007,* European Cetacean Society Special Publication Series *No. 49,* 1–68.

Evans, P.G.H. (Ed.) (2008b). Selection criteria for marine protected areas for cetaceans. *Proceedings of the ECS/ASCOBANS/ACCOBAMS Workshop, San Sebastián, Spain, 22 April 2007, European Cetacean Society Special Publication Series No. 48,* 1–104.

Evans, P.G.H. (Ed,) (2014). Chemical Pollution and Marine Mammals. *Proceedings of the ECS/ASCOBANS/ACCOBAMS Joint Workshop, held at the European Cetacean Society's 25th Annual Conference, Cadiz, Spain, 20 March 2011, ECS Special Publication Series No. 55,* 93 pp.

Evans, P.G.H. (Ed.) (2015). Introducing noise into the marine environment — what are the requirements for an impact assessment for marine mammals? *Proceedings of an ECS/ASCOBANS/ACCOBAMS Joint Workshop, the 28th Annual Conference of the European Cetacean Society, Liège, Belgium, 6 April 2014, ECS Special Publication Series No. 58,* 1–113.

Evans, P.G.H. & Simmonds, M. (2011). Report of the ECS/ASCOBANS/ACCOBAMS Workshop on Chemical Pollution and Marine Mammals. *20 March 2011, Cadiz, Spain. 18th ASCOBANS Advisory Committee Meeting, UN Campus, Bonn, Germany, 4–6 May 2011,* AC18/Doc.5-03 (S), 5 pp.

Evans, P.G.H. & Teilmann, J. (Eds.) (2009). *Report of ASCOBANS/HELCOM Small Cetacean Population Structure Workshop.* ASCOBANS/UNEP Secretariat, Bonn, Germany. 140 pp.

Evans, P. G. H., Harding, S., Tyler, G., & Hall, S. (1986). *Analysis of cetacean sightings in the British isles, 1958-1985.* Peterborough: Nature Conservancy Council, 71 pp.

Evans, P.G.H., Kinze, C.C., Kroger, R.H.H., & Smeenk, C. (1987) *Statement of Concern. The decline of the Harbour Porpoise in the North Sea.* European Cetacean Society. 5 pp.

Evans, P.G.H., Notarbartolo di Sciara, G., Hoyt, E., Tetley, M., Frisch, H., & Montiglio, C. (2013) Report of the Joint ECS/ASCOBANS/ACCOBAMS Workshop on the challenge of spatially managing cetaceans — a highly mobile animal group. *20th ASCOBANS Advisory Committee Meeting, Warsaw, Poland, 27–29 August 2013,* AC20/Doc.4.2.1 (O), 28 pp.

Evans, P.G.H., Siemensma, M., & Bräger, S. (2010) Report of the ASCOBANS/ECS Cetacean Bycatch Mitigation Workshop. *17th ASCOBANS Advisory Committee Meeting, Cornwall, United Kingdom, 21–23 April 2010,* AC17/Doc.4-07 (C), 8 pp.

Gaspari, S., Evans, P.G.H., & Montiglio, C. (2013) Report of the Joint ECS-ACCOBAMS–ASCOBANS Workshop on Cetacean Population Structure. *20th ASCOBANS Advisory Committee Meeting, Warsaw, Poland, 27–29 August 2013,* AC20/Doc.4.1.1 (O), 24 pp.

IJsseldijk, L.L. & Brownlow, A. (2018) Report of the Joint ECS-ASCOBANS Workshop on Cetacean Pathology: Necropsy Technique & Tissue Sampling. *30th Conference of the European Cetacean Society, Funchal, Madeira, 14–16 March 2016,* AC24/Inf.2.5.a, 27 pp.

Kayes, R. J. (1985). *The decline of porpoises and dolphins in the southern North Sea: a current status report.* Oxford: Political Ecology Research Group, 109 pp.

Kinze, C. C. (1987). Hvad ved vi om marsvinet? Okologisk status over Danmarks truede ynglehval. *Kaskelot, 75,* 14–23.

Kremer, H. (1987) *Untersuchungenzur Alterbestimmung am Schweinswal (Phocoena phocoena Linné 1758)* (MSc thesis, University of Kiel, 232 pp).

Kroger, R. H. H. (1986). *The decrease of harbour porpoise populations in the Baltic and North Sea.* Germany: University of Linköping, Sweden and University of Hamburg.

Lindstedt, I., & Lindstedt, M. (1988). Harbour porpoises deposited in Swedish museums in the years 1973-1986. *European Research on Cetaceans, 3,* 99–100.

Lindstedt, I., & Lindstedt, M. (1989). Incidental catch of harbour porpoises (*Phoceona phocoena*) in Swedish waters in the years 1973-1988. *European Research on Cetaceans, 3,* 96–98.

Smeenk, C. (1987). The harbour porpoise *Phocoena phocoena* (L., 1758) in the Netherlands: Stranding records and decline. *Lutra, 30,* 77–90.

Teixeira, A. M. (1979). Marine mammals of the Portuguese coast. *Sonderdruck aus Z.f. Saugetierkunde, 4,* 221–238.

Tetley, M. & Dolman, S. (2013) Towards a Conservation Strategy for White-beaked Dolphins in the Northeast Atlantic. *Report of the Joint ECS/ASCOBANS/WDC Workshop 20th ASCOBANS Advisory Committee Meeting, Warsaw, Poland, 27–29 August 2013,* AC20/Doc.4.1.c, 121 pp.

Chapter 2

Conservation agreements for the protection of whales, dolphins and porpoises

ASCOBANS is not the only piece of international legislation in the region that impacts upon cetaceans. The increased environmental awareness during the latter half of the 20th century spawned a number of treaties, conventions, directives and agreements.

European international agreements applicable to cetaceans

Habitats Directive

Probably the most important piece of conservation legislation in Europe has been the Council Directive on the Conservation of Natural Habitats and Wild Fauna and Flora [normally referred to as the European Union (EU) Habitats Directive], adopted in 1992 as the means by which the EU meets its obligations under the Bern Convention (1979). It is built around two pillars: a strict system for species protection and the Natura 2000 network of protected sites. All cetaceans occurring in Europe are listed in Annex IV of the Directive, requiring strict protection, and two species, the harbour porpoise and bottlenose dolphin, are additionally listed in Annex II, requiring the development of Special Areas of Conservation, where appropriate, as part of the Natura 2000 network.

Article 12(1) requires Member States to establish a system of strict protection, which includes prohibiting the deliberate capture or killing and disturbance of listed species. Additionally, Article 12(4) requires Member States to monitor incidental capture and killing to ensure that this does not have a significant negative impact on the species concerned.

Under Article 17 Member States must report on the status of each cetacean species in its waters every 6 years through the surveillance developed under Article 11. The first assessment of conservation status was undertaken in 2007, with the next report in 2013.

Marine Strategy Framework Directive

In June 2008 the EU Marine Strategy Framework Directive (MSFD) was adopted to provide 'a framework within which Member States shall take the necessary measures to achieve or maintain good environmental status within the marine environment by the year 2020 at the latest' [Article 1(1)]. It constitutes the environmental pillar of the EU's Integrated Maritime Policy. A 'good environmental status' involves the provision of 'ecologically diverse and dynamic oceans and seas which are clean, healthy and productive within their intrinsic conditions, and the use of the marine environment is at a level that is sustainable, thus safeguarding the potential for uses and activities by current and future generations' [Article 3(5)]. Member States are thus required to prepare national strategies to manage their seas, encompassing a clear assessment of their current environmental status and a targeted programme of measures to be introduced by 2016 at the latest [Article 5(1)], and to work together with one another at a wider regional level to achieve this [Article 5(2)]. It also places emphasis upon the use of existing regional structures.

In ascertaining the environmental status of seas within the European Community, a series of indicators and qualitative descriptors are established in the Annexes to the Directive. These include ocean noise, pollution, marine

European Whales, Dolphins, and Porpoises. DOI: https://doi.org/10.1016/B978-0-12-819053-1.00002-8

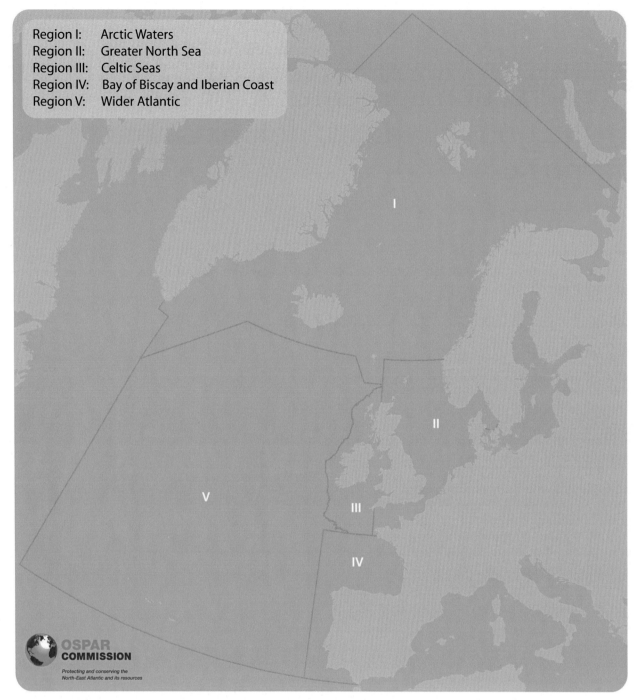

Region I: Arctic Waters
Region II: Greater North Sea
Region III: Celtic Seas
Region IV: Bay of Biscay and Iberian Coast
Region V: Wider Atlantic

FIGURE 2.1   Map of the region covered by OSPAR.

litter and shipping impacts. National assessments should examine essential features and characteristics of these areas, the predominant pressures upon them and their primary economic and social uses [Article 8(1)]. From this appraisal a series of environmental targets shall be identified (Article 10), as well as coordinated monitoring programmes for the ongoing assessment of these waters (Article 11). A detailed programme of measures is then to be developed [Article 13(1), including contributing to protected areas under the EU Nature Conservation Directives (Article 13(4))].

Advice to the European Commission is provided in part by ICES, the International Council for Exploration of the Sea, which meets annually, particularly focusing upon fisheries management. It has two working groups of relevance, one on fisheries bycatch (WGBYC, the working group on bycatch of protected species) and the other on marine mammals (WGMME, the working group on marine mammal ecology). Historically the latter has concentrated upon the management of seals, which are still exploited in some northern territories; however, in more recent years, cetaceans have been increasingly considered.

Two territorial conventions are also relevant

OSPAR (Oslo and Paris Convention)

The Convention for the Protection of the Marine Environment of the North-East Atlantic (open for signature in September 1992) entered into force in March 1998 (Fig. 2.1). It contains Annex V and Appendix 3 on the Protection and Conservation of the Ecosystem and Biological Diversity of the Maritime Area, and Criteria for Identifying Human Activities for the Purpose of Annex V, respectively. The OSPAR Strategy, in accordance with Annex V, promotes the selection and the establishment of a system of specific areas and sites that need to be protected, and the management of human activities in these areas and sites. Although OSPAR does not have competence over fisheries, it has prescribed a series of programmes of Ecological Quality Objectives (EcoQOs) including ones to reduce bycatch, and to tackle the problem of anthropogenic ocean noise. It has published two Quality Status Reports (OSPAR Commission, 2000, 2010), and a great number of shorter reports on specific topics of environmental concern.

Helsinki Convention (1974, 1992)

This was set up to protect the Baltic marine environment against all forms of pollution through the establishment of a system of coastal and marine Baltic Sea protected areas. In 1992 a new Convention was signed by all the countries bordering the Baltic Sea (Fig. 2.2) and by the European Economic Community, and incorporated marine biodiversity and conservation of habitats (Article 15). It entered into force in 2000. The governing body of the Convention is the Helsinki Commission — Baltic Marine Environment Protection Commission — also known as HELCOM. An environmental action programme aims to improve the quality of the Baltic environment with guidelines for offshore, protected areas in the Baltic Sea. It has produced major assessments of the environmental state of the Baltic (HELCOM, 2010, 2018), a review of its biodiversity (HELCOM, 2009) and several reports on specific environmental topics (e.g. hazardous substances).

ACCOBAMS

ASCOBANS has a sister Regional Agreement called ACCOBAMS, the Agreement on the Conservation of Cetaceans of the Black Sea, Mediterranean Sea and Contiguous Atlantic Area (Fig. 2.3). The final act was signed on 24 November 1996 in Monaco, and once seven Parties had ratified, it came into force on 1 June 2001. Its aims are similar to those of ASCOBANS, but with application to Southern Europe, being to reduce threats to cetaceans in Mediterranean and Black Sea waters and to improve our knowledge of these animals. Unlike ASCOBANS, however, this Agreement covers all cetaceans, and instead of having an Advisory Committee with meetings comprising scientific and administrative sessions, it has simply a Scientific Committee. Its Secretariat is based in the Principality of Monaco and operates relatively independently of the CMS Secretariat. There has always been some collaboration between the two Agreements, but this has become even more necessary since the Agreement Areas now overlap after Spain and Portugal decided to be a Party only to ACCOBAMS, for financial reasons, and in the case of Spain because ASCOBANS focuses only upon small cetaceans. The range states to ACCOBAMS are for the most part more economically challenged than those in the ASCOBANS Agreement Area. They also show a wider range of cultures, and several are outside the EU. This has made it more difficult for transnational collaboration, the extent of knowledge about Mediterranean cetaceans is weaker than in northern Europe, and tackling conservation pressures such as bycatch is even more challenging. Nevertheless, they are making great strides forward, with an initiative for a basin-wide abundance survey of cetaceans in the Mediterranean Sea in June—July 2018 (ACCOBAMS, 2017). The initial results have recently been made available (ACCOBAMS, 2018).

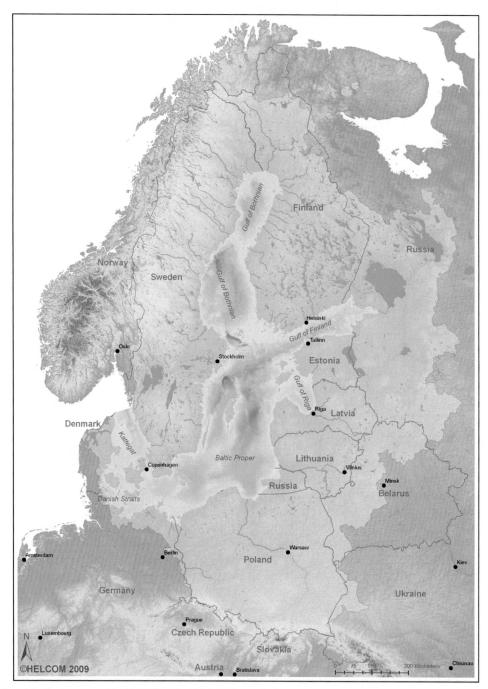

FIGURE 2.2 Map of the Region covered by HELCOM (the Helsinki Convention).

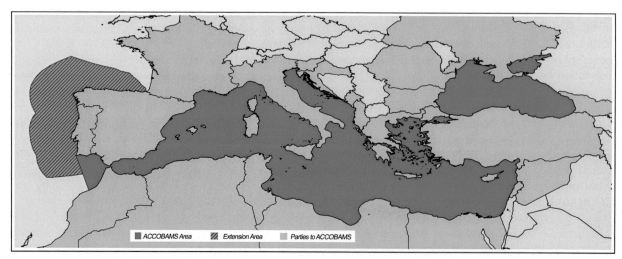

FIGURE 2.3 Map of agreement area of ACCOBAMS.

The relationship between ASCOBANS and its sister Agreement ACCOBAMS

The CMS Agreement for the Conservation of Cetaceans of the Black Sea, Mediterranean Sea and Contiguous Atlantic Area (ACCOBAMS) was created a couple of decades ago to implement activities and measures to address threats to cetaceans in the region in a cooperative fashion, in order to ensure their survival.

The collaboration with ASCOBANS started in the earliest year of the creation of ACCOBAMS. It addressed several issues of common interest since the two Agreements share common objectives for different connected geographical zones. This is clearly reflected in the similarities that exist between their respective Conservation Plans. In the beginning, the Agreements covered two neighbouring geographical areas. Since 2010, a geographical area came under the jurisdiction of the two sister Agreements, creating an interesting platform for consultation and cooperation — a base for exchanges for a better coordination emerges as a need to strengthen capabilities and knowledge (in order to reduce existing imbalances). Being entities of the CMS family, both Agreements have contributed jointly to develop concepts, activities and recommendations under the UNEP/CMS umbrella Convention. Furthermore, several scientists are active in both Agreements and contribute to the exchange of information and experience as well as to the harmonisation of the measures taken.

Even if the socioeconomic and cultural contexts of the two geographical areas differ, the migratory character of cetacean species brings together the international commitment of the countries.

The decision by the Parties of ACCOBAMS (November 2010) to extend the geographical scope of this Agreement to the marine waters of Spain and Portugal is expected to provide further opportunities for collaboration with ASCOBANS, strengthening the bridge for migratory species such as striped dolphin (Fig. 2.4) that can only benefit from a fertile collaboration through sharing experience and knowledge.

FIGURE 2.4 The striped dolphin is the commonest and most widely distributed species in the Mediterranean, but also occurs commonly in Atlantic waters around the Iberian Peninsula. *Photo: M. Reichelt/Mick E. Baines.*

(Continued)

(Continued)

In addition, for enhancing convergence and for optimising means and efforts deployed by both Secretariats, it appeared increasingly necessary to develop common activities. There are a wide range of conservation issues upon which ACCOBAMS and ASCOBANS share a common interest and where working together is of great mutual benefit. Among them, the following should be highlighted:

- The issue of anthropogenic noise on cetaceans where the collaboration was first materialised by the creation of guidelines addressed to governments for reducing the impact of noise, and secondly through the creation of a Joint Noise Working Group with ACCOBAMS, ASCOBANS and CMS in order to develop appropriate tools to assess the impact of anthropogenic noise on cetaceans, to further elaborate measures to mitigate such impacts and to coordinate efforts on this issue with other international bodies.
- The issue of unacceptable interactions and bycatch. The collaboration among ACCOBAMS and ASCOBANS helps in developing a number of practical conservation options, to assist the Parties of the two Agreements in meeting their conservation objectives, with respect to cetacean bycatch, by ensuring that bycatch management approaches are promulgated in working fisheries.
- Cetacean stranding issues present also clear opportunities for close collaboration and synergies. Necropsy protocols and best practice guidelines for stranding responses are being developed collaboratively by ACCOBAMS and ASCOBANS with the International Whaling Commission (IWC) and the European Cetacean Society. This initiative should lead to the development of principles and guidelines for handling live stranding events, including prevention, recognising the cultural, political and socioeconomic differences between countries.

Governance is part of the decision-making process and relies on political initiative and coordination of activities. The Mediterranean and the Atlantic regions are not lacking in tools for governance. However, their implementation articulation and complementarities need to be improved. In this spirit, all these items could be proposed to the relevant governments to be jointly implemented in the common area as key factors to achieve the CMS Convention objectives. This presupposes very close cooperation at the regional level among the states concerned. The contribution of their combined skills on scientific, legal and economic expertise is the *sine qua non*, for addressing marine and maritime governance issues.

Florence Descroix-Comanducci (Executive Secretary)
& Marie-Christine Grillo (Former Executive Secretary, ACCOBAMS, Monaco)

Barcelona Convention

Another important convention applying to the Mediterranean is the Barcelona Convention. In reviewing the international legal framework for marine mammal conservation in the Mediterranean, Scovazzi (2016) provides an excellent account of the Convention as it relates to cetaceans and much of the following is based upon that. Originally called the Convention for the Protection of the Mediterranean Sea against Pollution, it was opened for signature in 1976 in Barcelona. As a framework treaty, it was complemented by specific protocols. In subsequent years the Barcelona Convention and some of its protocols have been amended and new protocols adopted. The present Barcelona system includes eight instruments, most of which address issues relating to pollution. The Barcelona Convention itself was amended in 1995 to the Convention for the Protection of the Marine Environment and the Coastal Region of the Mediterranean. Of particular relevance to cetaceans has been the Protocol Concerning Specially Protected Areas and Biological Diversity in the Mediterranean (SPA Protocol; Barcelona, 1995). This applies to all the maritime waters of the Mediterranean, irrespective of their legal condition, to the seabed and its subsoil, and to the terrestrial coastal areas designated by each of the Parties. It also provides for the establishment of a List of Specially Protected Areas of Mediterranean Importance (SPAMI List). The emphasis here is on the inclusion of sites that are important for conserving biological diversity in the Mediterranean, ecosystems specific to the Mediterranean, and habitats of endangered species, or areas of special scientific, aesthetic, cultural or educational interest. One or more Parties may propose a SPAMI, and should then indicate the relevant protection and management measures, as well as how these would be implemented, taking into account the threats upon the site. If accepted into the list, then all Parties to the SPA Protocol are obliged to abide by it. Other States that are not a Party may also be invited to adopt appropriate measures, consistent with international law, to ensure that activities contrary to the principles and purposes of the Protocol do not take place.

The SPA Protocol has three annexes, which were adopted in 1996 in Monaco. These are the Common Criteria for the Choice of Protected Marine and Coastal Areas that Could be Included in the SPAMI List (Annex I), the List of Endangered or Threatened Species (Annex II) and the List of Species Whose Exploitation is Regulated (Annex III). In all, 19 species of marine mammals are listed in Annex II. Under Annex I the sites included in the SPAMI List must be 'provided with adequate legal status, protection measures and management methods and means', and must fulfil at least one of six general criteria (uniqueness, natural representativeness, diversity, naturalness, presence of habitats that are critical to endangered, threatened or endemic species, cultural representativeness).

The SPAMIs are required to have a management body, a management plan and a monitoring programme. However, only the Pelagos Sanctuary among the 34 SPAMIs so far established also covers areas beyond the limit of the territorial sea. In 1999 France, Italy and Monaco signed in Rome an Agreement for the establishment of a sanctuary for the protection of marine mammals, and this became known as the Pelagos Sanctuary. The Parties undertake to adopt measures to ensure a favourable state of conservation for every species of marine mammal within the sanctuary and to protect them and their habitat from negative impacts, both direct and indirect.

ASCOBANS and other international fora

'How successful has ASCOBANS been throughout the years?' This is a legitimate question, repeatedly asked, but very difficult to answer. On many occasions ASCOBANS has been criticised for only succeeding in presenting *soft law* to its Parties — nonlegally binding recommendations. One should, however, understand that most of the ASCOBANS Area is included within the geographical scope of other international fora, and also the subject ASCOBANS deals with is shared with other international conventions and with European Council Directives. Therefore an overlap exists between ASCOBANS activities and those undertaken in other international fora in which Parties of ASCOBANS participate or to which they are bound. Necessarily there is mutual influence, and action is taken at the most appropriate level. This level is not usually ASCOBANS, but the basis for action on small cetacean conservation can often be traced back to ASCOBANS, as this is the only framework with small cetaceans as its sole subject — and as such really digging deep into the problems that small cetaceans are facing.

ASCOBANS has been active on many fronts, and made recommendations on a number of issues related to small cetacean conservation: incidental catches in fisheries, disturbance from underwater noise, pollution, shipping, scientific research and public awareness.

Incidental catches

The limit of unacceptable anthropogenic removal of 1.7% of the population of harbour porpoises, as proposed jointly by ASCOBANS and the IWC, has found its way into the OSPAR Convention for the Protection of the Marine Environment of the North-East Atlantic. Within OSPAR it has become, at least for the North Sea, one of the EcoQOs for a healthy marine environment. This figure is also being used in practice as a concretisation of 'significant' in the 1992 Habitats Directive of the European Commission. Member States of the EU have the obligation to monitor the incidental capture and killing of the harbour porpoise, and they need to take further research or conservation measures to ensure that incidental capture and killing does not have a significant negative impact. Also in the European MSFD, a number of Member States of the EU use the figure of 1.7% as a limit for the bycatch rate to reach a 'Good Environmental Status' of European marine waters.

After the acceptance of a number of ASCOBANS resolutions on 'anthropogenic removal' (ASCOBANS has no mandate to deal with fisheries matters, and therefore is not supposed to deal with bycatch other than in an advisory capacity), the European Commission issued the 2004 Bycatch Regulation (EC/812/2004) in which measures were taken to better understand the level of bycatch in selected fisheries, to reduce the number of bycaught small cetaceans in selected fisheries, and to eliminate altogether certain types of fishing nets. Although not perfect, this Regulation provided a first concrete set of measures towards the reduction of bycatch. While not mentioned specifically in the preamble to the Bycatch Regulation, ASCOBANS undoubtedly influenced its development.

A step further than the Bycatch Regulation is the more detailed ASCOBANS Plans: The Recovery Plan for Baltic Harbour Porpoises (Jastarnia Plan), and the North Sea Conservation Plan for the Harbour Porpoise. These present a number of realistic actions and suggestions for measures which may find their way into national legislation or relevant international fora. These plans are not limited to bycatch; they also include proposals relevant to shipping, hydrocarbon extraction, discharges, construction and military activities and pay due attention to public awareness and research needs.

Underwater noise

ASCOBANS has been discussing underwater noise and its impact on small cetaceans for years. Certain activities, such as pile driving and seismic surveys, are known to potentially disturb cetaceans over large areas. Pile driving, necessary for constructing most types of wind farms, creates sound pressure levels so high that it can even cause physical injury to small cetaceans that are present at relatively close range. As the construction of offshore wind farms in the North Sea and the Channel — prime harbour porpoise habitat — is steadily increasing, it is likely that disturbance due to pile driving will remain an important subject on the agenda of ASCOBANS. ASCOBANS repeatedly issued recommendations concerning the reduction of excessive underwater noise, and during its Meeting of the Parties in 2009, a recommendation with detailed measures concerning pile driving and the disturbance of small cetaceans was adopted. These guidelines are now further developed, referred to in environmental impact assessments and in the permits for the construction of offshore wind farms. Also underwater noise is increasingly being

(Continued)

(Continued)

recognised as an important form of pollution, and maximum underwater (impulsive) noise levels to obtain a Good Environmental Status are being formulated in the framework of the MSFD. Seismic exploration has been on the agenda of ASCOBANS for many years, and several nations have now established guidelines – even if these are not taken up explicitly in legislation.

Public awareness

With all relevant partners around the table, governments, scientists and nongovernmental organisations, and with the ad hoc participation of the industry, of external experts and of other international bodies such as the European Commission, ASCOBANS is ideally placed to disseminate information to the public of the existence of small whales, dolphins and porpoises in the North Sea, the Baltic Sea and adjacent parts of the Northeast Atlantic Ocean, as well as the threats they face. Engaging the public and marine users is a key issue to ensure their conservation. ASCOBANS therefore has elaborated an education and public awareness plan.

Conclusion

The main ASCOBANS activities include those that should be dealt with jointly with other organisations and bodies, such as the European Commission, industry, nongovernmental organisations and other global and regional environmental agreements. Key to ASCOBANS is the bringing together of the expertise of both representatives of nations and of NGOs and IGOs – and discussing problems that small cetaceans are facing. So when answering the question about the success of ASCOBANS, one could reply with another question: is the forum in which decisions are taken really important if the aims of the Agreement are met?

Jan Haelters
Vice-Chair, Advisory Committee, 2007–10,
Head of Belgian delegation

Other international agreements applicable to cetaceans

International Whaling Commission

The most obvious piece of legislation applying globally specifically to cetaceans is the International Convention for the Regulation of Whaling. Its origins go back to 1931 when the League of Nations attempted to regulate the massive catches of whales that were occurring particularly in the Antarctic region. However, the convention was not ratified by some of the whaling nations (e.g. Germany and Japan). In 1937 another agreement, called the International Agreement for the Regulation of Whaling was signed in London by the whaling nations of Norway, United Kingdom and Germany (but not by Japan).

After a short respite during World War II, whaling resumed, involving several nations – Norway and the United Kingdom, followed by the Soviet Union, Japan, South Africa and The Netherlands. In December 1946 another attempt to regulate whaling activities internationally was made with the signing in Washington DC by representatives of 14 governments of the present day International Convention for the Regulation of Whaling, leading to the birth of the IWC. The preamble to the Convention was 'to provide for the proper conservation of whale stocks and thus make possible the orderly development of the whaling industry'. It incorporated the earlier agreements into its regulations, giving full protection worldwide to right whales and the gray whale, and to humpbacks in the Antarctic. It regulated the timing of the catching season and minimum sizes of individuals, as well as protecting certain areas in the Antarctic. However, it retained the concept of the Blue Whale Unit (BWU) which had been introduced in the 1930s largely to maximise oil production (i.e. to sustain the industry). This considered that one blue whale (Fig. 2.5) produced the same amount of oil as two fin, or two and a half humpbacks or six sei whales. Quotas set in the form of BWUs were expressed in biomass rather than numbers, took no account of the population status or biology of different whale species, and thus no account of the need to provide different levels of protection to different species. As a consequence these regulations did little to stop the further decline of some species whilst maintaining the industry on the more abundant ones. Furthermore, there were questions over the extent to which some nations observed these regulations. Blue whale catches went into sharp decline in the 1950s followed by fin whale catches in the early 1960s and sei whales in the late 1960s so that attention was turned to the smallest of the rorqual whales, the minke whale.

It was not until the 1972–73 season that the IWC abandoned the Blue Whale Unit in the Antarctic (it was never used in the North Pacific), and separate quotas were set for fin, sei, minke and sperm whales. In 1975 a new management procedure was introduced. It divided whale stocks into three categories: (1) Protection Stocks, which were more than 10% below the level giving the maximum sustainable yield (MSY), calculated on the basis of principles applied to the management of fish stocks; (2) Sustained Management Stocks, which were between 10% below and 20% above the

FIGURE 2.5 The concept of the Blue Whale Unit helped speed the dramatic decline in blue whale numbers in the Southern Ocean. *Photo: Peter GH Evans.*

level giving MSY; and (3) Initial Management Stocks, whose abundance exceeded 20% above the MSY level. No catching was allowed for Protection Stocks (estimated to have stock sizes <54% of the initial level); quotas were set for Sustained Management Stocks (estimated stock sizes 54%−72% of initial level) to keep them near the MSY level, and for Initial Management Stocks (>72% of initial level) to bring them gradually towards MSY levels. The concept of MSY is based on the idea that populations are regulated by density dependence. Each year, a surplus of animals is produced that can be harvested sustainably by humans indefinitely without causing its decline. Even if it were true, one of the main problems with this approach is that one never has full knowledge of population sizes, survival or reproductive rates. Populations are also affected by other pressures besides mortality from hunting, and their effects on demography are generally even more difficult to estimate. Small errors in estimates can have big effects, which were not necessarily observed even when catches appeared to be constant across years. The removal of particular individuals could also have important social effects.

In 1972 the UN Conference on the Human Environment held in Stockholm, Sweden, called for a complete moratorium on commercial whaling. Initially this was rejected by the IWC, but with increasing pressure of public opinion from several countries and greater scientific scrutiny over the decline of whale stocks, zero commercial catch-limits were finally agreed in 1982, and came into effect in 1986. In the meantime, a 'Comprehensive Assessment' was initiated − an in-depth evaluation of the status of all whale stocks in the light of management objectives and procedures. This included examination of current stock sizes, recent population trends, estimates of carrying capacity and productivity. The wording of the moratorium decision implied that with improved scientific knowledge in the future, it might be possible to set catch limits other than zero for certain stocks. Towards this objective, the IWC Scientific Committee spent 8 years developing a Revised Management Procedure (RMP). This was to be a scientifically robust method of setting safe catch limits for certain whale stocks, and was adopted by the Commission in 1994. With the development of computer modelling, it was possible to incorporate the large amounts of uncertainty that existed in the data used to assess the effects of varying catch limits. The objectives were that (1) catch limits should be as stable as possible; (2) catches should not be allowed on stocks below 54% of the estimated carrying capacity (the maximum number of whales that the environment can support); and (3) the highest possible continuing catch should be obtained from the stock. Computer simulations were used to test the performance of different approaches against various assumptions and uncertainties, before settling on one, known as the Catch Limit Algorithm (CLA), which it was felt best met the IWC's management objectives. This requires abundance estimates at regular (6-year) intervals along with the confidence limits (statistical uncertainty) around those estimates, and estimates of past catches (also allowing for the uncertainty in historic records) and the numbers of present catches.

The CLA does not allow catches from a stock estimated to be below 54% of its estimated preexploited population. The level of maximum productivity for whales is thought to be around 60%. Catches are set so that the population will stabilise at around 72% of its preexploitation level. This approach works as a feedback procedure: at each subsequent review of new abundance and catch information, the CLA learns more about the likely true status of the stock and narrows the range of potential catch limits. An implementation process was then developed to allow the CLA to be applied to particular whale species in specific regions, aiming for consideration of separate stocks where possible.

Although this approach is a significant improvement over past management procedures and could usefully also be applied in fisheries management, there remain a number of major challenges that can be difficult to overcome. First of all, rarely do we have adequate knowledge of how populations may be separated demographically. Then, our estimates of abundance often are of low precision, and uncertainty in their accuracy is created by the difficulty of fully accounting for the various inherent biases that exist in surveying cetacean abundance. In Chapter 6, Conservation research, I explore the various ways in which statisticians have attempted to deal with these issues.

Up to now the RMP has not been used in the regulation of whaling, so the moratorium on commercial whaling has remained in place. Nonetheless, the IWC regulations allow member governments to formally object to any decision reached, and thus to issue special permits to take whales for research purposes through what has been termed 'scientific whaling'. In the North Atlantic, Norway and Iceland have continued to take whales under objection, and Japan does so in the North Pacific and Antarctic. Also allowed are whales caught by indigenous communities, generally in remote regions such as Greenland, northern Alaska, and the far east of Russia, to provide for local needs. This is commonly termed aboriginal subsistence whaling, and catch limits are applied based upon the principles of the RMP. Since it is acknowledged that some animals may be struck but then lost, a Strike Limit Algorithm (SLA) rather than CLA was adopted in 2002 to the Bering–Chukchi–Beaufort Seas stock of bowhead whales, and in 2004 to gray whales in the North Pacific. There is ongoing research on fin and minke whales in Greenland, particularly addressing aspects of stock identity, but in the meantime, preliminary annual strike limits have been set.

One of the long-standing problems that the IWC has faced is the different attitudes towards whaling that different countries around the world currently possess. Some would like no whales to be intentionally killed anywhere; others believe they should be able to harvest whales as a sustainable resource (Fig. 2.6), and there are views in between. This impasse

FIGURE 2.6 Whaling for fin whales in Iceland, September 2018. *Photo: Sea Shepherd United Kingdom.*

has not really been overcome. In 1992 a separate body, the North Atlantic Marine Mammal Commission (NAMMCO), was formed for 'regional cooperation on conservation, management and study of marine mammals in the North Atlantic'. Its current members include Norway, Iceland, Greenland and the Faroe Islands. The organisation came about because the member nations were dissatisfied with the international management of cetaceans and other marine mammals by the IWC. They believe that decisions regarding whaling should be based on scientific principles regarding management of whale populations, taking into account both the complexity and vulnerability of the marine ecosystem, and the rights and needs of coastal communities to make a sustainable living from what the sea can provide. And at the time of writing Japan has announced that it will resume commercial whaling in its territorial seas in July 2019, and withdraw from the IWC.

ASCOBANS and IWC

I have been involved with ASCOBANS for two decades, attending the fourth Advisory Committee meeting and the Second Meeting of Parties in 1997. I was particularly pleased to foster the beginning of the development of close cooperation between ASCOBANS and the IWC, with a focus on scientific and conservation matters. I had the privilege and pleasure to work particularly closely with a Chair of the Advisory Committee, Dr Peter Reijnders on two key issues: the bycatch of harbour porpoises and chemical pollution. With respect to the former, of particular importance was the progress made towards setting conservation objectives for ASCOBANS and agreeing the need for urgent action to be taken when the estimated bycatch exceeded a percentage of current population size rather than waiting for evidence that a population had seriously declined.

It has to be said that on some topics the initial optimism has not always been matched by action and measurable improvements – that is the way for most intergovernmental bodies. However, the good progress that has been made (e.g. in our knowledge of cetaceans in the agreement area) should not be forgotten, and the vital role of ASCOBANS to provide a forum for discussion of the regional issues facing cetacean conservation remains as valid now as it was then.

It is important that as ASCOBANS goes forward it continues to learn the lessons from the past. Some obvious matters include strengthening the scientific basis for informed conservation actions and the monitoring of the success or otherwise of those actions; improving and clarifying the relationship between ASCOBANS and the EU on matters of cetacean conservation; improved communications *within* member governments to facilitate coordinated actions – on bycatch, for example the positions of environment and fisheries ministries or departments can sometimes be very different; the need for member governments to take ownership of implementing ASCOBANS resolutions by bringing them to the attention of other relevant intergovernmental or national bodies; the importance of working with all affected stakeholders especially those who may perceive their livelihoods are at risk. Many of these challenges are not unique to ASCOBANS and progress in ASCOBANS on them may provide a model for progress elsewhere.

Increasingly, we recognise the integral links between healthy ecosystems and healthy components of those including cetaceans. This requires increased collaboration at all levels from intergovernmental organisations through governments to industry and nongovernmental organisations and the general public – no single organisation can ensure healthy cetacean populations on its own – but ASCOBANS has an important role to play now and in the future.

Greg Donovan
Head of Science, International Whaling Commission

The US Marine Mammal Protection Act

Following concerns for the impact that a variety of human activities was having upon marine mammal populations, the United States Congress introduced a Marine Mammal Protection Act (MMPA) in 1972. It was one of the first pieces of legislation anywhere in the world to call specifically for an ecosystem approach to wildlife management. Its primary objective was to maintain the health and stability of the marine ecosystem and, when consistent with that objective, to obtain and maintain optimum sustainable populations of marine mammals. It prohibits the 'taking' of marine mammals, and enacts a moratorium on the import, export and sale of any marine mammal, along with any marine mammal part or product within the United States. The Act defines 'take' as 'the act of hunting, killing, capture, and/or harassment of any marine mammal, or, the attempt at such'. The MMPA defines harassment as 'any act of pursuit, torment or annoyance which has the potential to either (a) injure a marine mammal in the wild or (b) disturb a marine mammal by causing disruption of behavioural patterns, which includes, but is not limited to, migration, breathing, nursing, breeding, feeding, or sheltering'.

Under the Marine Mammal Protection Act, authority over conservation of marine mammals is shared between the Secretary of the Interior (through the US Fish and Wildlife Service) and the Secretary of Commerce (through the National Oceanic and Atmospheric Administration, NOAA). The US Fish and Wildlife Service has jurisdiction for the import and export of all marine mammals listed under CITES (the Convention on International Trade in Endangered Species of Wild Fauna and Flora). The National Oceanic and Atmospheric Administration's National Marine Fisheries Service (NMFS) has responsibility for the conservation and management of cetaceans and all pinnipeds other than walruses (which, along with sea and marine otters, polar bears, dugongs and manatees are the responsibility of the US Fish and Wildlife Service). Permits may be issued for scientific research, public display and the importation/exportation of marine mammal parts and products upon determination that the issuance is consistent with the MMPA's regulations. The two types of permits issued by the NMFS's Office of Protected Resources are incidental and directed. Incidental permits, which allow for some unintentional taking of small numbers of marine mammal, are granted to US citizens who engage in a specified activity other than commercial fishing in a specified geographic area. Directed permits are required for any proposed marine mammal scientific research activity that involves taking marine mammals. The US Marine Mammal Commission was set up as an independent government agency by the MMPA to further the conservation of marine mammals and their environment. Their stated aim is to work to ensure that marine mammal populations are restored and maintained as functioning elements of healthy marine ecosystems in the world's oceans, providing science-based oversight of domestic and international policies and actions of the federal agencies with mandates to address human impacts on marine mammals and their ecosystems.

Of particular relevance to the regulation of environmental impacts of ocean noise is the US Bureau of Ocean Energy Management (BOEM), within the US Department of the Interior. It is responsible for stewardship of US offshore energy and mineral resources, as well as protecting the environment that the development of those resources may impact. For more than 30 years this mandate has included studying, regulating and mitigating against the effects of industry-produced noise on marine life. BOEM also has a leadership role in the International Offshore Petroleum Environmental Regulators Marine Sound Working Group.

Sometime ago it was recognised that certain species and stocks of marine mammals were, or could be, in danger of extinction or depletion as a result of human activities, and that such species and stocks should not be permitted to diminish below the point at which they cease to be a significant functioning element in the ecosystem of which they were a part. In this context, a separate Endangered Species Act was passed in 1973 to prevent extinction, recover imperilled plants and animals, and protect the ecosystems on which they depend. This Act is administered by the same two agencies, the US Fish and Wildlife Service and the NMFS, the latter taking responsibility for the protection of cetaceans deemed to be endangered.

An important outcome of the MMPA has been government funded scientific research and monitoring, with regular Stock Assessment Reports providing updated information on stock sizes, trends and human-induced mortality for the populations of marine mammal species occurring in US waters.

The changing emphasis from MSYs as promoted within the IWC to an optimal sustainable population (OSP) as a goal for marine mammal populations represented a significant shift to a principal conservation objective. Recognising that 'takes' can occur incidentally (e.g. from fisheries bycatch), the Act was amended in 1988 and 1994, resulting in the key focus becoming the estimation of Potential Biological Removals (PBRs). The MMPA defines PBRs as 'the maximum number of animals, not including natural mortalities, that may be removed from a marine mammal stock while allowing that stock to reach or maintain its optimum sustainable population'. It is designed to ensure that stocks that are at or above OSP remain so, and that those below OSP increase to it within a reasonable amount of time. When human-caused 'take' exceeds PBR, then plans must be developed to reduce it below that level. The regular Stock Assessment Reports aim to determine whether or not stocks have exceeded PBR. As with the challenges that IWC have faced, this is not straightforward and often it has not been possible to robustly determine stock size trends.

The United States is a significant trading partner with a number of countries. One element of the MMPA that has an important potential consequence for conservation is that it requires the Secretary of the US Treasury to 'ban the importation of commercial fish or products from fish which have been caught with commercial fishing technology which results in the incidental kill or incidental serious injury of ocean mammals in excess of United States standards'. This clause has been used, for example to embargo yellowfin tuna catches from countries whose fleets caught large numbers of dolphins in the Eastern Tropical Pacific, and in 2017 a draft list was drawn up of foreign fisheries that export fish and fish products to the United States, classifying each according to their known frequency of interaction with marine mammals.

Under the Act Take Reduction Plans have been implemented to reduce marine mammal bycatch, including teams established to free large whales from fishing gear, and working with fishers to find technological and other solutions to monitor and mitigate this widespread threat. A goal has been set for zero mortality from bycatch, and towards that, the aim has been to attain insignificant levels equivalent to a threshold of 10% of the PBR level for any whale population. Fisheries experiencing mortality exceeding 1% of the PBR level must comply with any take reduction plan regulations applicable, including the placement of on-board fisheries observers if required.

The endangered North Atlantic right whale (Fig. 2.7) suffers serious mortality from ship strikes, and to minimise the overlap between whales and ships, the United States has moved and narrowed the shipping lanes crossing the Stellwagen National Marine Sanctuary, and established a seasonal 'Area To Be Avoided' in the Great South Channel

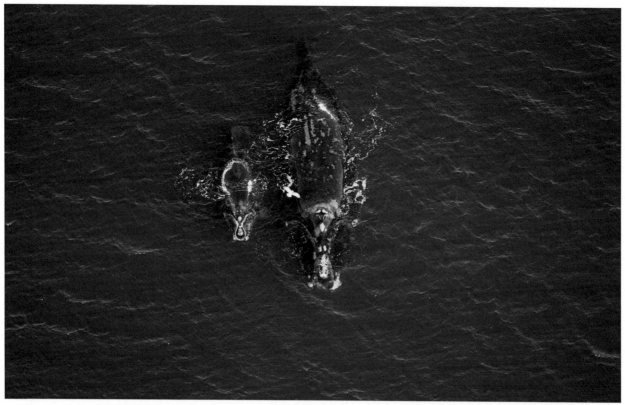

FIGURE 2.7 Northern right whale mother and calf. This most endangered species of whale in US waters has been the focus of much conservation action to reduce mortality from ship strikes and entanglement in fishing gear. *Photo: New England Aquarium, courtesy of Scott Kraus.*

southeast of Nantucket, Massachusetts, a critical spring foraging ground for the species. As a further conservation measure, a Ship Strike Rule was introduced by NMFS in 2008 imposing speed restrictions of no more than 10 knots for all vessels 65 ft. or longer in certain locations and at certain times of the year along the east coast of the US Atlantic seaboard. Voluntary speed restrictions around concentrations of right whales are also employed dynamically through email alerts to mariners, whilst commercial vessels of more than 300 gross tons are required to report to a shore-based station when they enter certain critical areas for right whales.

These are just some of the conservation actions that have been implemented as a result of the MMPA, which together have made a significant impact on the maintaining and restoring of marine mammal populations in US waters (Baur, Bean, & Gosliner, 1999; Roman et al., 2013). It was concluded that although information on population trends was unknown for most (71%) US marine mammal stocks, more stocks with known trends were improving than declining, with 19% increasing, 5% stable and 5% decreasing, and this situation compared favourably to that observed in other countries (Roman et al., 2013). Despite those successes, bycatch continues to be a problem for a number of marine mammal populations, not least for the North Atlantic right whale which experienced mortality of 3% of its entire population, with 17 confirmed mortalities in 2017 on the east coast of the United States and Canada (NOAA Fisheries, 2018).

An excellent recent review of the positive role the MMPA has played in marine mammal conservation has been published by IFAW (2017), following concerns that the US administration might weaken the Act for economic interests. I have drawn examples from that document in the above account. Two possible reasons why the MMPA has been relatively effective compared with legislation towards cetacean conservation in Europe are the greater resources that the United States have available to spend on these issues, and the fact that structurally it is rather more complex for unified action to take place across countries that differ culturally and socioeconomically, not to mention having separate political rulers.

Conservation legislation in Canada

Conservation legislation pertaining specifically to marine mammals was introduced relatively recently in Canada. For a long time, they have been largely regarded as a resource to be exploited. In the Fisheries Act of 1985, marine mammals were included in the definition of 'fish'. And until recently Fisheries and Oceans Canada, the administrative body for ocean and ocean resource issues, was focused primarily on conserving diverse fish stocks for industry (Campbell & Thomas, 2001). 'Marine Mammal Regulations', promulgated under the Fisheries Act in 1993, were designed to regulate the commercial seal and aboriginal seal and whale harvests. The regulations did not include requirements, or criteria, for determining at what levels harvest is sustainable in the long-term. Subsequent amendments have started to incorporate direct conservation measures. Currently, they apply to the following:

1. the management and control of fishing for marine mammals and related activities in Canada and in Canadian fisheries waters;
2. the management and control of fishing for marine mammals from Canadian fishing vessels in the Antarctic and
3. the conservation and protection of marine mammals in Canada and in Canadian fisheries waters.

As one can see, the emphasis on 'fishing' still remains. In this context licences may be issued for the direct exploitation of bowhead whales, belugas (Fig. 2.8) and narwhals in the aboriginal whale fisheries by Inuit in the Canadian arctic. These operate outside the remit of the IWC, Canada having left it in 1982 following disagreements over aboriginal whaling.

The federal government is responsible for protecting marine mammals. The responsibility to reduce threats from human activity is shared among several organisations:

1. Fisheries and Oceans Canada (DFO) is the lead organisation mandated to protect marine mammals in waters under the jurisdiction of Canada, according to the Fisheries Act and its Marine Mammal Regulations. It is responsible for developing and implementing policies and programmes to support Canada's economic, ecological and scientific interests in marine waters. It is also responsible for managing aquatic species listed under the Species at Risk Act (SARA) (2002) and has a leadership role in ocean management, as outlined in the Oceans Act (1996), which includes establishing marine protected areas. The Canadian Coast Guard, as a special operating agency of DFO, is responsible for providing aids to navigation, marine communications, marine traffic-management services and incident command during oil spills of unknown origin at sea.

FIGURE 2.8 Belugas in the St. Lawrence River Estuary, Canada. Aboriginal whaling is practised by the Inuit for beluga, narwhal and bowhead whale. *Photo: Caroline Weir.*

2. Environment and Climate Change Canada is responsible for listing aquatic and terrestrial species under the SARA, protecting terrestrial species at risk under the Act and administering the Act. It establishes marine and terrestrial national wildlife areas, under the Canada Wildlife Act, for wildlife interpretation, conservation and research. It provides scientific advice to responders and agencies, at the federal, provincial, territorial and municipal levels, on how best to reduce the environmental effects of spills of hazardous substances, such as oil.

3. Parks Canada is the agency responsible for establishing national marine conservation areas under the Canada National Marine Conservation Areas Act (2002), with the aim of protecting and conserving them for the benefit, education and enjoyment of the public. It is also responsible for establishing national parks, under the Canada National Parks Act (2000), which can include marine components, as well as the Saguenay–St. Lawrence Marine Park, established under its own separate legislation. Under the SARA, it is responsible for individuals of aquatic species on the federal lands that it administers.

4. Transport Canada is responsible for the government's transportation policies, plans and programs. The department develops policy and regulations for domestic marine vessels, with guidance from the International Maritime Organisation. It is mandated to protect the marine environment in navigation and shipping under the Canada Shipping Act (2001) and the Canada Marine Act (1998). It also leads Canada's National Oil Spill Preparedness and Response Regime.

An independent audit of Canada's progress in protecting marine mammals in its waters was undertaken in 2018 (Office of the Auditor General of Canada, 2018). It found that the government had done little to protect marine mammals from the risks of marine traffic and commercial fishing until recently. The four bodies had not fully applied existing policies and tools to proactively manage threats to marine mammals from commercial fishing and marine vessels, such as entanglements, bycatch, depletion of food sources, noise and disturbance, oil spills and collisions with marine vessels. These policies and tools included recovery strategies and action plans under the SARA, guidelines for planning and managing marine protected areas and integrated fisheries management plans to implement the Policy on Managing Bycatch. Part of the problem was the lack of national guidance and resources for Fisheries and Oceans Canada to effectively support the partners working to respond to incidents like gear entanglements.

The audit also found that marine protected areas (accounting for 3% of territorial seas) did not significantly contribute to the protection of marine mammals. Federal organisations did not have any criteria or guidance for considering the specific needs of marine mammals (such as large migratory ranges) when planning and managing marine protected areas. As a result, measures to protect marine mammals were not consistently considered when marine protected areas were being established.

Since 2017, when 12 endangered North Atlantic right whales were found dead in Canadian waters, federal organisations have begun to implement a variety of measures and actions aimed at protecting marine mammals by reducing threats from commercial fishing and marine vessels. They have put in place measures to reduce entanglements and bycatch from fisheries for the North Atlantic right whale; to introduce speed limits in some areas to protect both the North Atlantic right whale and the St. Lawrence Estuary beluga whale; and to limit commercial salmon fishing with the aim of increasing food sources for the southern resident killer whale. Although organisations have recently reacted to these threats for three endangered species, they have yet to apply sustained planning and management policies, tools and measures to reduce threats for all marine mammals.

Conservation legislation in Australia and New Zealand

In contrast to Canada where aboriginal whaling continues to be important for indigenous communities in Arctic regions, since the 1980s within the IWC, Australia and New Zealand have been strong voices against whaling. However, it was not long past that they themselves had been whaling. In Australia commercial whaling ceased in 1978 with the closure of Australia's last whaling station, the Cheynes Beach Whaling Company, in Western Australia. It was a relatively recent activity anyway with whaling in Australian waters only starting in 1791, although it became a major maritime industry during the 19th and early 20th centuries, initially targeting sperm and right whales and, later, particularly humpback whales. Whaling in New Zealand ended rather earlier, for economic reasons, in 1964. Nineteenth-century whaling in New Zealand was based on the southern right whale, and 20th-century whaling on the humpback whale.

During the mid-20th century both countries awakened to the serious threats to their endemic wildlife posed by introduced predators, and thus started strong conservation movements. Although focused upon the terrestrial environment, national legislation to protect marine mammals started soon after whaling ceased. The Marine Mammal Protection Act of New Zealand came into force in 1978 and the Whale Protection Act of Australia followed soon after, in 1980. Both prohibited their citizens from taking whales from any waters, even within the coastal zone of another state which permits whaling.

Those Acts initially focused upon protecting marine mammals from hunting. Then, during the 1990s, with a greater awareness of other environmental issues, more general legislation was introduced. Under the Environmental Protection and Biodiversity Act (EPSC) (1999), all cetaceans are protected in Australian waters:

- The Australian Whale Sanctuary includes all Commonwealth waters from the three nautical mile limit for state waters out to the boundary of the Exclusive Economic Zone (i.e. out to 200 nautical miles and further in some places).
- Within the Sanctuary it is an offence to kill, injure or interfere with a cetacean. Severe penalties apply to anyone convicted of such offences.
- Al states and territories also protect whales and dolphins within their waters.

Five whale species are currently listed under the EPSC Act as nationally threatened: the endangered blue whale and southern right whale, and the vulnerable sei, fin and humpback whales. Recovery plans identified whaling and habitat degradation as key threats, and established objectives and actions to ensure the ongoing recovery of the species. The recovery plans for these five species were reviewed in 2010, a recovery plan for the southern right whale was published in 2012 and a blue whale management plan was published in 2015.

Marine mammal protection in Australia is currently under the jurisdiction of the Australian government's Department of the Environment and Energy. In New Zealand it is the government's Department of Conservation (DOC) that is responsible for marine mammal protection.

In New Zealand the Marine Mammals Protection Act of 1978 was supplemented in 1992 by Marine Mammals Protection Regulations. The purpose of these regulations was to make provision for the protection, conservation and management of marine mammals and, in particular:

- To regulate human contact or behaviour with marine mammals either by commercial operators or other persons, in order to prevent adverse effects on and interference with marine mammals.
- To prescribe appropriate behaviour by commercial operators and other persons seeking to come into contact with marine mammals.

These regulations reflected concerns over growing evidence of disturbance of animals from the rapidly developing whale and dolphin watching industry and involved the introduction of a carefully regulated permitting scheme (Fig. 2.9).

FIGURE 2.9 A significant whale and dolphin watching industry has developed at Kaikoura, South Island, New Zealand, including swim-with-dolphin programmes. *Photo: Peter GH Evans.*

Two other Acts relate to marine mammal management in New Zealand, the Conservation Act (1987) and the Fisheries Act (1996), whilst also relevant are the New Zealand Marine Mammal Action Plan for 2005−10 produced by its DOC (Suisted & Neale, 2004) and the New Zealand Biodiversity Strategy (2000, with a new edition in 2016).

The Marine Mammal Action Plan served to underpin New Zealand's conservation legislation and policy, providing specific outputs with regard to the conservation of marine mammals that the DOC could systematically work to achieve. As such it aimed to be a significant guiding document, primarily for internal DOC use, to:

- Identify priorities for conservation management, according to species, issues and systems and administrative structures;
- Recommend and prioritise conservation actions to remedy the causative factors;
- Recommend priority scientific needs;
- Identify causative factors limiting recovery, or driving decline (including areas of conflict between marine mammals and human populations); and
- Recommend priority information collection areas.

The work of the DOC focused towards the following two broad aims:

1. Species protection.
   a. To actively protect marine mammal species and populations, and allow the recovery of those that are threatened with extinction or that have been depleted or otherwise adversely affected by human activities or unusual natural events;
   b. To build understanding of the main biological parameters for all marine mammals, and especially species threatened or affected by past or present human activities;
   c. To protect key sites in New Zealand waters that are of significance to marine mammals;
   d. To maintain and restore the distribution, abundance and diversity of marine mammals in New Zealand waters and beyond; and
   e. To achieve self-sustaining populations of all marine mammals throughout their natural range, and avoid extinctions of all marine mammal populations.
2. Management of human interactions and use.
   a. To manage human interactions with marine mammals in order to minimise adverse effects on their survival, welfare and recovery, and to ensure the appropriate management of both living and dead marine mammals;
   b. To identify and assess all significant threats to marine mammals (in general and as species, populations and individuals);
   c. To address and mitigate human-related threats to the welfare of marine mammals and the viability of their populations and habitats, and to progressively work towards eliminating human-related mortalities of marine mammals;
   d. To manage dead and distressed marine mammals, and the holding and taking of marine mammals (including body parts) and
   e. To address risks and uncertainty when making decisions and to ensure a precautionary approach is taken.

The New Zealand Biodiversity Strategy states the following desired outcomes for coastal and marine biodiversity for the year 2020:

- New Zealand's natural marine habitats and ecosystems are maintained in a healthy functioning state. Degraded marine habitats are recovering. A full range of marine habitats and ecosystems representative of New Zealand's indigenous marine biodiversity is protected.
- No human-induced extinctions of marine species within New Zealand's marine environment have occurred. Rare or threatened marine species are adequately protected from harvesting and other human threats, enabling them to recover.
- Marine biodiversity is appreciated, and any harvesting or marine development is done in an informed, controlled and ecologically sustainable manner.
- No new undesirable introduced species are established, and threats to indigenous biodiversity from established exotic organisms are being reduced and controlled.

Harcourt et al. (2015) have recently reviewed the present conservation status of marine mammals in Australia and New Zealand, emphasising the role both countries have played in foreign policy against whaling, not only within the IWC but also in the Southern Ocean through CCAMLR (the Commission for the Conservation of Antarctic Marine Living Resources). Whilst acknowledging the strong government and public focus on marine mammal conservation,

FIGURE 2.10 The South Island population of Hector's dolphin in New Zealand numbers around 10,000 individuals and supports an economically important wildlife watching industry in localities such as the Banks Peninsula. *Photo: Peter GH Evans.*

they note that species remain vulnerable to a number of threats including fisheries interactions, vessel disturbance, coastal and offshore development and climate change, and draw attention to the need for better understanding of their effects both individually and cumulatively.

One of the most threatened cetacean species, endemic to New Zealand waters, is the Hector's dolphin (Fig. 2.10), particularly its subspecies, the Māui's dolphin restricted to North Island and numbering little more than 50 individuals. Bycatch from commercial and recreational fishing is the main identified cause of death of Hector's and Māui's dolphins. Between 1921 and 2015, entanglement in fishing gear accounted for 71% of the 301 Hector's and Māui's dolphin deaths for which a cause of death was determined (New Zealand Department of Conservation, 2016). Of the 174 deaths where the type of fishing gear was recorded, 86% were caused by commercial and recreational set nets.

Bycatch of Hector's and Māui dolphins has decreased over the last decade. However, limitations to the data mean that it has not been possible to estimate the size of this decrease. In addition, numbers of Māui dolphins are now so low that the probability of a dolphin being caught has also diminished.

The need for evidence-based consensus at the broadest level

Meetings and workshops of the various conventions and agreements, advisory councils and commissions tend to involve many of the same people — as government representatives, advisors and NGO participants. Marine mammal science and conservation in Europe is a relatively small world. Frequently, the different fora discuss the same issues, and may even produce reports or reviews that largely overlap. Although there will be some differences particularly when the timing is different, one report then slightly updating the others, it does not seem to be the best use of human or financial resources. Although those fora can represent large international bodies, there are generally only a few individuals actually involved in discussions and report writing from each one, so the outputs may in fact have different emphases reflecting their personal perspectives or experience. That may not be a bad thing, but what might be more cost-effective would be for closer collaboration across the wider community of interested parties.[1] This applies not only to the collation and interpretation of scientific information but also in relation to management issues. At present, for example discussion of mitigation measures to reduce fisheries conflicts only marginally involves those directly engaged in fishing, and the same often applies to other human activities that may impact upon cetaceans.

It is human nature for each of us to imagine that the fora in which we have participated are the ones that have proved significant in executing change in a positive manner. In reality everyone plays their part by disseminating and

---

1. ASCOBANS and ACCOBAMS are already moving in this direction not only with jointly supported scientific workshops but also the establishment of joint working groups on noise and bycatch.

reinforcing information that can form the evidence base behind conservation action. Of course it is important that recommendations are sound, and in this modern world where views are formed on the basis of communications through social and other media, there is a danger that actions are taken that may be inappropriate, wasting resources and at times even proving counterproductive.

Fora such as the annual meetings of the Scientific Committee of the IWC (typically attended by 150–200 persons), the biennial conferences and workshops held by the Society of Marine Mammalogy (typically attended by 1000–3000 persons), and in Europe, the annual conferences and workshops held by the European Cetacean Society (typically attended by 400–500 persons), can serve as useful opportunities for interested parties to debate information on marine mammals, often in advance of it reaching a wider audience through peer-reviewed publications.

Notwithstanding the value of such routes for discussion and dissemination of information on cetaceans and the threats they face, there is no guarantee that they will lead to appropriate action. This is partly because those with concerns for marine mammal conservation form a small section of global society, and partly because other interests that could be conflicting may be deemed more pressing on economic, cultural or political grounds. The challenge therefore is to garner the support of public opinion and of policy makers that are responsible for effecting change.

Overview

Legislation that has been most effective in marine mammal conservation – such as the US Marine Mammal Protection Act, has reached a level of maturity that is still to be seen in much of Europe, where both national and regional legislation have been in existence for little more than a quarter of a century. This highlights one of the greatest blocks to conservation, that of the slowness to get action even when there is clear evidence to support it. A good example is the situation with respect to global attempts to halt man-induced climate change, with a number of people of influence still unprepared to accept the evidence let alone to respond at the necessary pace required.

Understandably scientists seek to undertake as much research as possible before they can feel confident about drawing conclusions. That is important so as not to reach the wrong conclusions. However, there are times when we have sufficient knowledge to take action, and delaying simply makes the situation worse, even costing lives. Fisheries bycatch is a good example. We have known this to be a major issue worldwide for more than 30 years (see, e.g. Allen, 1985; Evans, 1987; Northridge, 1984; Perrin, Smith, & Sakawaga, 1982), and yet it has been far from tackled. And discussions continue over how best to estimate bycatch mortality rates and at what point should we consider this unacceptable.

Environmental issues do reach public consciousness, taking prominence before waning again. In the 1960s to 1970s it was pollution, in the 1970s to 1980s it was whaling, in the 1990s to 2000s it was fisheries bycatch and in the 2000s to 2010s it has been noise disturbance along with climate change. And yet all these continue to be issues still to be properly addressed, alongside others such as prey depletion, ship strikes and plastic ingestion. The need for the establishment of marine protected areas came to the fore about 30 years ago, and since then national legislative actions have often centred upon this rather than tackling the more wide-scale environmental issues, probably because the latter are more difficult. The EU Habitats Directive is built largely upon the Natura 2000 network for selected species and habitats, whilst the system of strict protection which should be afforded to all cetacean species in EU waters is rarely pursued. In the meantime, few countries in Europe (or anywhere in the world) have fully developed management plans for protected areas with targets within specified time frames, routine monitoring to establish whether those are being met and then actions implemented where they are not.

In conclusion, despite various political and socioeconomic challenges, a case can be made for Europe to have its own cohesive Marine Mammal Protection legislation. This could be through the EU (with range states outside the EU invited to sign up), or through the United Nations, perhaps as an element of UNCLOS (the UN Convention on the Law of the Sea), as proposed in a global context by Jefferies (2017). The important ingredients should be ensuring (1) it forms 'hard' law rather than 'soft' law, linked to management; (2) there is consistency of policy across the target area; and (3) sufficient resources are made available for adequate monitoring and implementation.

References

ACCOBAMS (2017). *The ACCOBAMS Survey Initiative (ASI)*. Monaco: ACCOBAMS. <http://www.accobams.org/new_accobams/wp-content/uploads/2018/06/ASI-brochure.pdf>.

ACCOBAMS (2018). *ASI 2008 Summer Survey*. Monaco: ACCOBAMS. <http://www.accobams.org/main-activites/accobams-survey-initiative-2/asi-preliminary-results/>.

Allen, R. L. (1985). Dolphins and the purse-seine fishery for yellowfin tuna. In J. R. Beddington, R. J. H. Beverton, & D. M. Lavigne (Eds.), *Marin mammals and fisheries* (pp. 236−252). London: George Allen and Unwin.

Baur, D. C., Bean, M. J., & Gosliner, M. L. (1999). The laws governing marine mammal conservation in the United States. In J. R. Twiss, & R. R. Reeves (Eds.), *Conservation and management of marine mammals* (pp. 48−86). Washington and London: Smithsonian Institution Press, 471 pp.

Campbell, M. L., & Thomas, V. C. (2001). Protection and conservation of marine mammals in Canada: A case for legislative reform. *Ocean and Coastal Law Journal, 7*(2), 221−258.

Evans, P. G. H. (1987). *Natural history of whales and dolphins* (pp. 280−284). London: Christopher Helm.

Harcourt, R., Marsh, H., Slip, D., Chilvers, L., Noad, M., & Dunlop, R. (2015). Marine mammals, back from the brink? Contemporary conservation issues. In A. Stow, N. Maclean, & G. I. Holwell (Eds.), *Austral ark: The state of wildlife in Australia and New Zealand* (pp. 322−353). Cambridge: Cambridge University Press.

HELCOM (2009). Biodiversity in the Baltic Sea − An integrated thematic assessment on biodiversity and nature conservation in the Baltic Sea. In: *Baltic Sea environment proceedings No. 116B* (188 pp).

HELCOM (2010). Ecosystem health of the Baltic Sea 2003-2007: HELCOM initial holistic assessment. In: *Baltic Sea environment proceedings 122* (63 pp).

HELCOM (2018). State of the Baltic Sea − Second HELCOM holistic assessment 2011-2016. In: *Baltic Sea Environment Proceedings 155* (155 pp).

IFAW (International Fund for Animal Welfare) (2017). *A living success story: The Marine Mammal Protection Act* (32 pp). Washington, DC.

Jefferies, C. S. G. (2016). *Marine mammal conservation and the law of the sea*. Oxford: Oxford University Press, 401 pp.

New Zealand Department of Conservation (2016). *Bycatch of protected species: Hector's and Maui's Dolphins*. Auckland, New Zealand: Department of Conservation. <http://archive.stats.govt.nz/browse_for_stats/environment/environmental-reporting-series/environmental-indicators/Home/Marine/bycatch-hectors-mauis-dolphins/bycatch-hectors-mauis-dolphins-archived-27-10-16.aspx>.

NOAA Fisheries (2018). *2017-2018 North Atlantic right whale unusual mortality event*. <https://www.fisheries.noaa.gov/national/marine-life-distress/2017-2018-north-atlantic-right-whale-unusual-mortality-event>.

Northridge, S. (1984). *World review of interactions between marine mammals and fisheries*. Rome: FAO Fisheries Technical Paper 251, FAO.

Office of the Auditor General of Canada (2018). *Protecting marine mammals* (Report No. 2). Ottawa: Fall Reports of the Commissioner of the Environment and Sustainable Development. <http://www.oag-bvg.gc.ca/internet/English/parl_cesd_201810_02_e_43146.html>.

OSPAR Commission. (2000). *Quality Status Report 2000*. London: OSPAR Commission, 108 pp.

OSPAR Commission. (2010). *Quality Status Report 2010*. London: OSPAR Commission, 176 pp.

Perrin, W. F., Smith, T. D., & Sakawaga, G. T. (1982). Status of populations of spotted dolphin, Stenella attenuata, and spinner dolphin, S. longirostris, in the eastern tropical Pacific. In: *Mammals in the seas*, FAO Fisheries Series No. 5, (Vol. 4, pp. 67−84).

Roman, J., Altman, I., Dunphy-Daly, M. M., Campbell, C., Jasny, M., & Read, A. J. (2013). The Marine Mammal Protection Act at 40: Status, recovery, and future of U.S. marine mammals. *Annals of the New York Academy of Sciences, 1286*, 29−49.

Scovazzi, T. (2016). The international legal framework for marine mammal conservation in the Mediterranean Sea. *Advances in Marine Biology, 75*, 387−416.

Suisted, R., & Neale, D. (2004) Department of Conservation Marine Mammal Action Plan for 2005−2010. Marine Conservation Unit, New Zealand Department of Conservation, Wellington, New Zealand. 89 pp.

Chapter 3

Regional review of cetaceans in Northwest Europe

Northern Europe has a rich diversity of whales and dolphins inhabiting its seas. In all, 36 of the world's 90 cetacean species have been recorded within the ASCOBANS Agreement Area. Of these 27 species are small cetaceans within the InfraOrder Odontoceti, the toothed whales, for which the ASCOBANS Agreement currently applies. Table 3.1 lists all the species that have been found in the region, together with a summary of their overall status. Sixteen species occur regularly, of which 12 are small cetaceans (marked in bold). In the next chapter, I will focus attention upon those 12 species but will also briefly review the status, distribution and ecology of the others, including the large cetaceans that the Agreement does not cover. Before doing so, however, it might be of interest to the reader to take a short tour around the Agreement Area to see what he or she might expect to find, and to understand this in an oceanographic context (Fig. 3.1). I have based the latter partly upon descriptions of the region that I wrote in Reid, Evans, and Northridge (2003), updated here, where appropriate.

The Baltic

Let us start our tour in the semienclosed Baltic that is bordered by Denmark to the west, Sweden and Finland to the north, Germany and Poland to the south and Lithuania, Latvia, Estonia and the Russian Federation to the east. We refer to this as the inner Baltic Sea (or Baltic Proper). The species diversity here is low compared with open oceans and most freshwater systems, primarily owing to the brackish water that constitutes a stressful environment for many aquatic organisms, but also owing to its character as a geologically young sea (dating back less than 10,000 years) with a prehistory as a freshwater lake. Salinity has a particularly strong influence on the Baltic biodiversity and determines the distribution limit of many aquatic species, with more species able to live in the more saline southwestern waters than in the northern areas. The seabed is shaped into subbasins separated by shallow sills. Each subbasin is characterised by a different depth, volume and water exchange, resulting in subbasin-specific chemical and physical properties. In addition, the Baltic Sea has highly varied coastlines and large archipelago areas (Fig. 3.2).

Only one species of cetacean lives in the inner Baltic Sea and that is the harbour porpoise. Up to the 1940s, it was a common inhabitant here, particularly in the western portion (Tomilin, 1957). There followed a marked decline which could have been due to one or more of a number of factors — the unusually severe winter of 1940 when much of the Baltic froze, the introduction of a range of pollutants that certainly had their effects upon other marine top predators, and bycatch from increasingly intensive and technologically sophisticated fisheries (Berggren, Wade, Carlström, & Read, 2002; Koschinski, 2002). The population of the inner Baltic Sea appears to be genetically distinct whilst also being critically endangered, with numbers thought to be only around 500 individuals (Carlén et al., 2018; Evans & Teilmann, 2009; Lah et al., 2016; SAMBAH, 2016; Sveegaard et al., 2015; Wiemann et al., 2010). Although sightings, strandings and bycatch are still occasionally reported from Finland and the eastern Baltic States, most records come from the western portion — off the southeast coast of Sweden (Fig. 3.3), in German waters off Mecklenburg-Western Pomerania (Fig. 3.4), and around Puck Bay in Poland (Fig. 3.5) (Scheidat, Gilles, Kock, & Siebert, 2008; Skóra, Pawliczka, & Klinowska, 1988; Verfuß et al., 2007).

Porpoise densities in the German Baltic decline from west to east. Although porpoises are apparently present in this region throughout the year, there are seasonal peaks in spring to summer and a decline from autumn to winter, with a suggestion of an eastwards migration in spring (Kinze, Jensen, & Skov, 2003; Siebert et al., 2006).

TABLE 3.1 List of 36 cetacean species and their overall status in the ASCOBANS Agreement Area.

InfraOrder Odontoceti, the toothed whales

Family Phocoenidae

Phocoena phocoena	**Harbour porpoise**	COM/RAR

Family Delphinidae

Steno bredanesis	Rough-toothed dolphin	VAG
Tursiops truncatus	**Common bottlenose dolphin**	COM/RAR
Stenella frontalis	Atlantic spotted dolphin	VAG
Stenella coeruleoalba	**Striped dolphin**	COM/VAG
Delphinus delphis	**Common dolphin**	COM/VAG
Lagenodelphis hosei	Fraser's dolphin	VAG
Lagenorhynchus albirostris	**White-beaked dolphin**	COM/VAG
Lagenorhynchus acutus	**Atlantic white-sided dolphin**	COM/VAG
Grampus griseus	**Risso's dolphin**	REG/VAG
Peponocephala electra	Melon-headed whale	VAG
Feresa attenuata	Pygmy killer whale	VAG
Pseudorca crassidens	False killer whale	RAR/VAG
Orcinus orca	**Killer whale orca**	REG/VAG
Globicephala melas	**Long-finned pilot whale**	COM/VAG
Globicephala macrorhynchus	Short-finned pilot whale	VAG

Family Monodontidae

Monodon monoceros	Narwhal	VAG (RAR, Norway)
Delphinapterus leucas	White whale, beluga	VAG (RAR, Norway)

Family Ziphiidae

Ziphius cavirostris	**Cuvier's beaked whale**	REG/VAG
Hyperoodon ampullatus	**Northern bottlenose whale**	REG/VAG
Mesoplodon mirus	True's beaked whale	VAG
Mesoplodon europaeus	Gervais' beaked whale	VAG
Mesoplodon bidens	**Sowerby's beaked whale**	REG/VAG
Mesoplodon grayi	Grey's beaked whale	VAG
Mesoplodon densirostris	Blainville's beaked whale	VAG

Family Kogiidae

Kogia breviceps	Pygmy sperm whale	VAG/RAR
Kogia sima	Dwarf sperm whale	VAG

Family Physeteridae

Physeter microcephalus	Sperm whale	REG/RAR

InfraOrder Mysticeti, the baleen whales

Family Balaenidae (right whales)

Eubalaena glacialis	North Atlantic right whale	VAG
Balaenidae mysticetus	Bowhead whale	VAG

Family Balaenopteridae (rorquals)

Megaptera novaeangliae	Humpback whale	REG/VAG
Balaenoptera acutorostrata	Common minke whale	REG/VAG
Balaenoptera borealis	Sei whale	RAR/VAG
Balaenoptera brydei	Bryde's whale	VAG
Balaenoptera physalus	Fin whale	REG/VAG
Balaenoptera musculus	Blue whale	RAR/VAG

Notes: In bold are the regular species that are covered by the ASCOBANS Agreement. *COM*, Common; *REG*, regular; *RAR*, rare; *VAG*, vagrant; where two assessments are given, the second refers to the inner Baltic Sea.

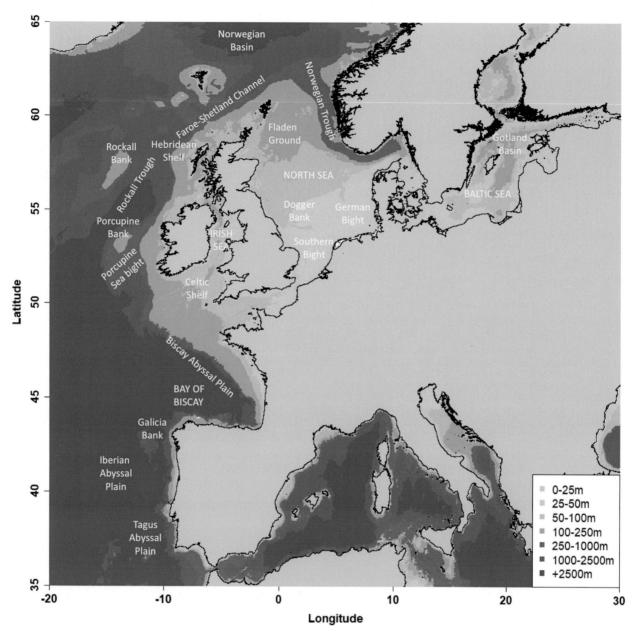

FIGURE 3.1   Map of the bathymetry of Northwest European seas.

Occasionally other species wander into the inner Baltic Sea. In recent years, these have included humpback whale, white-beaked dolphin, bottlenose dolphin, common dolphin and striped dolphin, but none can be regarded as native to the region.

The Belt Seas

The Inner Belt Seas (Fig. 3.6) around the Danish islands to the west have much higher densities of harbour porpoise, although since the 1990s numbers may have declined. The porpoises of inner Danish waters are distinct from those in the North Sea and the Baltic Proper (Lah et al., 2016; Sveegaard et al., 2015; Wiemann et al., 2010). Other species that occur regularly, although in small numbers, include the white-beaked dolphin and minke whale. White-beaked dolphins are quite common on Fisher Bank and Jutland Reef, and calves have been recorded here so they are probably breeding in the region. There has been a steady increase in white-beaked dolphin strandings over the last three decades, suggesting the species has become more regular in Danish waters (Kinze, Jensen, Tougaard, & Baagøe, 2010), assuming there has been no change in mortality rates. Similarly, minke whale strandings have also increased over the same period.

FIGURE 3.2 The coast of Finland is low-lying with many scattered wooded islands. Small numbers of porpoises occur offshore but in winter, sea ice forces animals to migrate farther south and west. *Photo: A.O. Laine/Parks and Wildlife Finland.*

FIGURE 3.3 In the Swedish Baltic, hotspots for the endangered Baltic harbour porpoise population occur around the Hoburgs and Northern and Southern Middle Banks. *Photo: Ida Carlén.*

FIGURE 3.4 The shallow seas of the German Baltic support good numbers of porpoises some of which seasonally migrate across the border into west Polish waters. *Photo: Michael Dähne/German Oceanographic Museum.*

FIGURE 3.5 Puck Bay off the Hel Peninsula in Poland is an important site for harbour porpoises. *Photo: Prof. Krzysztof Skóra Hel Marine Station, Department of Oceanography and Geography, University of Gdańsk.*

FIGURE 3.6   The Danish Belt Seas are home to a separate population of harbour porpoise. *Photo: Jonas Teilmann, Aarhus University, Denmark.*

Skagerrak and Kattegat

Porpoises are also the only regular species occurring in the Skagerrak and Kattegat (Fig. 3.7), waters that are shared by Sweden and Denmark, and in the case of the Skagerrak, also by Norway (Berggren & Arrhenius, 1995; Kinze et al., 2003, 2010). In those deeper areas that are closest to the North Sea, minke whales and white-beaked dolphins can be seen, and in recent years, common dolphin and humpback whale have also been reported.

Norwegian Atlantic

Moving westwards and northwards up the dissected mountainous coastline of Norway (Fig. 3.8), the deep fjords are occupied by numbers of porpoises and a few minke whales, whilst offshore in areas like the Norwegian Sea occur roving groups of Atlantic white-sided dolphin, killer whale, long-finned pilot whale and smaller numbers of sperm whale as well as a few fin and humpback whales. Andenes in Vesterålen supports a whale watching industry targeting sperm whales.

The general ocean circulation in the Northeast Atlantic is dominated by the northeastward extension of the Gulf Stream, known as the North Atlantic Current. This transports relatively warm, nutrient-rich and oxygen-rich water from the northwestern Atlantic towards the coasts of Europe, driven between 40° and 60°N by westerly winds.

North Sea

The main branch of the North Atlantic Current passes eastwards between Shetland and the Faroe Islands and then south into the northern North Sea, sweeping eastwards in the central North Sea. A separate current flows in a northeasterly direction along the coast of Belgium, the Netherlands and northern Germany. The North Sea can be divided into three distinct parts: (1) the Southern Bight (51−54°N) that borders northern Germany, The Netherlands and Belgium with water depths generally less than 40 m (Fig. 3.9); (2) the central North Sea (54−57°N) with water depths of 40−100 m

FIGURE 3.7 The Danish Kattegat coast at Anholt. Porpoises in the Kattegat are thought to form part of the North Sea population. *Photo: Anders Galatius.*

FIGURE 3.8 The waters around the islands of Lofoten and Vesterålen in western Norway are home to a wide variety of cetaceans from sperm whales, minke whales and humpbacks to killer whales and Atlantic white-sided dolphins. *Photo: Peter GH Evans.*

FIGURE 3.9 The German Wadden Sea is shallow with many offshore sandbanks providing suitable habitat for breeding harbour porpoise. *Photo: Anita Gilles/ITAW.*

(except for shallower areas on the Dogger Bank and coastal banks off western Denmark); and (3) the northern North Sea (north of 57°N), which includes an area of shelf water 100–200 m deep and the Norwegian Trench, with water depths from 200 m to more than 700 m in the Skagerrak between Denmark and Norway.

Some of the water moving northeastwards to the west and north of the British Isles turns southwards into the North Sea at several points. A relatively small flow enters via the Fair Isle Channel between Orkney and Shetland, with larger flows southwestwards to the east of Shetland and along the western edge of the Norwegian Trench.

In the North Sea the strongest tidal currents occur in the Southern Bight, the German Bight, off the east coast of Scotland, the Pentland Firth and between Orkney and Shetland. As elsewhere the combination of variations in water depth and in tidal currents leads to the development of distinct hydrographic regimes during the summer months when a seasonal thermocline extends over most of the central and northern North Sea in response to solar warming. The transitional or frontal zones are characterised by strong horizontal gradients in surface or bottom water temperatures. In contrast to the shelf waters, which are well mixed in the winter by tidal action and winds, the deeper waters of the Norwegian Trench exhibit stratification into layers of different densities and temperature throughout the year. Examples of frontal zones may be found between Orkney and Shetland; east of Buchan in Grampian, NE Scotland; in the outer Firths of Forth and Tay, Eastern Scotland; east of Spurn Head and Flamborough Head (Fig. 3.10) in Humberside; and around the Frisian Islands and Helgoland (Fig. 3.11) in North Germany (Holligan, Aarup, & Groom, 1989; Pingree & Griffiths, 1978). These biologically rich areas tend to be the ones to attract a variety of cetacean species — harbour porpoise, white-beaked dolphin, minke whale, occasional humpback and even fin whale.

FIGURE 3.10 Flamborough Head, in eastern England, has a productive frontal zone nearby that provides rich feeding grounds for a variety of seabirds and cetaceans. *Photo: Peter GH Evans.*

FIGURE 3.11 Helgoland, northern Germany, has high concentrations of harbour porpoise in neighbouring waters. *Photo: Helena Herr.*

Seasonal changes in surface temperature are most pronounced in the southern and eastern parts of the North Sea, where the water is relatively shallow and influenced by the more extreme continental climate. Seasonal variations in surface salinity are relatively small, the most significant being the decrease in salinity of the Norwegian coastal waters during summer as a result of relatively freshwater flowing out from the Baltic Sea.

Harbour porpoises are widely distributed around the North Sea (Hammond et al., 2002, 2013, 2017; Reid et al., 2003), but whereas in the 1980–90s, there were high densities in the northern sector from the Shetland Isles east to southern Scandinavia, now the species is more common in the central and southern North Sea (Camphuysen & Peet, 2006; Evans, 2010; Hammond et al., 2013, 2017). Although present year-round, the species tends to aggregate into larger groups in the late summer in the northwestern North Sea and in late winter or spring in the southernmost North Sea (Evans, 2010).

Although the harbour porpoise is the most common cetacean in the North Sea, other small cetacean species that occur regularly, in descending order of abundance, include white-beaked dolphin, bottlenose dolphin, short-beaked common dolphin and Risso's dolphin. White-beaked dolphins are the most widely distributed, occurring even in the southernmost North Sea, though they are most abundant in the northern sector. A population of bottlenose dolphins, numbering around 200 animals, ranges from the northeast coast of Scotland (Fig. 3.12) through coastal waters as far south as Whitby in Yorkshire, rarely venturing farther offshore. In the last two decades, the occurrence of common and Risso's dolphins in the northern and central North Sea has increased, in concert with increases in the region of warmer water fish like anchovy and sardine in the case of the former, and various squid species in the case of the latter. The Atlantic white-sided dolphin, which during the 1970s and 1980s was a regular visitor to the North Sea, particularly in winter, has in more recent times become relatively scarce here. Of the larger cetaceans in the North Sea, the minke whale is the commonest and most widely distributed, with summer aggregations around the Northern Isles, the Outer Moray Firth in NE Scotland, and at scattered localities from Fraserburgh in Aberdeenshire south to Whitby in Yorkshire. Offshore it is common around the Dogger Bank and farther north in the northwestern sector of the North Sea. The localities favoured by minke whales also at times attract other species like humpback, fin and even sei whale,

FIGURE 3.12 The North Sea population of bottlenose dolphins is concentrated around northeast Scotland, particularly the Moray Firth where a special area of conservation has been established for its protection. *Photo: Peter GH Evans.*

FIGURE 3.13 Fair Isle, the southernmost outlier of the Shetland Isles, is regularly visited by killer whales and minke whale, with Atlantic white-sided dolphin, Risso's dolphin and sperm whale also occurring occasionally. *Photo: Peter GH Evans.*

possibly associated with shoals of herring, sprat or sandeel. With the influence of the North Atlantic Current, the northernmost parts of the North Sea also hold the highest species diversity, and other whales that come into these waters on a seasonal basis include killer whale, long-finned pilot whale and sperm whale (Fig. 3.13).

English Channel

Moving south, the eastern sector and coastal areas of the English Channel (Figs. 3.14 and 3.15) are shallow, with depths rarely exceeding 50 m. Depths are greater in the central zone and generally slope from east to west reaching 100 m along the western edge, although a trough to the northwest of the Channel Islands, the Hurd Deep, reaches a depth of more than 170 m. Currents flow eastwards, bringing more saline water from the Atlantic. A frontal system (termed the Ushant Front) develops in summer in the transitional zone between cooler, tidally mixed Channel water and the warmer stratified water of the Atlantic (Pingree, Holligan, & Mardell, 1978).

The easternmost part of the Channel has experienced increased numbers of harbour porpoise in recent years whilst small groups of bottlenose dolphin move back and forth in summer along the south coast of England. Farther offshore the common dolphin is the most abundant small cetacean until one reaches the Channel waters of northern France (Fig. 3.16) and the Channel Islands, which are occupied by resident groups of bottlenose dolphin, with seasonal appearances of minke whale, Risso's dolphin and long-finned pilot whale.

Irish Sea

The Irish Sea bordered by Scotland in the north, Ireland to the west, and England and Wales to the east is a relatively shallow sea, mostly less than 100 m deep with shallow embayments in the east but a deeper channel in the west, reaching depths of 275 m in the Beaufort Dyke within the North Channel. It is largely sheltered from the winds and currents

FIGURE 3.14 The Isle of Wight off the coast of southern England. *Photo: Peter GH Evans.*

FIGURE 3.15 Sidmouth in south Devon. A small population of white-beaked dolphins inhabit neighbouring Lyme Bay. *Photo: Peter GH Evans.*

FIGURE 3.16 The coast of Normandy, northern France. One of the largest coastal populations of bottlenose dolphin in Europe ranges between the Normandy coast and the Channel Islands of Jersey and Guernsey. *Photo: Peter GH Evans.*

of the North Atlantic, although its relatively high salinity reflects the influence of oceanic water from the south. In the Irish Sea, the inshore Coastal Current carries water from St George's Channel northwards through the North Channel, mixing with water from the outer Clyde. Southerly winds can strengthen this current, increasing the northward transport of water from the Irish Sea into the Sea of Hebrides, with northerlies retarding the current.

Surface tidal currents reach a speed of more than four knots in St George's Channel, near the Irish coast, and are locally strong elsewhere. Currents are weakest in the west-central Irish Sea. The greatest ranges of tide occur on the northwest English coast in and out of Liverpool Bay and the Solway Firth, leaving an area of almost permanently slack water off the Irish coast north of Dublin (Pingree & Griffiths, 1978). Tidal streams enter the Irish Sea from both the north and the south, meeting near latitude 54°N, just south of the Isle of Man.

The annual mean temperature decreases northward from a little over 11°C at the southern end of St George's Channel to 10°C in the North Channel and it also decreases towards the sides. The water is coolest in February or March with temperature decreasing from the deeper channel towards the coasts, and warmest in August where it increases closest to the coasts, exceeding 16°C in Liverpool and Cardigan Bays. The annual mean salinity also decreases from south to north and from the centre of the channel to the sides.

Throughout most of the region tidal mixing is sufficiently intense to ensure that the water column remains well mixed throughout the year. However, in the regions of weak tidal currents to the east and west of the Isle of Man and in Cardigan Bay, stratification (warm, lighter water on top of cool, denser water) occurs in summer. Near to estuaries and especially in Liverpool Bay, the water column can also stratify because freshwater is lighter than salty water. The transitions between mixed and stratified regions are in many cases marked by sharp fronts, which may be manifest at the surface and/or near the seabed, and have an important influence on the dynamics, inhibiting mixing across the fronts. There are important fronts in summer between St George's Channel and the Celtic Sea in the south (called the Celtic Sea front), and running southwestwards from the Isle of Man towards the Irish coast of Co. Dublin in the north (the Western Irish Sea front). Such tidal mixing fronts are often zones of high biological activity (Pingree et al., 1978), where plankton growth and activity can be much higher than in adjacent stratified and mixed zones, due to elevated nutrient levels. The Irish Sea Front exhibits little variability in either position or structure and is particularly well developed in August (Huang, Cracknell, Vaughan, & Davies, 1991; Simpson, 1981). These are areas where short-beaked common dolphins and minke whales are mainly found, along with small numbers of fin whale and occasional humpbacks (Baines & Evans, 2012).

As with most other parts of the northwest European continental shelf, the most common and widely distributed species in the Irish Sea is the harbour porpoise (Fig. 3.17). The bottlenose dolphin is the next most frequently recorded species, with a predominantly coastal distribution, the main summer concentrations of sightings being in southern Cardigan Bay (Fig. 3.18) extending north into Tremadog Bay, and in winter, the north coast of Wales, including Anglesey, north to the Isle of Man and beyond. The common dolphin has a largely offshore distribution centred upon the Celtic Deep at the southern end of the Irish Sea, where water depths range from 50 to 150 m. This high-density area extends eastwards towards the coast and islands of west Pembrokeshire (Fig. 3.19) and northwards over into the central channel towards the Isle of Man (Baines & Evans, 2012).

The other small cetacean species that is regular in the Irish Sea is Risso's dolphin. It has a relatively localised distribution, forming a wide band running southwest to northeast that encompasses the southeast coast of Ireland, west of Pembrokeshire, the western end of the Llŷn Peninsula and off north Anglesey in Wales, along with the waters around the Isle of Man (Fig. 3.20) in the north (Baines & Evans, 2012).

FIGURE 3.17 Point Lynas on the north coast of Anglesey is one of the best locations in the United Kingdom to see harbour porpoise. Aggregations of up to 100 animals have been recorded here. *Photo: Peter GH Evans.*

FIGURE 3.18 Morfa Harlech in West Wales. Cardigan Bay is home to Britain's largest coastal population of bottlenose dolphins, numbering approximately 200–300 individuals. *Photo: Peter GH Evans.*

FIGURE 3.19 The waters around the Pembrokeshire coast and islands such as Skomer in southwest Wales support populations of harbour porpoise and common dolphin whilst, farther offshore, minke whales, fin whales and humpbacks may be present in summer. *Photo: David Ord.*

FIGURE 3.20   The coastal waters around the Isle of Man are visited regularly by Risso's dolphin, common dolphin and minke whale. *Photo: Peter GH Evans.*

Atlantic Scotland and Ireland

Moving westwards to the Atlantic shores of Scotland and Ireland, the influence of the North Atlantic Current becomes very obvious. The open ocean is exposed to the winds, and gales are frequent with big seas built up by the prevailing southwesterlies, resulting in the regular mixing of the surface layers and the dispersal of plankton and fish. Gelatinous animals and planktonic crustaceans dominate the fauna of the surface waters, whereas at depths of approximately 300−600 m, various species of squid and pelagic fish may aggregate, forming prey patches for several species of cetacean, such as sperm whale, the beaked whales (mainly Cuvier's and Sowerby's but also True's, and in the north, northern bottlenose whale), long-finned pilot whales, and offshore dolphins such as the striped dolphin farther south, Atlantic white-sided dolphin farther north, and common dolphin towards the shelf edge. Fish are less abundant in the deep ocean than over the shelf although they may form seasonal aggregations or concentrate over banks. Along the shelf slope, the Atlantic Ocean water masses meet the less saline waters of the continental shelf, which receive freshwater inputs from rivers in Britain, Ireland and continental Europe. Although there is relatively little mixing along this boundary, currents of slightly warmer water move northwards along this shelf edge from the Celtic and Irish Seas to Shetland carrying plankton, including fish eggs and larvae, from south to north. Turbulence brings deepwater up the shelf slope to within 200 m of the surface, resulting in enhanced productivity of plankton and associated aggregations of cephalopods and fish such as blue whiting and mackerel. These in turn attract concentrations of pelagic seabirds and species of cetacean such as fin whale, killer whale, striped, common and Atlantic white-sided dolphins, with SW Ireland a particularly rich area (Fig. 3.21) (Evans, 1990).

Most of the continental shelf on the Atlantic side is exposed to the prevailing southwesterly winds, and saline oceanic water crosses the shelf edge between Malin Head in northwest Ireland and Barra Head in the Outer Hebrides, often intruding over other parts of the shelf in winter.

Along the west coast of Ireland, the Irish Shelf Current flows northwards and then eastwards along its north coast (Figs. 3.22 and 3.23). The main front in the Atlantic region is the Irish Shelf Front that occurs to the south and west of Ireland (at approximately 11°W) around the 150-m isobath and which exists year-round (Huang et al., 1991). This front

FIGURE 3.21   The Skellig Rocks off Co. Kerry with a 6th century Celtic monastery and large colonies of seabirds are located in one of the most westerly parts of Europe. Several species of cetaceans, particularly minke whale, but also harbour porpoise, Risso's dolphin, humpback whale and killer whale occur here. *Photo: Peter GH Evans.*

FIGURE 3.22   Cliffs of Moher, Co. Galway. The west coast of Ireland has a rich cetacean fauna with pelagic species coming onto the continental shelf and mixing with more coastal species. *Photo: Peter GH Evans.*

FIGURE 3.23 Ramore Head, Co. Antrim along the coast of Northern Ireland, frequented by minke whale and numbers of harbour porpoise. *Photo: Peter GH Evans.*

marks the boundary between water over the Irish shelf (often mixed vertically by the tide) and offshore North Atlantic water. To the west of this, waters governed by the North Atlantic Current flow north off the northern half and south off the southern half of Ireland (McMahon, Raine, Titov, & Boychuk, 1995).

In Ireland the shelf edge comes closest off the southwest and northwest coasts and this is where schools of common dolphin and small numbers of minke whale can most frequently be seen as well as the more oceanic species such as fin whale, sperm whale, northern bottlenose whale, long-finned pilot whale and killer whale. The southern and southwestern Irish coasts have some of the highest densities of harbour porpoise, and this region is one of the areas where fin whale, humpback and occasional sei whale are most likely to be seen. A resident bottlenose dolphin population of 100−150 animals occupies the Shannon Estuary, which seem to keep distinct from other groups that range up and down the shelf edge and west coast and bays. Small groups of Risso's dolphin range all along the west coast of Ireland, but particularly off Counties Kerry and Cork (Berrow, Whooley, O'Connell, & Wall, 2010; Wall et al., 2013).

Farther north off the west coast of Scotland and around the Hebrides (Fig. 3.24), nutrient upwellings around headlands and islands provide suitable conditions for plankton concentrations to develop, and upon these feed large numbers of fish, which then fall prey to seabirds, seals and cetaceans. The waters are often relatively calm, being protected from the Atlantic by the long chain of islands that form the Outer Hebrides (Fig. 3.25). Tidal currents passing over uneven bottom topography can have a considerable mixing effect, as for example in the Sea of the Hebrides. This region has a diversity of cetacean species. High densities of harbour porpoise occur particularly near the west mainland coasts and Inner Hebrides, with aggregations in late summer that may number 100 or more individuals. The second most frequently observed cetacean is the minke whale, which tends to occur in the same areas as porpoises. Other species regularly found in the region include common dolphin (throughout), Atlantic white-sided dolphin (particularly in the southwestern Sea of the Hebrides), white-beaked dolphin (mainly in the North Minch), Risso's dolphin, bottlenose dolphin and killer whale. West and north of the Outer Hebrides deeper water species such as fin whale, humpback whale, sperm whale, northern bottlenose whale and long-finned pilot whale occur regularly, and sometimes these enter shelf seas and coastal waters.

FIGURE 3.24 The Small Isles (Rum, Eigg, Canna and Muck) in the Inner Hebrides support important populations of harbour porpoise, common dolphin and minke whale, whilst killer whales are seen here annually. *Photo: Peter GH Evans.*

FIGURE 3.25 Tiumpan Head, Isle of Lewis, Outer Hebrides. One of the best sites to see Risso's dolphin from the coast. Minke whales, humpbacks and schools of white-beaked dolphins also regularly occur here. *Photo: Peter GH Evans.*

FIGURE 3.26 Bottlenose dolphins feeding amongst the algal forests around Ile de Sein, in the Iroise Sea Marine Park, France, situated within the Armorican Shelf between the Celtic Sea and the Bay of Biscay. *Photo: Yves Gladou/Iroise Natural Marine Park, courtesy of Philippe Le-Niliot.*

Bay of Biscay

South of the British Isles lies the Bay of Biscay bordered by France to the east and mainly by Spain to the south. This open bay has sea depths ranging from the shallow continental shelf (less than 100 m) to the abyssal plain (greater than 4000 m), with many underwater features such as submarine canyons, seamounts and a steep continental slope. The Armorican shelf in the north of the bay (Fig. 3.26) is up to 180 km wide, whilst in the south the continental shelf is narrow, only 30–40 km in width (Koutsikopoulos & Le Cann, 1996).

The Capbreton Canyon, in the southeast corner of the Bay of Biscay, is a major morphological feature that cuts into the continental slope in an east to west direction with the 1000 m contour only 3 km from the coast. It is the deepest submarine canyon in the world, the longest off Europe, and its head is located just 250 m from the coastline (Cirac, Bourillet, Griboulard, Normand, & Mulder, 2001). The canyon runs westward parallel to the north coast of Spain for 160 km, then turns northward, widens and abruptly disappears in the continental rise at a water depth of 3500 m (Gaudin et al., 2003).

The Bay of Biscay is located between the eastern part of the subpolar gyre and the subtropical gyre, and may be affected by both gyres depending on latitude and the general circulation. For the oceanic part of the Bay, it is weak and anticyclonic in direction (Koutsikopoulos & Le Cann, 1996). In the southern part, east-flowing shelf and slope currents are common in autumn and winter due to southerly and westerly winds (Valencia, Franco, Borja, & Fontan, 2004). On the continental shelf the circulation is governed by the combined effects of tides, river inputs and wind so that shelf waters are colder in winter and warmer and less saline in summer (Valencia, Borja, Fontan, Perez, & Rios, 2003).

In spring and summer, northeasterly winds of medium to low intensity are prevalent, causing frequent coastal upwelling events (Koutsikopoulos & Le Cann, 1996). In autumn and winter, southerly and westerly winds are dominant, causing frequent downwelling events (Borja, Uriarte, Valencia, Motos, & Uriarte, 1996). Upwelling is observed along the French and Spanish coasts, and the strength of the upwelling corresponds to the winds and water masses (Gil & Sanchez, 2003).

FIGURE 3.27 The Basque coast of northern Spain is close to deep canyons occupied by striped dolphin, long-finned pilot whale and Cuvier's beaked whale. *Photo: Peter GH Evans.*

The extension of the warm water Iberian poleward current is also observed in the Bay and is a common feature of winter circulation, coinciding with the spawning season of various pelagic and demersal fish species (Gil, 2003).

The Bay of Biscay is one of the world's strongest generation sites for internal tides which, along with surface tides, are amplified by the interaction of bottom topography and result in high phytoplankton abundance and cool water at the surface above the continental shelf break (New, 1988). The prevailing winds and sea currents make the waters along the continental slope productive and attractive to marine life, with over 30 species of cetaceans recorded here (Kiszka, Macleod, Van Canneyt, Walker, & Ridoux, 2007; Smith, 2010).

In the French sector of the Bay of Biscay, common dolphins are the most abundant species, with smaller numbers of bottlenose dolphin and harbour porpoise (Certain, Ridoux, Van Canneyt, & Bretagnoile, 2008; Certain et al., 2011). Along the north Spanish coasts (Fig. 3.27) are groups of bottlenose dolphin, and offshore in the deeper areas around the shelf break and within canyons, common and striped dolphin, long-finned pilot whale, Cuvier's beaked whale, sperm whale and fin whale all occur (Kiszka et al., 2007; Lambert et al., 2017; Laran et al., 2017; Smith, 2010). Small numbers of harbour porpoise, Risso's dolphin, northern bottlenose whale and minke whale can also be seen. There is evidence that locally distinct populations of bottlenose dolphin and harbour porpoise exist around the Iberian Peninsula (Fernández, García-Tiscar, et al., 2011; Fernández, Santos, et al., 2011; Fontaine et al., 2007, 2010).

Iberian Peninsula

The northern coastline of Galicia (Fig. 3.28) is mostly rocky and shallow, with many rias — flooded tectonic valleys of moderate depth, the oceanography being dominated by winds particularly in summer when freshwater input is at a minimum (Alvarez, de Castro, & Gomez-Gesteira, 2005; Fraga, 1981). The rias form a semiclosed system

FIGURE 3.28   The shelf edge comes close to the coast of Galicia in northwest Spain, providing favourable habitats for deepwater species such as striped dolphin, long-finned pilot whale, sperm whale and fin whale. *Photo: M Begoña Santos.*

because of the downwelling winds, the presence of the poleward flow and the upwelling that occurs inside them (Torres & Barton, 2007). To the south of Silleiro Cape, a rectilinear sandy coast extends to just north of the Nazaré Canyon, interrupted only by Cape Mondego. Farther south beaches are replaced by cliffs which extend to Cape Raso, at the latitude of Lisbon.

As with northern Spain, the continental shelf is relatively narrow west of Portugal, with depths of 500 m or more frequently within 50 km of the coast. The prevailing current is a southward branch of the warm North Atlantic Current (referred to as the Portuguese Current) that initially sweeps eastwards in a wide arc from the central Atlantic. Beyond the shelf break, Atlantic water interacts with salty Mediterranean water, which then moves northward along the continental slope, forming the Navidad Current. Off the Iberian Peninsula, northerly winds cause an upwelling of cold and nutrient-rich deeper water to the surface during summer, with coastal upwelling interacting with strong outflow from the rias generating eddies in the poleward slope flow which may contribute to breakdown of the Iberian Pole Current (IPC) during the start of the upwelling regime (Alvarez, de Castro, Prego, & Gomez-Gesteira, 2003; Alvarez, Gomez-Gesteira, Lorenzo, Crespo, & Dias, 2011; Figueiras, Labarta, & Fernández Reiriz, 2002; Torres & Barton, 2007). This region corresponds to the Lusitanian zone and is highly diverse with many different types of coastal habitat from rocky cliffs to sandy and muddy shores, from coastal lagoons to open bays and estuaries. Areas of upwelling off the Iberian coast give rise to spring blooms that occur earlier than farther north, and are highly productive, supporting large populations of pelagic fish such as sardine, at least 22 species of cetaceans in Galicia (Covelo et al., 2009; Covelo, Martínez-Cedeira, Llavona, Díaz, & López, 2016; López, 2003; López el al., 2004, Penas-Patiño & Piñeiro-Seage, 1989) and 16 species of cetaceans in Portugal (Brito, Vieira, Sá, & Carvalho, 2009; Ferreira et al., 2012; Llavona, 2017).

The most frequent small cetacean species occurring off the Portuguese mainland coast (Fig. 3.29) is the common dolphin, followed by bottlenose dolphin, striped dolphin (with higher densities in deeper waters) and small numbers of minke whale and Risso's dolphin (Santos et al., 2012). Fin whale, sperm whale, long-finned pilot whale, Cuvier's beaked whale and killer whale occur farther offshore, the first two being the target of whaling activities in the first half of the 20th century (Brito et al., 2009). A small isolated population of bottlenose dolphin (numbering less than 30 individuals) occupies the Sado Estuary (Fig. 3.30) but appears to be in long-term decline (Dos Santos, Couchinho, Luís, & Gonçalves, 2010; Lacey, 2015).

FIGURE 3.29 The rocky coast of Portugal provides good vantage points for observing a range of deepwater species including common dolphins. *Photo: Graham .J. Pierce.*

FIGURE 3.30 The small isolated bottlenose dolphin population in the Sado Estuary, Portugal lives alongside a very busy and industrialised area. *Photo: Ines Carvalho.*

FIGURE 3.31 The Strait of Gibraltar looking across to the North African coast of Morocco from the Rock of Gibraltar. *Photo: Peter GH Evans.*

And so we approach the southeastern boundary of the ASCOBANS Agreement at the western edge of the Gulf of Cadiz where the shelf is comparatively wide, and the cetacean community is dominated by bottlenose and common dolphins with small numbers of harbour porpoise (Cañadas, Sagarminaga, De Stephanis, Urquiola, & Hammond, 2009). East of here the relatively deep and narrow Strait of Gibraltar (Fig. 3.31) is an important location for fin whale, sperm whale, long-finned pilot whale, killer whale, common and striped dolphin (De Stephanis et al., 2008).

References

Alvarez, I., de Castro, M., & Gomez-Gesteira, M. (2005). Inter- and intra-annual analysis of the salinity and temperature evolution in the Galician Rías Baixas—ocean boundary (northwest Spain). *Journal of Geophysical Research, 110*, 14.

Alvarez, I., de Castro, M., Prego, R., & Gomez-Gesteira, M. (2003). Hydrographic characterization of a winter-upwelling event in the Ria of Pontevedra (NW Spain). *Estuarine Coastal and Shelf Science, 56*, 869—876.

Alvarez, I., Gomez-Gesteira, M., Lorenzo, M. N., Crespo, A. J. C., & Dias, J. M. (2011). Comparative analysis of upwelling influence between the western and northern coast of the Iberian Peninsula. *Continental Shelf Research, 31*(5), 388—399. <http://ephyslab.uvigo.es/publica/documents/file_248CSR_Comp_IP_Ines10.pdf>.

Baines, M. E., & Evans, P. G. H. (2012). *Atlas of the marine mammals of wales* (2nd ed.). CCW Monitoring Report No. 68, 143 pp.

Berggren, P., & Arrhenius, F. (1995). Sightings of harbour porpoises (*Phocoena phocoena*) in Swedish waters before 1990. *Reports of the International Whaling Commission* (Special issue 16), 99—108.

Berggren, P., Wade, P. R., Carlström, J., & Read, A. J. (2002). Potential limits to anthropogenic mortality for harbour porpoises in the Baltic region. *Biological Conservation, 103*, 313—322.

Berrow, S., Whooley, P., O'Connell, M., & Wall, D. (2010). *Irish cetacean review (2000-2009)*. Kilrush: Irish Whale and Dolphin Group, 60 pp.

Borja, A., Uriarte, A., Valencia, V., Motos, L., & Uriarte, A. (1996). Relationships between anchovy (*Engraulis encrasicolus*) recruitment and the environment in the Bay of Biscay. *Scientia Marina, 60*, 179—192.

Brito, C., Vieira, N., Sá, E., & Carvalho, I. (2009). Cetaceans' occurrence off the west central Portugal coast: A compilation of data from whaling, observations of opportunity and boat-based surveys. *Journal of Marine Animals and Their Ecology, 2*, 1—4.

Camphuysen, C. J., & Peet, G. (2006). *Whales and dolphins of the North Sea*. Kortenhoef: Fontaine Uitgevers, 160 pp.

Cañadas, A., Sagarminaga, R., De Stephanis, R., Urquiola, E., & Hammond, P. S. (2005). Habitat preference modelling as a conservation tool: proposals for marine protected areas for cetaceans in southern Spanish waters. *Aquatic Conservation: Marine and Freshwater Ecosystems, 15,* 495−521.

Carlén, I., Thomas, L., Carlström, J., Amundin, M., Teilmann, J., Tregenza, N., & Tougaard, J. (2018). Basin-scale distribution of harbour porpoises in the Baltic Sea provides basis for effective conservation actions. *Biological Conservation, 226,* 42−53.

Certain, G., Masse, J., Van Canneyt, O., Petitgas, P., Doremus, G., Santos, M. B., & Ridoux, V. (2011). Investigating the coupling between small pelagic fish and marine top predators using data collected from ecosystem-based surveys. *Marine Ecology Progress Series, 422,* 23−39.

Certain, G., Ridoux, V., Van Canneyt, O., & Bretagnoile, V. (2008). Delphinid spatial distribution and abundance estimates over the shelf of the Bay of Biscay. *ICES Journal of Marine Science, 65,* 656−666.

Cirac, P., Bourillet, J.-F., Griboulard, R., Normand, A., & Mulder, T. (2001). Le canyon de Capbreton: nouvelles approaches morphostructurales et morphosedimentaires. Premiers résultats de la campagne ITSAS. (Canyon of capbreton: New morphostructural and morphosedimentary approaches. First results of the ITSAS cruise). *Earth and Planetary Sciences, 332,* 447−455.

Covelo, P., Martínez-Cediera, J., Caldas, M., Díaz, J. I., Palacios, G., Ferreira, A., López, A. (2009). Cetáceos y pinnípedos en Galicia. 10 años de funcionamiento de la red oficial de varamientos (1999-2008). In: *IX Congreso de la Sociedad Española para la Conservación y Estudio de los Mamíferos, 4-7 diciembre. Bilbao, Spain.*

Covelo, P., Martínez-Cedeira, J. A., Llavona, Á., Díaz, J. I., & López, A. (2016). Strandings of Beaked Whales (Ziphiidae) in Galicia (NW Spain) between 1990 and 2013. *Journal of the Marine Biological Association of the United Kingdom, 96*(4), 925−931.

De Stephanis, R., Cornulier, T., Verborgh, P., Salazar Sierra, J., Perez Gimeno, N., & Guinet, C. (2008). Summer spatial distribution of cetaceans in the Strait of Gibraltar in relation to the oceanographic context. *Marine Ecology Progress Series, 353,* 272−288.

Dos Santos, M. E., Couchinho, M. N., Luís, A. R., & Gonçalves, E. J. (2010). Monitoring underwater explosions in the habitat of resident bottlenose dolphins. *Journal of the Acoustical Society of America, 128*(6), 3805−3808.

Evans, P. G. H. (1990). European cetaceans and seabirds in an oceanographic context. *Lutra, 33,* 95−125.

Evans, P. G. H. (2010). *Review of trend analyses in the ASCOBANS area* (68 pp.). ASCOBANS AC17/Doc. 6-08 (S).

Evans, P. G. H., & Teilmann, J. (Eds.), (2009). *Report of ASCOBANS/HELCOM small cetacean population structure workshop.* Bonn: ASCOBANS/UNEP Secretariat, 140 pp.

Fernández, R., García-Tiscar, S., Santos, M. B., López, A., Martínez-Cedeira, J. A., Newton, J., & Pierce, G. J. (2011). Stable isotope analysis in two sympatric populations of bottlenose dolphins *Tursiops truncatus*: Evidence of resource partitioning? *Marine Biology, 158,* 1043−1055.

Fernández, R., Santos, M. B., Pierce, G. J., Llavona, A., López, A., Silva, M. A., . . . Piertney, S. B. (2011). Fine-scale genetic structure of bottlenose dolphins, *Tursiops truncatus*, in Atlantic coastal waters of the Iberian Peninsula. *Hydrobiologia. Ecosystems and Sustainability, 610,* 211. Available from https://doi.org/10.1007/s10750-011-0669-5.

Ferreira, M., Marçalo, A., Nicolau, L., Araújo, H., Santos, J., Pinheiro, C., . . . Vingada, J. (2012). *Estado actual das redes de arrojamentos e de reabilitação em Portugal continental.* Anexo do Relatório intercalar do projecto LIFE MarPro PT/NAT/00038.

Figueiras, F., Labarta, U., & Fernández Reiriz, M. (2002). Coastal upwelling primary production and mussel growth in the Rías Baixas of Galicia. *Hydrobiologia, 484,* 121−131.

Fontaine, M. C., Baird, S. J., Piry, S., Ray, N., Tolley, K. A., Duke, S., . . . Michaux, J. R. (2007). Rise of oceanographic barriers in continuous populations of a cetacean: The genetic structure of harbour porpoises in Old World waters. *BMC Biology, 5,* 1−16.

Fontaine, M. C., Tolley, K. A., Michaux, J. R., Birkun, A., Jr., Ferreira, M., Jauniaux, T., & Llavona, Ń. (2010). Genetic and historic evidence for climate-driven population fragmentation in a top cetacean predator: The harbour porpoises in European waters. *Proceedings of the Royal Society B: Biological Sciences, 277*(1695), 2829−2837.

Fraga, F. (1981). Upwelling of the Galician coast, Northwest Spain. In F. Richards (Ed.), *Coastal upwelling.* (pp. 176−182). Washington, DC: American Geophysical Union.

Gaudin, M., Cirac, P., Trainer, J., Cremer, M., Schneider, J.-L., Bourillet, J.-F., . . . Griboulard, R. (2003). Morphology and sediment dynamics of the Capbreton canyon (Bay of Biscay, SW France). *Geophysical Research Abstracts, 5,* 13900.

Gil, J. (2003). Changes in the pattern of water masses resulting from a poleward slope current in the Cantabrian Sea (Bay of Biscay). *Estuarine, Coastal and Shelf Science, 57,* 1139−1149.

Gil, J., & Sanchez, R. (2003). Aspects concerning the occurrence of summer upwelling along the southern Bay of Biscay during 1993-2003. *ICES Marine Science Symposia, 219,* 337−339.

Hammond, P. S., Berggren, P., Benke, H., Borchers, D. L., Collet, A., Heide-Jørgensen, M. P., . . . Øien, N. (2002). Abundance of harbour porpoise and other cetaceans in the North Sea and adjacent waters. *Journal of Applied Ecology, 39,* 361−376.

Hammond, P. S., Macleod, K., Berggren, P., Borchers, D. L., Burt, M. L., Cañadas, A., . . . Vázquez, J. A. (2013). Cetacean abundance and distribution in European Atlantic shelf waters to inform conservation and management. *Biological Conservation, 164,* 107−122.

Hammond, P. S., Lacey, C., Gilles, A., Viquerat, S., Borjesson, P., Herr, H., . . . Øien, N. (2017). *Estimates of cetacean abundance in European Atlantic waters in summer 2016 from the SCANS-III aerial and shipboard surveys.* Available at <https://synergy.standrews.ac.uk/scans3/files/2017/05/SCANS-III-design-based-estimates-2017-05-12-final.revised.pdf>.

Holligan, P. M., Aarup, T., & Groom, S. B. (1989). The north sea satellite colour atlas. *Continental Shelf Research, 9,* 665−765.

Huang, W. G., Cracknell, A. P., Vaughan, R. A., & Davies, P. A. (1991). A satellite and field view of the Irish Shelf Front. *Continental Shelf Research, 11,* 543−562.

Kinze, C. C., Jensen, T., & Skov, R. (2003). Fokus på hvaler i Danmark 2000-2002. *Biologiske Skrifter, 2,* 1 47.

Kinze, C. C., Jensen, T., Tougaard, S., & Baagøe, H. J. (2010). Danske hvalfund i perioden 1998-2007 [Records of cetacean strandings on the Danish coastline during 1998-2007]. *Flora og Fauna, 116*(4), 91—99.

Kiszka, J., Macleod, K., Van Canneyt, O., Walker, D., & Ridoux, V. (2007). Distribution, encounter rates, and habitat characteristics of toothed cetaceans in the Bay of Biscay and adjacent waters from platform-of-opportunity data. *ICES Journal of Marine Science, 64*, 1033—1043.

Koschinski, S. (2002). Current knowledge on harbour porpoises (*Phocoena phocoena*) in the Baltic Sea. *Ophelia, 55*(3), 167—197.

Koutsikopoulos, C., & Le Cann, B. (1996). Physical processes and hydrological structures related to the Bay of Biscay anchovy. *Scientia Marina, 60*, 9—19.

Lacey, C. (2015). *Current status of the resident bottlenose dolphin population in the Sado Estuary, Portugal* (Unpublished MSc dissertation). University of Edinburgh.

Lah, L., Trense, D., Benke, H., Berggren, P., Gunnlaugsson, Þ, Lockyer, C., ... Tiedemann, R. (2016). Spatially explicit analysis of genome-wide SNPs detects subtle population structure in a mobile marine mammal, the harbor porpoise. *PLoS One, 11*(10), e0162792. Available from https://journals.plos.org/plosone/article?id = 10.1371/journal.pone.0162792.

Lambert, C., Pettex, E., Dorémus, G., Laran, S., Stephan, E., Van Canneyt, O., & Ridoux, V. (2017). How does ocean seasonality drive habitat preferences of highly mobile top predators? Part II: The eastern North-Atlantic. *Deep-Sea Research II, 14*, 133—154.

Laran, S., Authier, M., Blanck, A., Dorémus, G., Falchetto, H., Monestiez, P., ... Ridoux, V. (2017). Seasonal distribution and abundance of cetaceans within French waters: Part II: The Bay of Biscay and the English Channel. *Deep-Sea Research II, 14*, 31—40.

Llavona, Á. (2017). *Population parameters and genetic structure of the harbour porpoise (Phocoena phocoena, L. 1758) in the northwest coast of the Iberian Peninsula* (Ph.D. thesis) (373 pp.). Universidade de Aveiro.

López, A. (2003). *Estatus dos pequenos cetáceos da plataforma de Galicia* (Ph.D. thesis) (337 pp). Universidade de Santiago de Compostela.

López, A., Pierce, G. J., Valeiras, X., & Santos, M. B. (2004). Distribution patterns of small cetaceans in Galician waters. *Journal of the Marine Biological Association UK, 84*, 283—294.

McMahon, T., Raine, R., Titov, O., & Boychuk, S. (1995). Some oceanographic features of north-eastern Atlantic waters west of Ireland. *ICES Journal of Marine Science, 52*, 221—232.

New, A. L. (1988). Internal tidal mixing in the Bay of Biscay. *Deep-Sea Research, 35*, 691—709.

Penas-Patiño, X. M., & Piñeiro, A. (1989). *Cetáceos, focas e tartarugas mariñas das costas Ibéricas*. Santiago de Compostela: Ed. Consellería de pesca. Xunta de Galicia, 379 pp.

Pingree, R. D., & Griffiths, D. K. (1978). Tidal fronts on the shelf seas around the British Isles. *Journal of Geophysical Research, 83*, 4615—4622.

Pingree, R. D., Holligan, P. M., & Mardell, G. T. (1978). The effects of vertical stability on phytoplankton distribution in the summer on the northwest European Shelf. *Deep-Sea Research, 25*, 1011—1028.

Reid, J. B., Evans, P. G. H., & Northridge, S. P. (2003). *Atlas of cetacean distribution in north-west European waters*. Peterborough: Joint Nature Conservation Committee, 76 pp.

SAMBAH (2016). *Final report for LIFE + project SAMBAH LIFE08 NAT/S/000261 covering the project activities from 01/01/2010 to 30/09/2015* (80 pp.). Reporting date 29/02/2016.

Santos, J., Araújo, H., Ferreira, M., Henriques, A., Miodonski, J., Monteiro, S., ... & Vingada, J. (2012). Chapter I: Baseline estimates of abundance and Distribution of target species. In: *MARPRO Conservação de especies marinhas protegidas em Portugal continental* (80 pp.). Project LIFE09/NAT/PT/000038.

Scheidat, M., Gilles, A., Kock, K.-H., & Siebert, U. (2008). Harbour porpoises (*Phocoena phocoena*) abundance in the south western Baltic Sea. *Endangered Species Research, 5*, 215—223.

Siebert, U., Gilles, A., Lucke, K., Ludwig, M., Benke, H., Kock, K.-H., & Scheidat, M. (2006). A decade of occurrence of harbour porpoises in German waters: Aerial surveys, incidental sightings and strandings. *Journal of Sea Research, 56*, 65—80.

Simpson, J. H. (1981). The shelf-sea fronts: Implications of their existence and behaviour. *Philosophical Transactions of the Royal Society of London, A, 302*, 531—546.

Skóra, K. E., Pawliczka, I., & Klinowska, M. (1988). Observations of the harbour porpoise (*Phocoena phocoena*) on the Polish Baltic coast. *Aquatic Mammals, 14*(3), 113—119.

Smith, J. (2010). *The ecology of Cuvier's beaked whale, Ziphius cavirostris (Cetacea, Ziphiidae), in the Bay of Biscay* (Ph.D. thesis). University of Southampton.

Sveegaard, S., Galatius, A., Dietz, R., Kyhn, L., Koblitz, J. C., Amundin, M., ... Teilmann, J. (2015). Defining management units for cetaceans by combining genetics, morphology, acoustics and satellite tracking. *Global Ecology & Conservation, 3*, 83—850. Available from https://doi.org/10.1016/j.gecco.2015.04.002.

Tomilin, A. G. (1957). *Zveri SSSR i Prilezhasfchikh Stran. Zveri VostochnoiEvropy i Severnoi Azii*. Izdatel' stvo Akademi Nauk SSSR, Moscow. 756 pp. (translated in 1967 as V. G. Heptner (Ed.), *Mammals of the USSR and adjacent countries. Mammals of eastern Europe and adjacent countries* (717 pp.). Vol. IX, Cetacea by the Israel Program for Scientific Translations, Jerusalem).

Torres, R., & Barton, E. D. (2007). Onset of the Iberian upwelling along the Galician coast. *Continental Shelf Research, 27*(13), 1759—1778.

Valencia, V., Borja, A., Fontan, A., Perez, F. F., & Rios, A. F. (2003). Temperature and salinity fluctations along the Basque Coast (southern Bay of Biscay), from 1986-2000, related to climatic factors. *ICES Marine Science Symposia, 219*, 340—342.

Valencia, V., Franco, J., Borja, A., & Fontan, A. (2004). Hydrography of the southeastern Bay of Biscay. In A. Borja, & M. E. Collins (Eds.), *Oceanography and the marine environment of the Basque Country* (pp. 159—194). Elsevier B.V.

Verfuß, U. K., Honnef, C. G., Medling, A., Dähne, M., Mundrry, R., & Benke, H. (2007). Geographical and seasonal variation of harbour porpoise (*Phocoena phocoena*) in the German Baltic Sea revealed by passive acoustic monitoring. *Journal of the Marine Biological Association of the UK*, *87*, 165–176.

Wall, D., Murray, C., O'Brien, J., Kavanagh, L., Wilson, C., Ryan, C., ... Berrow, S. (2013). *Atlas of the distribution and relative abundance of marine mammals in Irish offshore waters 2005-2011*. Merchants Quay, Kilrush, Co. Clare: Irish Whale and Dolphin Group.

Wiemann, A., Andersen, L. W., Berggren, P., Siebert, U., Benke, H., Teilmann, J., ... Tiedemann, R. (2010). Mitochondrial control region and microsatellite analyses on harbour porpoise (*Phocoena phocoena*) unravel population differentiation in the Baltic Sea and adjacent waters. *Conservation Genetics*, *11*, 195–211.

Chapter 4

Systematic list of European cetacean species

Let us now consider in a little more detail the 36 species of cetacean that have been recorded in the ASCOBANS Agreement Area, with an emphasis upon the 16 species that occur regularly in the region, including 12 species of small cetacean covered by the Agreement. The list does not cover all species recorded in European seas. One species, Antarctic minke whale, has been identified in the Norwegian Arctic (see common minke whale species account). Two other species have been recorded as vagrants in the Mediterranean Sea but not yet in northwest Europe: a grey whale (presumed to have migrated from the North Pacific via the Arctic Ocean) was sighted in May 2010 off Barcelona, Spain, having been recorded 3 weeks earlier off Israel; and the Indian Ocean humpback dolphin has been recorded off the island of Crete, Greece in November 2017, as well as in Turkey in April 2016.

In the species accounts below, small cetaceans are treated first since they are the subject of the ASCOBANS Agreement. Taxonomy follows that of the Society of Marine Mammalogy's Committee on Taxonomy (2018), although within families, species are arranged as in Würsig, Thewissen, and Kovacs (2018), rather than alphabetically by scientific name (see Table 3.1). Readers should note that abundance estimates given in the species accounts come from publications but there is uncertainty around them and methods are continually evolving so they may get revised subsequently (see, e.g. boxed feature by Philip Hammond in Chapter 6, Conservation research, which also shows the precise survey areas for which the abundance estimates apply). For discussion of the various issues related to obtaining abundance estimates, see Chapter 6, Conservation research.

References

Committee on Taxonomy (2018). *List of marine mammal species and subspecies*. Society for Marine Mammalogy <www.marinemammalscience.org> consulted on December 31, 2018.

Würsig, B., Thewissen, J. G. M., & Kovacs, K. (Eds.), (2018). *Encyclopedia of marine mammals* (3rd ed.). New York: Academic Press, 1157 pp.

ORDER Cetartiodactyla
SUBORDER Cetacea
INFRAORDER Odontoceti, the toothed whales

Family Phocoenidae (porpoises)

Harbour porpoise *Phocoena phocoena*

This, the commonest and most widely distributed species of cetacean in northern European seas, is also the smallest, with adults averaging a length of around 1.5 m. Porpoises only occasionally leap clear of the water, and one's usual view is of a small, centrally placed triangular dorsal fin and a glimpse of the back (Fig. 4.1). It has a small rotund body with a small head with no forehead or beak, and short rounded flippers (Fig. 4.2). Its back is dark grey with a paler

FIGURE 4.1 Harbour Porpoise. On surfacing, the porpoise rarely shows more than its back, upper flanks and small triangular dorsal fin. *Photo: Pia Anderwald.*

FIGURE 4.2 The Harbour Porpoise has a small rotund body with a small head and no forehead or beak. *Photo: Ian Birks.*

patch on the flanks extending up the sides in front of the dorsal fin and from the flippers a dark grey line extends to the jawline.

The harbour porpoise is restricted to temperate and subarctic (mainly 11°C–14°C) seas of the northern hemisphere, occurring in both the Atlantic and Pacific. In the North Atlantic, the species occurs mainly from central west Greenland and Novaya Zemlya in the north to North Carolina and Senegal in the south, with a geographically isolated population in the Black Sea (Fig. 4.3) (Evans, Lockyer, Smeenk, Addink, & Read, 2008).

In European seas it is common and widely distributed over the continental shelf (mainly at depths of 20–200 m) from the Barents Sea and Iceland south to the coasts of France and Spain, although in the 1970s it became scarce in the southernmost North Sea, English Channel and Bay of Biscay. Nevertheless, it remains the most widely distributed and frequently observed cetacean in Northwest European shelf seas, and since the 1990s, has returned to the southernmost North Sea, English Channel and French Biscay coast (Camphuysen, 2004; Evans, 2010; Hammond et al., 2013; Kiszka, Hassani, & Pezeril, 2004; Kiszka, Macleod, Van Canneyt, & Ridoux, 2007).

Porpoises tend to be relatively solitary although aggregations in the tens or even low hundreds have been observed on occasions, and may be associated with concentrations of prey such as sandeel, sprat, herring or whiting. Numbers

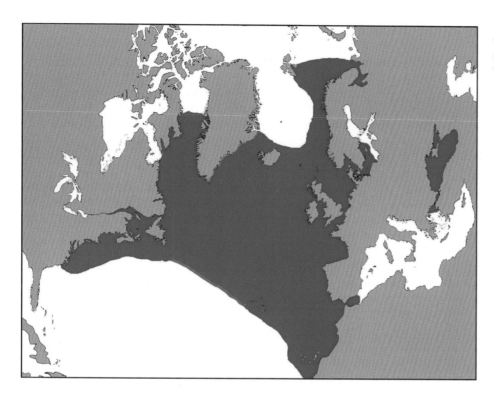

FIGURE 4.3 North Atlantic distribution of Harbour Porpoise (depicting those areas where the species is thought to regularly occur).

and densities of porpoise tend to be higher during spring or summer months, particularly in shelf seas and nearer the coast (see, e.g. Evans et al., 2003; Gilles et al., 2016; Laran et al., 2017).

Up to 16 separate Management Units (MU), thought to be demographically if not genetically distinct were proposed for the harbour porpoise in the North Atlantic, using a precautionary approach of division rather than lumping where there was evidence to suggest this (Evans & Teilmann, 2009). These included up to 11 MUs within the ASCOBANS Agreement Area which encompasses most of northwest Europe. These were: (1) the inner Baltic Sea (with its western boundary being around the Darss/Gedser underwater ridge or Rügen); (2) the southern Belt Sea; (3) the northern Belt Sea and southern Kattegat; (4) the northern Kattegat and Skagerrak; (5) the northeast sector of the North Sea; (6) the western sector of the North Sea including the Eastern Channel; (7) the Celtic Sea along with southwest Ireland, the Irish Sea and Western Channel; (8) northwest Ireland and western Scotland; (9) the waters around the Faroe Islands south towards the Faroe–Shetland Channel; (10) the west French part of the Bay of Biscay and (11) the Iberian Peninsula (coasts of Portugal and Atlantic Spain). The ICES Working Group on Marine Mammal Ecology (ICES, 2014) combined some of these into five assessment units. Further research is likely to refine and modify these subpopulations. The main populations distinguished genetically appear to occur in the following five regions: inner Baltic Sea; the Belt Seas/southern Kattegat; northern Kattegat, Skagerrak and North Sea; Celtic Seas from western Scotland south to the Bay of Biscay; and around the Iberian Peninsula (with some admixing between those in the Irish Sea, western English Channel and Bay of Biscay and those around the Iberian Peninsula) (Fontaine et al., 2017, 2010; Lah et al., 2016; Sveegaard et al., 2015). Re-visiting the genetic basis for these divisions, an IMR/NAMMCO workshop (NAMMCO & IMR, 2019) has proposed nine areas in the ASCOBANS Agreement Area with some adjustments over those indicated earlier: (1) the inner Baltic Sea; (2) the Belt Seas & southern Kattegat; (3) an area of overlap between the two in the western inner Baltic; (4) Greater North Sea (including Skagerrak, northern Kattegat, and eastern English Channel); (5) Celtic Seas (including western English Channel, Bristol Channel, and north-eastern Bay of Biscay); (6) Irish Sea; (7) West Scotland & Ireland; (8) the waters around the Faroe Islands; and (9) the Iberian Peninsula (including Spanish coast of the Bay of Biscay). Elsewhere in the North Atlantic, seven areas were distinguished: (1) Norwegian & Russian Coasts (north of the Greater North Sea); (2) Iceland; (3) East Greenland; (4) West Greenland; (5) Newfoundland & Labrador; (6) Scotian Shelf; and (7) Eastern United States. Further research is likely to refine and modify these various proposed divisions.

Although porpoises can be found in deep waters off the edge of the continental shelf (e.g. within the Faroe Bank Channel), they are comparatively rare in waters exceeding depths of 200 m. However, recent tagging of porpoises in west Greenland has shown they move out seasonally into the deep waters of the central North Atlantic (Nielsen et al., 2018). The species frequently uses tidal conditions for foraging (see e.g. Evans, 1997; Isojunno, Matthiopoulos, & Evans, 2012; Johnston, Westgate, & Read, 2005; Jones et al., 2014; Marubini, Gimona, Evans, Wright, & Pierce, 2009; Pierpoint, 2008; Waggitt, Dunn, Evans, Hiddink et al., 2017).

Population estimates do not exist for the entire North Atlantic range of the harbour porpoise, or even for the European range. However, the widest scale surveys were SCANS undertaken in 1994, followed by SCANS-II in 2005, and SCANS-III in 2016. Estimates of abundance have been published for SCANS and SCANS-II (Hammond et al., 2002, 2013), but these have been superseded by revised estimates calculated as part of the SCANS-III project (Hammond et al., 2017) and it is these revised estimates that are presented below and included in Table 6.1.

From the SCANS survey in July 1994 the overall population estimate was 407,177 porpoises (CV 0.18, 95% CI: 289,000−574,000) (revised from Hammond et al., 2002), with the following regional estimates: North Sea including Shetland and Orkney (289,000), NW Scotland (c. 9,150), Skagerrak/Kattegat/Belt Seas/Western Baltic Sea (51,700), Channel (0) and Celtic Shelf (57,200). Only a small portion of the Baltic Proper was surveyed.

The repeat survey in July 2005 (SCANS-II), covering a wider area (continental shelf seas from SW Norway, south to Atlantic Portugal), gave an estimate of 519,864 (CV 0.21, 95% CI: 344,000−797,000) (revised from Hammond et al., 2013), with regional estimates: North Sea including the eastern Channel (355,000), E. Skagerrak/Kattegat/Belt Seas/Western Baltic Sea (27,900), Celtic and Irish Seas, including the western Channel (107,000), Atlantic Ireland and Scotland (26,300) and the Iberian Peninsula (2,880).

Comparing these two surveys, the number of harbour porpoise estimated for the North Sea overall increased slightly from 289,000 (CV 0.14) in 1994 to 355,000 (CV 0.22) in 2005. However, the estimated number in the northern North Sea declined from 234,000 to 189,000, whereas in the central and southern North Sea it increased from 54,700 to 167,000. This is thought to represent a southward shift in distribution rather than actual changes in population size (Hammond et al., 2013), at least for the month of July.

In July 2016 the SCANS-III survey covered the same continental shelf area surveyed in 2005 (except Irish waters covered by the ObSERVE survey), and also waters along the coast of Norway as far north as Vestfjorden. The overall estimate from SCANS-III was 466,569 (CV 0.15, 95% CI: 345,000−630,000) harbour porpoise (Hammond et al., 2017), with regional estimates: North Sea (345,000), Kattegat/Belt Seas/Western Baltic Sea (42,300), Celtic and Irish Seas (49,200), Atlantic Ireland and Scotland (29,000), and the Iberian Peninsula (2,900). The southerly shift in distribution seen from 1994 to 2005 was maintained in 2016.

The ObSERVE survey of Irish waters in summer 2016 (Rogan et al., 2018) complemented SCANS-III and together these surveys provided area coverage equivalent to SCANS-II 2005. The estimated abundance of harbour porpoise from ObSERVE (excluding the Irish Sea which was included in SCANS-III) was 26,636 (CV 0.23, 95% CI: 17,100−41,500) and the total estimate for summer 2016 (SCANS-III + ObSERVE) is thus 493,205 (CV 0.15, 95% CI: 371,000−656,000).

A comparison of abundance estimates for an equivalent area (North Sea) in 1994, 2005 and 2016 showed no significant trend, nor any in inner Danish waters (Hammond et al., 2017).

In Norwegian waters, estimates of 11,000 porpoises (95% CI: 4790−25,200) for the Barents Sea and Norwegian waters north of 66°N, and 82,600 (95% CI: 52,100−131,000) for southern Norway and the northern North Sea, were made during July 1989 (Bjørge & Øien, 1995). A more recent abundance assessment for Norwegian waters in 2016 indicated 166,700 porpoises (CV 0.18), split more or less equally between the Norwegian North Sea and the Barents Sea and eastern Norwegian Sea north of 62°N (A. Bjørge, personal communication). These abundance estimates exclude the Norwegian fjords where densities appear to be higher than in the open North Sea, so the overall numbers will be higher.

The population of harbour porpoises in the inner Baltic Sea (the Baltic Proper) was too small to estimate by conventional visual surveys, so an alternative approach using a large number of static acoustic devices (C-PODs), deployed between 2011 and 2013, has been used. This has resulted in an abundance estimate of 497 individuals (95% CI: 80−1091) (SAMBAH, 2016).

The main threats to harbour porpoise in the ASCOBANS Agreement Area are thought to be bycatch (mainly from bottom set gill nets and pelagic trawls), attacks by bottlenose dolphins, depletion of prey resources, physiological effects on reproduction and infectious disease potentially arising from high contaminant levels (particularly of PCBs − see, e.g. Jepson et al., 2016). From postmortem examinations of 1692 porpoises stranded in the United Kingdom between 1991 and 2010, 23% were thought to have died from infectious disease, 19% as a result of bottlenose dolphin attacks, 17% were diagnosed as bycatch, 15% died from starvation and 4% live-stranded (Deaville & Jepson, 2011).

References

Bjørge, A., & Øien, N. (1995). Distribution and abundance of harbour porpoise, *Phocoena phocoena*, in Norwegian waters. *Reports of the International Whaling Commission* (Special issue 16), 89–98.

Camphuysen, C. J. (2004). The return of the harbour porpoise (*Phocoena phocoena*) in Dutch coastal waters. *Lutra, 47*, 113–122.

Deaville, R., & Jepson, P. D. (compilers) (2011). *UK Cetacean strandings investigation programme* (98 pp). Final Report to Defra for the period 1st January 2005–31st December 2010. London: Institute of Zoology. http://randd.defra.gov.uk/Document.aspx?Document = FinalCSIPReport2005-2010_finalversion061211released[1].pdf.

Evans, P. G. H. (1997). *Ecological studies of the harbour porpoise in Shetland, North Scotland* (106 pp). Report for WWF-UK. Oxford: Sea Watch Foundation.

Evans, P. G. H. (2010). *Review of trend analyses in the ASCOBANS Area* (68 pp). ASCOBANS AC17/Doc. 6-08 (S).

Evans, P. G. H., Anderwald, P., & Baines, M. E. (2003). *UK Cetacean Status Review. Report to English Nature and Countryside Council for Wales.* Oxford: Sea Watch Foundation, 160 pp.

Evans, P. G. H., Lockyer, C. H., Smeenk, C., Addink, M., & Read, A. J. (2008). Harbour porpoise *Phocoena phocoena*. In S. Harris, & D. W. Yalden (Eds.), *Mammals of the British Isles* (4th ed., pp. 704–709). Southampton: The Mammal Society, 800 pp, Handbook.

Evans, P.G.H. and Teilmann, J. (editors) (2009) *Report of ASCOBANS/HELCOM Small Cetacean Population Structure Workshop.* ASCOBANS/UNEP Secretariat, Bonn, Germany. 140 pp.

Fontaine, M. C., Thatcher, O., Ray, N., Piry, S., Brownlow, A., Davison, N. J., . . . Goodman, S. J. (2017). Mixing of porpoise ecotypes in southwestern UK waters revealed by genetic profiling. *Royal Society Open Science, 4*, 160992. Available from https://doi.org/10.1098/rsos.160992.

Fontaine, M. C., Tolley, K. A., Michaux, J. R., Birkun, A., Jr., Ferreira, M., Jauniaux, T., . . . Baird, S. J. E. (2010). Genetic and historic evidence for climate-driven population fragmentation in a top cetacean predator: The harbour porpoises in European waters. *Proceedings of the Royal Society B: Biological Sciences, 277*(1695), 2829–2837.

Gilles, A., Viquerat, S., Becker, E. A., Forney, K. A., Geelhoed, S. C. V., Haelters, J., . . . Aarts, G. (2016). Seasonal habitat-based density models for a marine top predator, the harbour porpoise, in a dynamic environment. *Ecosphere, 7*(6), e01367. Available from https://doi.org/10.1002/ecs2.1367.

Hammond, P. S., Berggren, P., Benke, H., Borchers, D. L., Collet, A., Heide-Jørgensen, M. P., . . . Øien, N. (2002). Abundance of harbour porpoise and other cetaceans in the North Sea and adjacent waters. *Journal of Applied Ecology, 39*, 361–376.

Hammond, P.S., Lacey, C., Gilles, A., Viquerat, S., Borjesson, P., Herr, H., . . . and Øien, N. (2017) Estimates of cetacean abundance in European Atlantic waters in summer 2016 from the SCANS-III aerial and shipboard surveys. Available at https://synergy.standrews.ac.uk/scans3/files/2017/05/SCANS-III-design-based-estimates-2017-05-12-final revised.pdf.

Hammond, P. S., Macleod, K., Berggren, P., Borchers, D. L., Burt, M. L., Cañadas, A., . . . Vázquez, J. A. (2013). Cetacean abundance and distribution in European Atlantic shelf waters to inform conservation and management. *Biological Conservation, 164*, 107–122.

ICES (2014). *Report of the Working Group on Marine Mammal Ecology (WGMME), 10-13 March, Woods Hole, Massachusetts, USA* (232 pp). ICES CM2014/ACOM: 27.

Isojunno, S., Matthiopoulos, J., & Evans, P. G. H. (2012). Harbour porpoise habitat preferences: Robust spatio-temporal inferences from opportunistic data. *Marine Ecology Progress Series, 448*, 155–170.

Jepson, P. D., Deaville, R., Barber, J. L., Aguilar, A., Borrell, A., Murphy, S., . . . Law, R. J. (2016). PCB pollution continues to impact populations of orcas and other dolphins in European waters. *Nature Scientific Reports, 6*, 18573. Available from https://doi.org/10.1038/srep18573.

Johnston, D. W., Westgate, A. J., & Read, A. J. (2005). Effects of fine-scale oceanographic features on the distribution and movements of harbour porpoises *Phocoena phocoena* in the Bay of Fundy. *Marine Ecology Progress Series, 295*, 279–293. Available from http://www.int-res.com/abstracts/meps/v295/p279-293/.

Jones, A. R., Hosegood, P., Wynn, R. B., De Boer, M. N., Butler-Cowdry, S., & Embling, C. B. (2014). Fine-scale hydrodynamics influence the spatio-temporal distribution of harbour porpoises at a coastal hotspot. *Progress in Oceanography, 128*, 30–48. Available from http://www.sciencedirect.com/science/article/pii/S0079661114001256.

Kiszka, J., Macleod, K., Van Canneyt, O., & Ridoux, V. (2007). Distribution, relative abundance and bathymetric preferences of toothed cetaceans in the English Channel and Bay of Biscay. *ICES Journal of Marine Science*, 1033–1043.

Kiszka, J. J., Hassani, S., & Pezeril, S. (2004). Distribution and status of small cetaceans along the French Channel coasts: Using opportunistic records for a preliminary assessment. *Lutra, 47*, 33–46.

Lah, L., Trense, D., Benke, H., Berggren, P., Gunnlaugsson, Þ, Lockyer, C., . . . Tiedemann, R. (2016). Spatially explicit analysis of genome-wide SNPs detects subtle population structure in a mobile marine mammal, the harbor porpoise. *PLoS One, 11*(10), e0162792. Available from https://journals.plos.org/plosone/article?id = 10.1371/journal.pone.0162792.

Laran, S., Authier, M., Blanck, A., Dorémus, G., Falchetto, H., Monestiez, P., . . . Ridoux, V. (2017). Seasonal distribution and abundance of cetaceans within French waters: Part II: The Bay of Biscay and the English Channel. *Deep-Sea Research II, 14*, 31–40.

Marubini, F., Gimona, A., Evans, P. G. H., Wright, P. J., & Pierce, G. J. (2009). Habitat preferences and interannual variability in occurrence of the harbour porpoise *Phocoena phocoena* in the north-west of Scotland. *Marine Ecology Progress Series, 381*, 297–310.

Nielsen, N. H., Teilmann, J., Sveegaard, S., Hansen, R. G., Sinding, M., Dietz, R., & Heide-Jørgensen, M. P. (2018). Oceanic movements, site fidelity and deep diving in harbour porpoises from Greenland show limited similarities to animals from the North Sea. *Marine Ecology Progress Series, 597*, 259–272.

North Atlantic Marine Mammal Commission and the Norwegian Institute of Marine Research. (2019) *Report of Joint IMR/NAMMCO International Workshop on the Status of Harbour Porpoises in the North Atlantic.* Tromsø, Norway. 235 pp.

Pierpoint, C. (2008). Harbour porpoise (*Phocoena phocoena*) foraging strategy at a high energy, near-shore site in south-west Wales, UK. *Journal of the Marine Biological Association UK, 88*, 1167−1173.

Rogan, E., Breen, P., Mackey, M., Cañadas, A., Scheidat, M., Geelhoed, S., & Jessopp, M. (2018). *Aerial surveys of cetaceans and seabirds in Irish waters: Occurrence, distribution and abundance in 2015-2017*. Dublin: Department of Communications, Climate Action & Environment and National Parks and Wildlife Service (NPWS), Department of Culture, Heritage and the Gaeltacht, 297 pp.

SAMBAH (2016) Final report for LIFE + project SAMBAH LIFE08 NAT/S/000261 covering the project activities from 01/01/2010 to 30/09/2015. Reporting date 29/02/2016, 80 pp.

Sveegaard, S., Galatius, A., Dietz, R., Kyhn, L., Koblitz, J. C., Amundin, M., ... Teilmann, J. (2015). Defining management units for cetaceans by combining genetics, morphology, acoustics and satellite tracking. *Global Ecology and Conservation, 3*, 839−850. Available from https://doi.org/10.1016/j.gecco.2015.04.002.

Waggitt, J. J., Dunn, H., Evans, P. G. H., Hiddink, J., Holmes, L., Keen, E., ... Veneruso, G. (2017). Regional-scale patterns in harbor porpoise occupancy of tidal stream environments. *ICES Journal of Marine Science, 75*(2), 701−710. Available from https://doi.org/10.1093/icesjms/fsx164.

Winship, A. J. (2009). *Estimating the impact of bycatch and calculating bycatch limits to achieve conservation objectives as applied to harbour porpoise in the North Sea* (Ph.D. thesis). UK: University of St Andrews.

Family Delphinidae

Rough-toothed dolphin *Steno bredanensis*

The rough-toothed dolphin is a tropical species occurring in all oceans of the world. In the North Atlantic, it ranges from equatorial waters north to the Gulf of Mexico in the west, and the Mediterranean in the east. It is regular in the Windward Islands, Canaries and Cape Verdes, occurring also less commonly in the Azores and Madeira. It has the typical delphinid body shape, with adults reaching a length of 2.5−2.8 m. Its most distinctive feature is the long, conical head without any demarcation between the gently sloping melon and the fairly long beak. The lips and much of the lower jaw are white, sometimes with a pinkish hue. The flippers are relatively large and set further back than most dolphins. The dorsal fin is fairly tall and erect, and slightly recurved. Body coloration is dark grey with a distinct narrow cape over the back that dips down onto the upper flanks below the dorsal fin, and a white or pinkish belly (Fig. 4.4).

FIGURE 4.4 The Rough-toothed Dolphin has a long, conical head, gently sloping melon, white or pinkish lips to its long beak, and often pinkish underparts. *Photo: Caroline R Weir.*

There are two, possibly three, records of this species thought to be from within the ASCOBANS Agreement Area. Around 1825 Prof. Van Breda made drawings of a fresh animal and its skull, assumed to have come locally from the Ghent area — the Scheldt Estuary on the border between Belgium and the Netherlands. He then published these drawings 4 years later, after showing them to the French anatomist, Prof. Georges Cuvier, but neither included any description nor the location from which it came. Since then the assumption that it was local has entered into various publications (see, e.g. Miyazaki & Perrin, 1994; Rice, 1998; West, Mead, & White, 2011). However, there is no firm evidence for this (De Smet, 1974), although in 1862 Schlegel specified the locality as in or near the mouth of the Scheldt, information that was presumed to come from Van Breda (Smeenk & Camphuysen, 2016). Subsequent to that report, there was a lower jaw of the species allegedly found in 1877 in a ditch near Bruinisse, on the former island of Schouwen-Duiveland, province of Zeeland and preserved in the Zeeuws Museum in Middelburg. There has been some speculation that it might have belonged to Van Breda (who came from, and later returned to, nearby Zierikzee on the same island), and so could have been of his specimen of *Steno*, the whereabouts of which are unknown, or it may have been from an animal caught or stranded in the area in the period 1822—55 (Smeenk & Camphuysen, 2016). Smeenk (2018) provides a chronological review of the nomenclature of this species in an attempt to elucidate the, often confused, history of its nomenclature.

In France there is a sighting report of rough-toothed dolphin from near Brest (Brittany), but the accompanying drawing is insufficient to identify the specimen so the record remains uncertain (Duguy, 1983). Another case of this species was described near Paimpol, northern Brittany, in an old reference (Van Beneden, 1889), but this relates to a mass stranding of pilot whales in 1812 (Cuvier, 1812), with no reference to *Steno*. Van Beneden also mentions the presence of skulls from *Steno* in the collection of the National Museum of Paris; however, no such skull has been found, suggesting that he may have been confused with another species. We therefore have no definite record of the species from northern Europe, although it is quite possible that some of those records from the region are indeed of this species. Interestingly a small population exists in the eastern Mediterranean (from the Sicily Channel eastwards) with genetic affinities to the Atlantic (Kerem et al., 2016).

References

Cuvier, G. (1812). Description des cétacés échoués dans la baye de Paimpol. *Nouveau Bulletin des Sciences, par la Société Philomatique de Paris, 3* (56), 69—91.

De Smet, W. M. A. (1974). Inventaris van de walvisachtigen (Cetacea) van de Vlammse kust en de Schelde. *Bulletin van het Kominklijk Belgisch Instituut voor Natuurwetenschappen, 50*(1), 1—156.

Duguy, R. (1983). Les Cétacés des Côtes de France. *Annalès de la Société des Sciences Naturelles de la Charente-Maritime* (Suppl.), 112 pp.

Kerem, D., Goffman, O., Elasar, M., Hadar, N., Scheinin, A., & Lewis, T. (2016). The rough-toothed dolphin *Steno bredanensis*, in the eastern Mediterranean Sea: A relict population? *Advances in Marine Biology, 75*, 233—258.

Miyazaki, N., & Perrin, W. F. (1994). Rough-toothed dolphin *Steno bredanensis*. In S. H. Ridgway, & R. Harrison (Eds.), *Handbook of marine mammals. Vol. 5: The first book of dolphins* (pp. 1—22). London: Academic Press.

Rice, D. W. (1998). *Marine mammals of the world. Systematics and distribution* (pp. 1—231). The Society for Marine Mammalogy Special Publication No. 4.

Smeenk, C. (2018). A chronological review of the nomenclature of *Delphinus rostratus*, Shaw, 1801 and *Delphinus bredanensis* (Lesson, 1828). *Lutra, 61*(1), 189—195.

Smeenk, C., & Camphuysen, C. J. (2016). Snaveldolfojn *Steno bredanensis*. In S. Broekhuizen, K. Spoelstra, J. B. M. Thissen, K. J. Canters, & J. C. Buys (Eds.), *Atlas van de Nederlandse Zoogdieren — Natuur van Nederland 12* (pp. 354—355). Leiden, The Netherlands: Naturalis Biodiversity Center & EIS Kniscntrum Insecten en andere ongerwervelden.

Van Beneden, P.-J. (1889). *Histoire naturelle des cétacés des mers d'Europe*. Brussels: F. Hayez.

West, K. L., Mead, J. G., & White, W. (2011). *Steno bredanensis* (Cetacea: Delphinidae). *Mammalian Species, 43*(886), 177—189.

Common bottlenose dolphin *Tursiops truncatus*

Probably the most familiar of all whales and dolphins through TV appearances and displays in dolphinaria is the bottlenose dolphin, now frequently referred to as the common bottlenose dolphin to distinguish it from the smaller Indo-Pacific bottlenose dolphin *Tursiops aduncus*. Particularly in cool temperate North Atlantic waters, it is a relatively large, stout dolphin, reaching adult lengths of 3.0—3.8 m. In warmer tropical seas the species is distinctly smaller and slimmer. Geographic variation also commonly occurs between animals occupying coastal areas and those living offshore. Unlike many other dolphin species, it lacks distinctive markings, being dark grey on the back, lighter grey on the

FIGURE 4.5 The Bottlenose Dolphin is generally grey above grading to white below wth a rounded forehead and short beak. *Photo: Peter GH Evans.*

flanks and grading to white on the belly. When breaching backwards, it displays its white throat and belly and bottle-shaped nose, which gives the species its name (Fig. 4.5).

The rounded head has a distinct short beak that is often tipped with white on the lower jaw. The flippers are fairly long and pointed; the centrally placed dorsal fin is tall, slender and recurved, although its size and shape can be very variable. Nicks along the edge of the dorsal fin and scratches on the fin and back are commonly used to identify individuals.

The bottlenose dolphin has a worldwide distribution in tropical and temperate seas in both hemispheres. In the North Atlantic, it occurs from Nova Scotia in the west and the Faroe Islands in the east (occasionally as far north as northern Norway and Svalbard), southwards to the Equator and beyond (Fig. 4.6).

Along the Atlantic seaboard of Europe, the species is locally fairly common near-shore off the coasts of Spain, Portugal, northwest France, western Ireland (particularly the Shannon Estuary and Connemara), northeast Scotland (particularly the Moray Firth south to the Firth of Forth), southwest Scotland, in the Irish Sea (particularly north and west Wales, including all of Cardigan Bay) and in the English Channel (Arso Civil et al., 2019; Baines & Evans, 2012; Berrow et al., 2010; Cheney et al., 2013; Evans et al., 2003; Louis et al., 2015; ICES, 2016; Nykanen et al., 2015; Reid et al., 2003). Smaller groups of bottlenose dolphins collectively numbering about 45 individuals have also taken up residence at other localities — for example around the Outer Hebridean island of Barra, and in the Inner Hebrides (Islay, Mull, Coll, Tiree and southern Isle of Skye) in west Scotland (Cheney et al., 2013; Evans et al., 2003).

The species also occurs offshore in the eastern North Atlantic, especially along the shelf edge (where it may occur in association with long-finned pilot whales), as far north as the Faroe Islands and even Svalbard (Reid et al., 2003). In the Bay of Biscay there are particularly high numbers over the outer shelf and shelf break, with abundance increasing during the winter months (Certain, Ridoux, Van Canneyt, & Bretagnoile, 2008; Laran et al., 2017). During the summer some may enter near-shore waters around the Faroe Islands, northern and western Scotland, western Ireland, in the Bay of Biscay and around the Iberian Peninsula (Galicia and coast of Portugal) (Certain et al., 2008; Evans et al., 2003; Reid et al., 2003).

In coastal waters bottlenose dolphins often favour river estuaries, headlands or sandbanks where there is uneven bottom relief and/or strong tidal currents (Ingram & Rogan, 2002; Lewis & Evans, 1993; Liret, 2001; Liret, Allali, Creton, Guinet, & Ridoux, 1994; Rogan, Ingram, Holmes, & O'Flanagan, 2000; Wilson, Thompson, & Hammond, 1997). Those locations often combine favourable feeding conditions with shallow seas that enable calves to learn to forage off the seabed (Fig. 4.7). The species is rare in the southern North Sea, in Danish waters and in the Baltic Sea.

Genetic studies using mitochondrial DNA indicated some population substructuring between animals from east and west Scotland and from Wales, although for the latter two regions sample sizes were low; there was also a low level of genetic diversity in the east Scottish sample (Parsons et al., 2002). A more recent and wider genetic analysis, using both

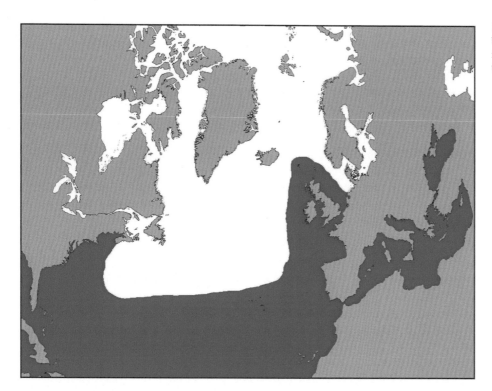

FIGURE 4.6 North Atlantic distribution of bottlenose dolphin (depicting those areas where the species is thought to regularly occur).

FIGURE 4.7 Bottlenose Dolphin adult and calf off the coast of Wales. *Photo: Pia Anderwald.*

mtDNA and microsatellites, found that coastal and pelagic bottlenose dolphins were highly differentiated in the Northeast Atlantic, with some finer-scale population structure within the two groups (Louis et al., 2014). Using both mtDNA and microsatellite techniques, fine-scale population structure was revealed also among three distinct populations in Ireland – one in the Shannon Estuary, another from the Connemara–Mayo region and a third from strandings of unknown origin but thought to be part of a large offshore population (Mirimin et al., 2011). This study found moderate (microsatellite) to low (mitochondrial) gene diversity in dolphins using the Shannon Estuary and the Connemara—Mayo region, while dolphins that stranded along the Atlantic coast showed much higher levels of gene diversity at both classes of markers (Mirimin et al., 2011).

Further south a genetic study by Fernández, Santos, et al., (2011) of stranded bottlenose dolphins in Galicia, northwest Spain, using mtDNA and microsatellites, supported by a stable isotope study (Fernández, García-Tiscar, et al., 2011) found significant differences between animals in northern and southern Galicia, as well as from Portugal. There were also significant differences in microsatellite frequencies between southern Galicia and the northeastern corner of Spain. However, most of these sample sizes were very small. These studies, as well as one by Nichols et al., (2007) using ancient DNA that found significant differentiation in a now extinct population from the Humber Estuary, eastern England, suggest that local adaptation leading to isolation and potential extinction of coastal populations of bottlenose dolphin in Europe may be a feature of this species, exposing them to long-term conservation risk. Offshore populations, on the other hand, seem to exhibit much higher levels of gene flow (e.g. Madeira and the Azores — see Querouil et al., 2007) and be more diverse genetically (Louis et al., 2014).

SCANS-II surveys of Northwest European shelf waters in July 2005 gave an overall abundance estimate of 16,641 (CV 0.42, 95% CI: 7600—36,400) (revised from Hammond et al., 2013), whereas, offshore, the CODA survey (July 2007) yielded an abundance estimate, uncorrected for g(0) and responsive movement, of 19,295 (CV 0.25, 95% CI: 11,900—31,400) (Hammond et al., 2009). A combined estimate for the entire region surveyed in 2005/07 is 35,936 (CV 0.24, 95% CI: 22,800—56,800) (see Table 6.1).

The SCANS-III survey in July 2016 (covering the same waters as SCANS-II and CODA combined but excluding Irish waters surveyed by the ObSERVE project) yielded an abundance estimate of 27,697 (CV 0.23, 95% CI: 17,700—43,400) bottlenose dolphins (see Table 6.1). Greatest numbers occurred in deep waters west of the Outer Hebrides and in the Bay of Biscay (Hammond et al., 2017).

The ObSERVE survey covering Irish waters complemented SCANS-III and together these surveys provided equivalent coverage to SCANS-II + CODA 2005/07. Greatest numbers occurred offshore south and west of Ireland, and, in both years (2015 and 2016), they were more frequently seen in winter than in summer, with a maximum count of 212,600 (CV 0.15, 95% CI: 157,000—288,000) in winter 2016—17 (Rogan et al., 2018). The estimated abundance of bottlenose dolphins from ObSERVE in summer 2016 was 87,330 (CV 0.21, 95% CI: 58,000—131,000). The total estimate for summer 2016 (SCANS III + ObSERVE) is thus 115,027 (CV 0.17, 95% CI = 83,100—159,000). All these surveys highlight the significant sizes of populations offshore compared with coastal areas.

Regional inshore studies indicate a population of 195 bottlenose dolphins (95% PI: 162—253) in eastern Scotland from the Moray Firth to the Fife coast (Arso Civil et al., 2019; Cheney et al., 2013), whilst the population occupying Cardigan Bay between 2005 and 2016 reached a maximum of 318 bottlenose dolphins (CV 0.39; 95% CI: 251—440) in 2009, but has been declining since then (Feingold & Evans, 2014; Lohrengel et al., 2017). Neither population is closed, and individuals may join up for periods of time from elsewhere. In western Ireland estimates of between 107 and 140 individuals have been reported as occupying the Shannon Estuary (Berrow, O'Brien, Groth, Foley, & Voigt, 2010; Berrow, O'Brien, Groth, Foley, & Voigt, 2012; Englund, Ingram, & Rogan, 2007, 2008; Ingram, 2000; Ingram & Rogan, 2003; Rogan, Nykanen, Gkaragkouni, & Ingram, 2015). Elsewhere in Ireland a population of around 150 animals occurs particularly around the west Connacht coast (Fig. 4.8), although it may range widely from Donegal Bay to the south coast (Ingram, Englund, & Rogan, 2001, 2003; Ingram & Rogan, 2003; Nykanen, Ingram, & Rogan, 2015).

A resident population of 319—391 animals inhabits the Gulf of St Malo, ranging between the French coast of Normandy and the Channel Islands (Couet, 2015; Louis et al., 2015). Two small populations inhabit the Iroise Sea, with 33 in the vicinity of Île de Sein, and around 70 around the island of Ouessant and the Achipelago of Molène (S. Hassani, *pers. comm.*).

Scattered sightings occur south to the Bay of Biscay, with regular groups along the coasts of Cantabria and Asturias. From the SCANS-II Survey, the overall abundance estimate for the Iberian bottlenose dolphin shelf sea population (from La Rochelle to the Strait of Gibraltar) was 5061 (CV = 0.57) individuals (Hammond et al., 2013). López et al., (2013) provides an estimate of 10,687 individuals (CV 0.26) for the shelf and deep waters of North and Northwest Spain, obtained using spatial modelling and survey data (surveys that took place between 2003 and 2011, including CEMMA surveys and also the SCANS-II Survey, but this estimate has not been corrected for perception and availability biases). The estimate for the shelf waters of Cantabria and Galicia was 4,592 individuals.

For the coastal bottlenose dolphin population in the nearshore waters of Southern Galicia (including all five Galician Rias), an estimate of 285 indivduals (95% CI: 199—330) has been obtained recently from photo-ID studies

FIGURE 4.8 Bottlenose Dolphin in west Connacht, West Ireland. *Photo: Pia Anderwald.*

(Methion, 2019), with between 56 and 144 estimated during the years 2014−16 in the biggest of the Galician rías (Ría of Arousa) (Methion & Díaz López, 2018). The bottlenose dolphin that inhabits the Sado Estuary is the only resident population in mainland Portugal. Around 25 individuals live in the estuary, although the population has been in decline for some years (Dos Santos, Couchinho, Luís, & Gonçalves, 2010; Lacey, 2015).

The main threats to bottlenose dolphins in northern Europe are probably high contaminant loads (Jepson et al., 2016), as well as sound disturbance, and potential vessel strikes particularly from high-speed recreational craft. From postmortem examinations of 63 bottlenose dolphins stranded in the United Kingdom between 1991 and 2010, 13% died as a result of infectious disease, 8% live-stranded, 7% were the result of possible infanticide, 6% were diagnosed as bycatch and 3% died from starvation (Deaville & Jepson, 2011).

References

Arso Civil, M., Quick, N. J., Cheney, B., Pirotta, H., Thompson, P. M., & Hammond, P. S. (2019). Changing distribution of the east coast of Scotland bottlenose dolphin population and the challenges of area-based management. *Aquatic Conservation.*

Baines, M.E. and Evans, P.G.H. (2012) *Atlas of the Marine Mammals of Wales.* 2nd Edition. CCW Monitoring Report No. 68. 143 pp.

Berrow, S., O'Brien, S., Groth, L., Foley, A., & Voigt, K. (2010). Bottlenose Dolphin SAC Survey (pp. 1−24). Report to the National Parks and Wildlife Service.

Berrow, S., O'Brien, S., Groth, L., Foley, A., & Voigt, K. (2012). Abundance estimate of bottlenose dolphin in the lower shannon candidate special area of conservation, Ireland. *Aquatic Mammals, 38*(2), 136−144. Available from https://doi.org/10.1578/AM.38.2.2012.136.

Certain, G., Ridoux, V., Van Canneyt, O., & Bretagnoile, V. (2008). Delphinid spatial distribution and abundance estimates over the shelf of the Bay of Biscay. *ICES Journal of Marine Science, 65,* 656−666.

Cheney, B. J., Thompson, P. M., Ingram, S. N., Hammond, P. S., Stevick, P. T., Durban, J. W., ... Wilson, B. (2013). Integrating multiple data sources to assess the distribution and abundance of bottlenose dolphins (*Tursiops truncatus*) in Scottish waters. *Mammal Review, 43,* 71−88.

Couet, P. (2015). *De l'identification des animaux aux modèles mathématiques: une remise en question des méthodes usuelles de suivi des populations. Le cas de la population de grands dauphins (Tursiops truncatus) en mer de la Manche* (M.Sc. thesis). Centre d'Ecologie Fonctionnelle and Evolutive.

Deaville, R. and Jepson, P.D. (compilers) (2011) *UK Cetacean Strandings Investigation Programme Final report for the period 1st January 2005 - 31st December 2010.* Zoological Society of London, UK. (http://randd.defra.gov.uk/Document.aspx?Document = FinalCSIPReport2005-2010_finalversion061211released[1].pdf).

Dos Santos, M. E., Couchinho, M. N., Luís, A. R., & Gonçalves, E. J. (2010). Monitoring underwater explosions in the habitat of resident bottlenose dolphins. *Journal of the Acoustical Society of America, 128*(6), 3805–3808.

Englund, A., Ingram, S. N., & Rogan, E. (2007). *Population status report for bottlenose dolphins using the Shannon Estuary 2006-7* (35 pp). Report to the National Parks and Wildlife Service.

Englund, A., Ingram, S. N., & Rogan, E. (2008). *Population status report for bottlenose dolphins using the Shannon Estuary 2008* (20 pp). Report to the National Parks and Wildlife Service.

Evans, P. G. H., Anderwald, P., & Baines, M. E. (2003). *UK Cetacean Status Review. Report to English Nature and Countryside Council for Wales.* Oxford: Sea Watch Foundation, 160 pp.

Feingold, D., & Evans, P. G. H. (2014). *Bottlenose dolphin and harbour porpoise monitoring in Cardigan Bay and Pen Llyn a'r Sarnau special areas of conservation 2011-2013* (124 pp). Natural Resources Wales Evidence Report Series No. 4.

Fernández, R., García-Tiscar, S., Santos, M. B., López, A., Martínez-Cedeira, J. A., Newton, J., & Pierce, G. J. (2011). Stable isotope analysis in two sympatric populations of bottlenose dolphins *Tursiops truncatus*: Evidence of resource partitioning? *Marine Biology, 158*, 1043–1055.

Fernández, R., Santos, M. B., Pierce, G. J., Llavona, A., López, A., Silva, M. A., & Piertney, S. B. (2011). Fine-scale genetic structure of bottlenose dolphins, Tursiops truncatus, in Atlantic coastal waters of the Iberian Peninsula. *Hydrobiologia, 670*(1), 111–125. Ecosystems and Sustainability (Published online doi:10.1007/s10750-011-0669-5).

Hammond, P.S., Burt, M.L., Cañadas, A., Castro, R., Certain, G., Gillespie, D., ... Winship, A. (2009). Cetacean Offshore Distribution and Abundance in the European Atlantic (CODA). Report available from SMRU, Gatty Marine Laboratory, University of St Andrews, St Andrews, Fife KY16 8LB, UK. http://biology.st-andrews.ac.uk/coda/.

Hammond, P.S., Lacey, C., Gilles, A., Viquerat, S., Borjesson, P., Herr, H., ... and Øien, N. (2017). Estimates of cetacean abundance in European Atlantic waters in summer 2016 from the SCANS-III aerial and shipboard surveys. Available at https://synergy.standrews.ac.uk/scans3/files/2017/05/SCANS-III-design-based-estimates-2017-05-12-final-revised.pdf.

Hammond, P. S., Macleod, K., Berggren, P., Borchers, D. L., Burt, M. L., Cañadas, A., ... Vázquez, J. A. (2013). Cetacean abundance and distribution in European Atlantic shelf waters to inform conservation and management. *Biological Conservation, 164*, 107–122.

ICES. (2016). *OSPAR request on indicator assessment of coastal bottlenose dolphins. ICES special request advice. Northeast Atlantic ecoregion.* Copenhagen: ICES, 14 pp.

Ingram, S. N. (2000). The ecology and conservation of bottlenose dolphins in the Shannon Estuary, Ireland (Ph.D. thesis) (213 pp). Ireland: University College Cork.

Ingram, S. N., Englund, A., & Rogan, E. (2001). *An extensive survey of bottlenose dolphins (Tursiops truncatus) on the west coast of Ireland* (17 pp). Heritage Council Report no. WLD/2001/42.

Ingram, S. N., Englund, A., & Rogan, E. (2003). *Habitat use, abundance and site-fidelity of bottlenose dolphins (Tursiops truncatus) in Connemara coastal waters, Co. Galway* (27 pp). Heritage Council Report No. 12314.

Ingram, S. N., & Rogan, E. (2002). Identifying critical areas and habitat preferences of bottlenose dolphins *Tursiops truncatus*. *Marine Ecology Progress Series, 244*, 247–255.

Ingram, S. N., & Rogan, E. (2003). Estimating abundance, site fidelity and ranging patterns of bottlenose dolphins (Tursiops truncatus) in the Shannon Estuary and selected areas of the west-coast of Ireland (28 pp). Report to the National Parks and Wildlife Service.

Jepson, P. D., Deaville, R., Barber, J. L., Aguilar, A., Borrell, A., Murphy, S., ... Law, R. J. (2016). PCB pollution continues to impact populations of orcas and other dolphins in European waters. *Nature Scientific Reports, 6*, 18573. Available from https://doi.org/10.1038/srep18573.

Lacey, C. (2015). *Current status of the resident bottlenose dolphin population in the Sado Estuary, Portugal* (Unpublished M.Sc. dissertation). University of Edinburgh.

Lewis, E., & Evans, P. G. H. (1993). Comparative ecology of bottle-nosed dolphins (*Tursiops truncatus*) in Cardigan Bay and the Moray Firth. *European Research on Cetaceans, 7*, 57–62.

Liret, C. (2001). *Domaine vital, utilization de l'espace et des ressources: les grands dauphins, Tursiops truncatus, de l'île de Sein* (155 pp.) (Thèse de doctorat de l'Université de Bretagne Occidentale). Brest.

Liret, C., Allali, P., Creton, P., Guinet, C., & Ridoux, V. (1994). Foraging activity pattern of bottlenose dolphins around Ile de Sein, France, and its relationships with environmental parameters. *European Research on Cetaceans, 8*, 188–192.

Lohrengel, K., Evans, P.G.H., Lindenbaum, C., Morris, C., and Stringell, T. (2017) *Bottlenose dolphin and harbour porpoise monitoring in Cardigan Bay and Pen Llŷn a'r Sarnau Special Areas of Conservation, 2014-16.* NRW Evidence Report No: 191, 1-154.

López, A., Vázquez, J.A., Martínez-Cedeira, J., Cañadas, A., Marcos, E., Maestre, I., ... Evans, P. (2013). Estimas de abundancia, mediante modelización espacial, de las poblaciones de marsopa común (*Phocoena phocoena*), delfín mular (*Tursiops truncatus*), cachalote (*Physeter macrocephalus*) y rorcual común (*Balaenoptera physalus*) en el norte Peninsular. XI Congreso de la Sociedad Española para la Conservación y Estudio de los Mamíferos (SECEM). Avilés, 5-8 de diciembre de 2013.

Louis, M., Gally, F., Barbraud, C., Béesau, J., Tixier, P., Simon-Bouhet, B., ... Guinet, C. (2015). Social structure and abundance of coastal bottlenose dolphins, *Tursiops truncatus*, in the Normano-Breton Gulf, English Channel. *Journal of Mammalogy, 96*, 481–493. Available from https://doi.org/10.1093/jmamma/gyv053.

Louis, M., Viricel, A., Lucas, T., Peltier, H., Alfonsi, E., Berrow, S., ... Simon-Bouhet, B. (2014). Habitat-driven population structure of bottlenose dolphins, *Tursiops truncatus*, in the North-east Atlantic. *Molecular Ecology, 23*, 857−874.

Methion, S. (2019). Behavioural ecology of common bottlenose dolphins *(Tursiops truncatus)* in a highly affected coastal ecosystem. France: University of Bordeaux, PhD thesis.

Methion, S., & Díaz López, B. (2018). Abundance and demographic parameters of bottlenose dolphins in a highly affected coastal ecosystem. *Marine and Freshwater Research, 69*(9), 1355−1364. Available from https://doi.org/10.1071/MF17346.

Mirimin, L., Miller, R., Dillane, E., Berrow, S. D., Ingram, S., Cross, T. F., & Rogan, E. (2011). Fine-scale population genetic structuring of bottlenose dolphins using Irish coastal waters. *Animal Conservation, 14*, 342−353.

Nichols, C., Herman, J., Gaggiotti, O. E., Dobney, K. M., Parsons, K., & Hoelzel, A. R. (2007). *Genetic isolation of a now extinct population of bottlenose dolphins (Tursiops truncatus)*. Proceedings of the Royal Society, B, 247, 1611−1616.

Nykanen, M., Ingram, S. D., & Rogan, E. (2015). *West coast dolphins (Tursiops truncatus): Abundance, distribution, ranging patterns and habitat use* (33 pp). Report to the National Parks and Wildlife Service.

Querouil, S., Silva, M. A., Freitas, L., Prieto, R., Magalhaes, S., Dinis, A., ... Santos, R. S. (2007). High gene flow in oceanic bottlenose dolphins *(Tursiops truncatus)* of the North Atlantic. *Conservation Genetics, 8*, 1405−1418.

Reid, J. B., Evans, P. G. H., & Northridge, S. P. (2003). *Atlas of Cetacean Distribution in North-west European Waters*. Peterborough: Joint Nature Conservation Committee, 76 pp.

Rogan, E., Breen, P., Mackey, M., Cañadas, A., Scheidat, M., Geelhoed, S., & Jessopp, M. (2018). *Aerial surveys of cetaceans and seabirds in Irish waters: Occurrence, distribution and abundance in 2015-2017*. Dublin, Ireland: Department of Communications, Climate Action & Environment and National Parks and Wildlife Service (NPWS), Department of Culture, Heritage and the Gaeltacht, 297 pp.

Rogan, E., Ingram, S., Holmes, B., & O'Flanagan, C. (2000). *A survey of bottlenose dolphins (Tursiops truncatus) in the Shannon Estuary*. Dublin: Marine Resource Series.

Rogan, E., Nykanen, M. Gkaragkouni, M., & Ingram, S. N. (2015). *Bottlenose dolphin survey in the lower River Shannon SAC, 2015* (pp. 1−21). Report to the National Parks and Wildlife Service.

Wilson, B., Thompson, P. M., & Hammond, P. S. (1997). Habitat use by bottlenose dolphins: Seasonal distribution and stratified movement patterns in the Moray Firth, Scotland. *Journal of Applied Ecology, 3*, 1365−1374.

Atlantic spotted dolphin *Stenella frontalis*

Two species of spotted dolphin are now recognised in the tropical Atlantic: one endemic to the region, referred to as the Atlantic spotted dolphin (and occurring up to about 40°N), and the other called the Pan-tropical spotted dolphin (apparently rarer and occurring up to 25°N). The Atlantic spotted dolphin (Fig. 4.9) is the only species recorded in the

FIGURE 4.9 Atlantic Spotted Dolphin adult and young. *Photo: Maren Reichelt/Mick E Baines.*

ASCOBANS Agreement Area, with a single confirmed stranding at Finistère, France in 2005, and a number of sightings off the coast of Portugal (Santos et al., 2012). Further south in the central and eastern North Atlantic the species occurs seasonally in Madeira and the Azores (being less common in winter), whereas in the Canaries it occurs mainly during winter months.

The Atlantic spotted dolphin resembles bottlenose and spinner dolphins with a slender body, 1.6−2.3 m in length, a relatively short but distinct beak and centrally placed recurved dorsal fin. However, a distinctive feature is the patterning of black and white spots, most apparent in adults, as well as a pale blaze below the dorsal fin. The Pan-tropical spotted dolphin lacks this blaze, the tail-stock is divided into upper dark and lower light halves, and in adults the dark ventral spots merge to form a slightly mottled or uniform grey underside.

The species varies geographically in body size and coloration, tending to be larger and more heavily spotted in coastal waters compared with those found in offshore regions. Confusion is most likely to occur with young spotted dolphins that may have few or no spots on their flanks and back. Before puberty, spotting develops around the abdomen and as the animal gets older, these extend to cover the entire body as well as enlarging and coalescing in some areas.

Reference

Santos, J., Araújo, H., Ferreira, M., Henriques, A., Miodonski, J., Monteiro, S., ... Vingada, J. (2012). Chapter I: Baseline estimates of abundance and distribution of target species. *MARPRO Conservação de especies marinhas protegidas em Portugal continental* (80 pp). Project LIFE09/NAT/PT/000038.

Striped dolphin *Stenella coeruleoalba*

The striped dolphin is a small social delphinid superficially resembling the common dolphin but with a more robust body, a shorter beak (maximum length 10−12 cm), and two distinct black stripes on the flanks one from the eye to the flipper and the other to the anus (Fig. 4.10). A distinct groove separates the beak from the forehead. Also evident is a white or light grey V-shaped blaze, one branch of this narrowing to a point below the dorsal fin, the other extending backwards towards the tail. The species lacks the yellow patches of the common dolphin. Adults are between 1.9 and 2.4 m in length. The coloration of the body is variable, dark grey or bluish grey on the back, with lighter grey flanks, the posterior part of which is light grey and sometimes extends upwards over the dorsal surface of the tail-stock.

FIGURE 4.10 Striped Dolphin showing its two distinct black stripes, from the eye to the flipper and the eye to the anus. *Photo: Caterina Lanfredi, Tethys Research Institute.*

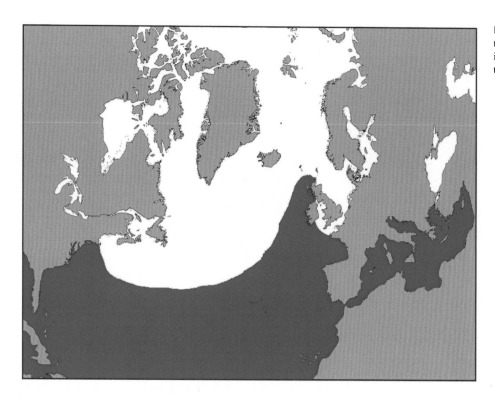

FIGURE 4.11 North Atlantic distribution of Striped Dolphin (depicting those areas where the species is thought to regularly occur).

The belly is white. Tapering, black flippers are inserted in this white region, although some individuals show rather pale flippers towards the base. The dorsal fin is slender, recurved and centrally placed. The tail-stock is narrow with no obvious keel; the dark tail flukes have a median notch in them.

The striped dolphin has a worldwide distribution, occurring mainly in tropical and warm temperate waters. The species is the most common and widely distributed delphinid in the Mediterranean. In the eastern North Atlantic, it generally occurs further offshore than the common dolphin, with highest densities in the deep waters of the western Bay of Biscay beyond the continental shelf of Spain, Portugal and France (Fig. 4.11). Further north, around the British Isles and Ireland, it is an occasional visitor, recorded mainly in the southwest, although there has been a sharp increase in records in this region since the 1980s, with the species occurring occasionally also in the North Sea and even straying into the western Baltic (Evans et al., 2003), possibly reflecting warming sea temperatures in the region. Striped dolphins occur in the mid-Atlantic to at least 62°N, suggesting that its distribution offshore may be extended northward by the Gulf Stream. The species has also been recorded recently casually from Icelandic, Danish, Swedish and Norwegian waters (with sightings up to 66.5°N) (Bloch, Petersen and Sigurjónsson, 1996; Isaksen & Syvertsen, 2002).

At latitudes between 30° and 45°N, the striped dolphin is one of the most abundant delphinids in the eastern North Atlantic (Fig. 4.12). In the ASCOBANS Agreement Area, the SCANS-II survey in July 2005 only detected striped dolphins off the coast of Iberia, giving an abundance estimate of 1846 (CV 0.69). In July 2007 the CODA survey yielded an overall abundance estimate of c. 61,364 (CV 0.93, 95% CI: 12,300−306,000) (Hammond et al., 2009) (see Table 6.1). The abundance estimate in the southern Bay of Biscay was 20,045 (CV 0.56), and 7546 (CV 0.62) in offshore Galicia. An overall revised estimate of striped dolphins combining SCANS-II and CODA surveys is 63,210 (CV 0.90, 95% CI: 13,900−287,000). As noted above, striped and common dolphins can be difficult to distinguish and several sightings were classified as unidentified common of striped dolphins. The estimate of the abundance of common and striped dolphins combined, including unidentified common/striped dolphins, from SCANS-II and CODA in 2005/07 was 305,460 (CV 0.36, 95% CI: 154,000−606,000) (see Table 6.1). Almost all confirmed striped dolphin sightings were in the Bay of Biscay.

The SCANS-III survey in July 2016 covering an area equivalent to SCANS-II + CODA combined but excluding Irish waters surveyed by the ObSERVE project gave a total abundance estimate of 372,340 (CV 0.33, 95% CI: 199,000−698,000) striped dolphins; in addition, there were an estimated 158,167 (CV 0.19, 95% CI: 110,000−228,000) unidentified common/striped dolphins (Hammond et al., 2017; see also Table 6.1). The estimate of the abundance of common and striped dolphins combined, including unidentified common/striped dolphins from SCANS-III was 998,180 (CV 0.18, 95% CI: 703,000-1,416,000).

FIGURE 4.12 Striped Dolphin is abundant in warmer waters of the North Atlantic and is the most common dolphin in the Mediterranean. *Photo: Enrico Pirotta/Tethys.*

The ObSERVE survey covering Irish waters in summer 2016 complemented SCANS-III and together these surveys provided area coverage equivalent to SCANS-II + CODA in 2005/07. There were too few sightings to estimate the abundance of striped dolphins from ObSERVE but a combined estimate for common and striped, including unidentified common/striped dolphins was 33,215 (CV 0.44, 95% CI: 19,300−57,100). The estimate of the abundance of common and striped dolphins combined, including unidentified common/striped dolphins for summer 2016 (SCANS-III and ObSERVE) is thus 1,031,395 (CV 0.18, 95% CI = 734,000−1,449,000.

In the ASCOBANS Agreement Area, the main threats have been from fisheries bycatch (see, e.g. Goujon, 1996; Rogan & Mackey, 2007, in relation to the driftnet fishery for albacore tuna). Other potential threats include prey depletion and infectious disease as a result of high contaminant levels (see, e.g. Jepson et al., 2016). From postmortem examinations of 111 striped dolphins stranded in the United Kingdom between 1991 and 2010, 34% live-stranded, 18% died as a result of infectious disease, 11% died from starvation and 7% were diagnosed as bycatch (Deaville & Jepson, 2011).

References

Bloch, D., Petersen, A., & Sigurjönsson, J. (1996). Strandings of striped dolphins (*Stenella coeruleoalba*) in Iceland and the Faroe Islands and sightings in the northeast Atlantic, north of 50°N latitude. *Marine Mammal Science*, *12*, 125−132.

Deaville, R. and Jepson, P.D. (compilers) (2011) *UK Cetacean Strandings Investigation Programme Final report for the period 1st January 2005 - 31st December 2010*. Zoological Society of London, UK. (http://randd.defra.gov.uk/Document.aspx?Document = FinalCSIPReport2005-2010_finalversion061211released[1].pdf).

Evans, P. G. H., Anderwald, P., & Baines, M. E. (2003). *UK Cetacean Status Review. Report to English Nature and Countryside Council for Wales*. Oxford: Sea Watch Foundation, 160 pp.

Goujon, M. (1996). *Captures accidentelles du filet maillant derivant et dynamique des populations de dauphins au large du Golfe de Gascogne* (Ph.D. thesis) (239 pp). France: Laboratoire Halieutique, D.E.E.R.N., Ecole Nationales Superieure Agronomique de Rennes.

Hammond, P.S., Burt, M.L., Cañadas, A., Castro, R., Certain, G., Gillespie, D., ... Winship, A. (2009). Cetacean Offshore Distribution and Abundance in the European Atlantic (CODA). Report available from SMRU, Gatty Marine Laboratory, University of St Andrews, St Andrews, Fife KY16 8LB, UK. http://biology.st-andrews.ac.uk/coda/.

Hammond, P.S., Lacey, C., Gilles, A., Viquerat, S., Borjesson, P., Herr, H., ... and Øien, N. (2017). Estimates of cetacean abundance in European Atlantic waters in summer 2016 from the SCANS-III aerial and shipboard surveys. Available at https://synergy.standrews.ac.uk/scans3/files/2017/05/SCANS-III-design-based-estimates-2017-05-12-final-revised.pdf.

Isaksen, K., & Syvertsen, P. O. (2002). Striped dolphins, *Stenella coeruleoalba*, in Norwegian and adjacent waters. *Mammalia*, *66*(1), 33−41.

Jepson, P. D., Deaville, R., Barber, J. L., Aguilar, A., Borrell, A., Murphy, S., ... Law, R. J. (2016). PCB pollution continues to impact populations of orcas and other dolphins in European waters. *Nature Scientific Reports*, *6*, 18573. Available from https://doi.org/10.1038/srep18573.

Rogan, E., Breen, P., Mackey, M., Cañadas, A., Scheidat, M., Geelhoed, S., & Jessopp, M. (2018). *Aerial surveys of cetaceans and seabirds in Irish waters: Occurrence, distribution and abundance in 2015-2017*. Dublin, Ireland: Department of Communications, Climate Action & Environment and National Parks and Wildlife Service (NPWS), Department of Culture, Heritage and the Gaeltacht, 297 pp.

Rogan, E., & Mackey, M. (2007). Megafauna bycatch in drift nets for albacore tuna (*Thunnus allunga*) in the NE Atlantic. *Fisheries Research, 86,* 6−14.

Common dolphin *Delphinus delphis*

Common dolphins are amongst the smallest of the true dolphins, being around 1.7−2 m in length. They are active, fast-moving species, frequently bow-riding boats and jumping clear of the water (Fig. 4.13). Two species of common dolphin had been recognised in the North Atlantic − a short-beaked form, *Delphinus delphis*, and a long-beaked form, *D. capensis* of more tropical waters (Heyning & Perrin, 1994). However, the latter is now considered simply a regional variant of *D. delphis* (Perrin, 2018). In Europe, a distinct form occurs in the Black Sea, currently thought to be a separate subspecies, *D. d. ponticus* (Amaha, 1994).

Common dolphins have a slender body with a falcate to erect centrally placed dorsal fin. In temperate waters they are the only dolphin species likely to be seen with a long, narrow beak − dark in colour, though sometimes tipped white. On its flanks there is an hourglass pattern of tan or yellowish tan becoming pale grey behind the dorsal fin. This pale patch may reach the dorsal surface. Underwater the hourglass pattern shows up very clearly, as does the creamy white belly contrasting with the brownish black back and upper flanks (see Fig. 4.14).

The common dolphin has a worldwide distribution in oceanic and shelf-edge waters of tropical, subtropical and temperate seas, occurring in both hemispheres. It is abundant and widely distributed in the eastern North Atlantic, mainly occurring in deeper waters from the Iberian Peninsula north to approximately 65°N latitude (though rare north of 62°N), west of Norway and the Faroe Islands (Murphy, Evans, & Collet, 2008; Reid et al., 2003). It occurs westwards at least to the mid-Atlantic ridge (Cañadas, Desportes, Borchers, & Donovan, 2009; Doksaeter, Olsen, Nottestad, & Ferno, 2008), and eastwards into the Mediterranean where they have a patchy distribution, with a distinct isolated population in the Black Sea (Bearzi et al., 2003; Bearzi et al., 2016) (Fig. 4.15). In the offshore North Atlantic it seems to favour waters over 15°C and shelf-edge features at depths of 400−1000 m between 49 and 55°N especially between 20 and 30°W (Cañadas et al., 2009).

FIGURE 4.13 The Common Dolphin frequently jumps clear of the water displaying its relatively long beak and patterned flanks. *Photo: Mike J Tetley.*

FIGURE 4.14 The Common Dolphin has an hourglass pattern of tan or yellowsh tan on its flanks, bcoming pale grey behind the dorsal fin, contrasting with the brownish back and creamy white belly. *Photo: Peter GH Evans.*

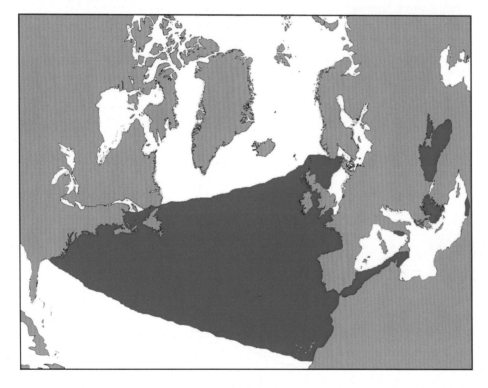

FIGURE 4.15 North Atlantic distribution of Common Dolphin (depicting those areas where the species is thought to regularly occur).

On the United Kingdom continental shelf, the species is common in the western half of the English Channel and deeper waters of the Irish Sea, and further north in the Sea of the Hebrides and southern part of the Minch (Evans, Anderwald, & Baines, 2003; Reid et al., 2003). It is also common south and west of Ireland. In some years, the species occurs further north and east in shelf seas — in the northern Hebrides, around Shetland and Orkney and in the northern North Sea. It is generally rare in the central and southern North Sea and eastern portion of the English Channel (Camphuysen & Peet, 2006; Evans et al., 2003), but is abundant in the Bay of Biscay (Evans et al., 2003; Kiszka et al., 2007; Laran et al., 2017; Reid et al., 2003). Range shifts may be related to changing oceanographic conditions (warmer sea temperatures and extensions of the range of potential prey species like anchovy and sardine). Since the

FIGURE 4.16 Common Dolphins can be seen frequently in groups during summer in the Celtic and Irish Seas. *Photo: Katrin Lohrengel.*

1990s, the species has become regular in the North Sea and even entered the Baltic (Evans & Bjørge, 2013; Evans et al., 2003; Kinze et al., 2010, 2018; Murphy et al., 2008).

In shelf waters off the west coasts of Ireland and Scotland and in the Irish Sea, common dolphin abundance tends to be greatest in the summer months (Berrow et al., 2010; Evans et al., 2003; see Fig. 4.16), whereas offshore and further south, although the species is more abundant in summer, the western English Channel and Bay of Biscay remain important winter habitats for the species (Brereton, Williams, & Martin, 2005; De Boer, Leaper, Keith, & Simmonds, 2008; Laran et al., 2017) and west of Ireland, higher numbers have been recorded in winter in some years (Rogan et al., 2018). Common dolphins also occur all along the coast of Portugal and southwest Spain.

Whereas common dolphins can be found over large parts of the North Atlantic and are currently considered a panmictic population (Amaral et al., 2012; Mirimin et al., 2009), it is possible there is some substructuring between those occurring offshore and those over the continental shelf, although this has yet to be fully explored (Caurant et al., 2009; Evans & Teilmann, 2009; Forcada, Aguilar, Evans, & Perrin, 1990).

There have been several abundance surveys of common dolphin in various parts of the eastern North Atlantic though none spanning the entire region (e.g. MICA survey in 1993 — Goujon, Antoine, Collet, & Fifas, 1993; ATLANCET aerial survey in 2001 — Ridoux, van Canneyt, Doremus, & Certain, 2003; MARPRO surveys in 2007—12 — Santos et al., 2012). In the ASCOBANS Agreement Area the SCANS survey (July 1994), covering an area from the Celtic shelf to approximately 11°W and 48°S, produced an estimate of 75,449 individuals (CV 0.67; 95% CI: 23,900—248,900) (Hammond et al., 2002). However, as Hammond et al., (2002) pointed out, this survey did not have sufficient double-platform data to use in the analysis, they were not able to correct for animals missed on the track-line, nor, perhaps most importantly, responsive movement — common dolphins notoriously respond positively to the presence of vessels that they can bow-ride.

In 2005 SCANS-II resurveyed the same area as SCANS, but extended this to include also the Irish Sea, the waters off western and Northern Ireland, west Scotland, and continental shelf waters off France, Spain and Portugal. The total summer abundance for those Northeast Atlantic shelf waters was 54,995 (CV 0.21, 95% CI: 36,500—82,800) (revised from Hammond et al., 2013). This was supplemented by the CODA survey conducted in July 2007 in offshore waters of the ASCOBANS Agreement Area, which estimated a total abundance of 118,264 (CV 0.38, 95% CI: 56,900—247,000) (Hammond et al., 2009). These estimates are corrected for animals missed along the track-line and also for responsive movement. An overall revised estimate combining SCANS-II and CODA surveys (2005/07) is 173,219 (CV 0.27, 95% CI: 103,000—290,000) common dolphins.

The SCANS-III survey in July 2016 (covering the same waters as SCANS-II + CODA combined but excluding Irish waters surveyed by the ObSERVE project) yielded an abundance estimate of 467,673 (CV 0.26, 95% CI: 281,000—778,000) common dolphins, with an additional estimate of unidentified common/striped dolphins of 158,167 (CV 0.19, 95% CI: 110,000—228,000) (Hammond et al., 2017). Highest densities of identified common dolphins were along the edge of shelf seas in the Bay of Biscay and around the Iberian Peninsula. Although survey areas vary between years, it is clear that numbers counted in summer 2016 were considerably higher than in earlier years, suggesting that there is extensive annual movement of common dolphins in and out of the areas surveyed.

The ObSERVE survey covering Irish waters in summer 2016 complemented SCANS-III and together these surveys provided area coverage equivalent to SCANS-II + CODA in 2005/07. The estimated abundance of common dolphins from ObSERVE was 13,633 (CV 0.86, 95% CI: 5,210−35,600) and the total estimate for summer 2016 (SCANS-III + ObSERVE) is thus 481,306 (CV 0.26, 95% CI: 293,000−791,000). The estimate of the abundance of common and striped dolphins combined, including unidentified common/striped dolphins from SCANS-III was 998,180 (CV 0.18, 95% CI: 703,000−1,416,000).

The Irish ObSERVE surveys were undertaken in the summer and winter of 2015 and 2016, allowing both annual and seasonal comparisons. In winter 2015/16 almost twice as many common dolphins were recorded compared with summer 2015, although the following year, numbers were comparable between seasons; highest densities occurred to the south and west of Ireland (Rogan et al., 2018). Further south, in the Bay of Biscay, densities and overall abundance were higher in summer than in winter, although this applied collectively to common and striped dolphins, since the two could not be readily distinguished during aerial surveys (Laran et al., 2017).

An analysis in the wider region that not only corrected for animals missed along the track-line but also for responsive movement was for the western block of the 1995 North Atlantic sightings survey (NASS) that covers part of the central North Atlantic (52−57.5°N, 18−28°W), resulting in an estimate of 273,000 common dolphins (CV 0.26, 95% CI: 153,000−435,000) (Cañadas et al., 2009).

The main threat to the species in the ASCOBANS Agreement Area appears to be fisheries bycatch (Murphy et al., 2013). In spring 2019 more than 1000 common dolphins with evidence of by-catch stranded on the French Biscay coast, and around 500 two years earlier. From postmortem examinations of 537 common dolphins stranded in the United Kingdom between 1991 and 2010, 51% were diagnosed as bycatch, 18% live-stranded, 7% from infectious disease and 4% died from starvation (Deaville & Jepson, 2011).

References

Amaha, A. (1994) *Geographic variation of the common dolphin Delphinus delphis (Odontoceti Delphinidae)*. Ph.D. thesis, Tokyo University of Fisheries, Tokyo.

Amaral, A. R., Luciano, B. B., Bilgmann, K., Dmitri Boutov, D., Freitas, L., Robertson, K. M., . . . Møller, L. M. (2012). Seascape genetics of a globally distributed, highly mobile marine mammal: The short-beaked common dolphin (Genus *Delphinus*). *PLoS One*, *7*(2). Available from https://doi.org/10.1371/journal.pone.0031482.

Bearzi, G., Bonizzoni, S., Santostasi, N. L., Furey, N. B., Eddy, L., Valavanis, V. D., & Gimenez, O. (2016). Dolphins in a scaled-down Mediterranean: The Gulf of Corinth's odontocetes. *Advances in Marine Biology*, *75*, 297−331.

Bearzi, G., Reeves, R. R., Notarbartolo-di-Sciara, G., Politi, E., Cañadas, A., Frantzis, A., & Mussi, B. (2003). Ecology, status and conservation of short-beaked common dolphins *Delphinus delphis* in the Mediterranean Sea. *Mammal Review*, *33*, 224−252.

Berrow, S., Whooley, P., O'Connell, M., & Wall, D. (2010). *Irish Cetacean Review (2000−2009)*. Kilrush: Irish Whale and Dolphin Group, 60 pp.

Brereton, T., Williams, A., & Martin, C. (2005). Ecology and status of the common dolphin *Delphinus delphis* in the English Channel and Bay of Biscay 1995−2002. In K. Stockin, et al. (Eds.), *Proceedings of the workshop on common dolphins: Current research, threats and issues* (pp. 15−22). Cambridge: European Cetacean Society, Special issue April 2005, Kolmärden, 1 April 2004.

Cañadas, A., Desportes, G., Borchers, D., & Donovan, G. (2009). A short review of the distribution of short-beaked common dolphin (*Delphinus delphis*) in the central and eastern North Atlantic with an abundance estimate for part of this area. *NAMMCO Scientific Publications*, *7*, 201−220.

Caurant, F., Chouvelon, T., Lahaye, V., Mendez-Fernandez, P., Rogan, E., Spitz, J., & Ridoux, V. (2009). *The use of ecological tracers for discriminating dolphin population structure: The case of the short-beaked common dolphin Delphinus delphis in European Atlantic waters* (11 pp). IWC SC/61/SM34.

Deaville, R. and Jepson, P.D. (compilers) (2011) *UK Cetacean Strandings Investigation Programme Final report for the period 1st January 2005 - 31st December 2010*. Zoological Society of London, UK. (http://randd.defra.gov.uk/Document.aspx?Document = FinalCSIPReport2005-2010_finalversion061211released[1].pdf).

De Boer, M. N., Leaper, R., Keith, S., & Simmonds, M. P. (2008). Winter abundance estimates for the common dolphin (*Delphinus delphis*) in the western approaches of the English Channel and the effect of responsive movement. *Journal of Marine Animals and Their Ecology*, *1*(1), 15−21.

Doksaeter, L., Olsen, E., Nottestad, L., & Ferno, A. (2008). Distribution and feeding ecology of dolphins along the Mid-Atlantic Ridge between Iceland and the Azores. *Deep Sea Research Part II: Topical Studies in Oceanography*, *55*, 243−253.

Evans, P. G. H., & Bjørge, A. (2013). Impacts of climate change on marine mammals. Marine Climate Change Impacts Partnership (MCCIP). *Science Review*. Available from https://doi.org/10.14465/2013.arc15.134-148.

Evans, P. G. H., Anderwald, P., & Baines, M. E. (2003). *UK Cetacean Status Review. Report to English Nature and the Countryside Council for Wales*. Oxford: Sea Watch Foundation, 160 pp.

Forcada, J., Aguilar, A., Evans, P. G. H., & Perrin, W. F. (1990). Distribution of common and striped dolphins in the temperate waters of the eastern North Atlantic. *European Research on Cetaceans, 4*, 64−66.

Goujon, M., Antoine, L., Collet, A., & Fifas, S. (1993). *Approche de l'mpact de la ecologique de la pecherie thoniere au filet maillant derivant en Atlantique nord-est*. Rapport internese al Direction des Resources Vivantes de l'IFREMER. Plouzane: Ifremer, Centre de Brest.

Hammond, P. S., Berggren, P., Benke, H., Borchers, D. L., Collet, A., Heide-Jørgensen, M. P., . . . Øien, N. (2002). Abundance of harbour porpoise and other cetaceans in the North Sea and adjacent waters. *Journal of Applied Ecology, 39*, 361−376.

Hammond, P.S., Burt, M.L., Cañadas, A., Castro, R., Certain, G., Gillespie, D., . . . Winship, A. (2009). Cetacean Offshore Distribution and Abundance in the European Atlantic (CODA). Report available from SMRU, Gatty Marine Laboratory, University of St Andrews, St Andrews, Fife KY16 8LB, UK. http://biology.st-andrews.ac.uk/coda/.

Hammond, P.S., Lacey, C., Gilles, A., Viquerat, S., Borjesson, P., Herr, H., . . . and Øien, N. (2017). Estimates of cetacean abundance in European Atlantic waters in summer 2016 from the SCANS-III aerial and shipboard surveys. Available at https://synergy.standrews.ac.uk/scans3/files/2017/05/SCANS-III-design-based-estimates-2017-05-12-final-revised.pdf.

Hammond, P. S., Macleod, K., Berggren, P., Borchers, D. L., Burt, M. L., Cañadas, A., . . . Vázquez, J. A. (2013). Cetacean abundance and distribution in European Atlantic shelf waters to inform conservation and management. *Biological Conservation, 164*, 107−122.

Heyning, J. E., & Perrin, W. F. (1994).) Evidence for two species of common dolphins (genus *Delphinus*) from the eastern North Pacific. *Contributions in Science (Los Angeles County Museum), 442*, 1−35.

Kinze, C. C., Jensen, T., Tougaard, S., & Baagøe, H. J. (2010). Danske hvalfund i perioden 1998-2007 [Records of cetacean strandings on the Danish coastline during 1998-2007]. *Flora og Fauna, 16*(4), 91−99.

Kinze, C. C., Thøstesen, C. B., & Olsen, M. T. (2018). Cetacean stranding records along the Danish coastline: records for the period 2008-2017 and a comparative review. *Lutra, 61*(1), 87−105.

Laran, S., Authier, M., Blanck, A., Dorémus, G., Falchetto, H., Monestiez, P., . . . Ridoux, V. (2017). Seasonal distribution and abundance of cetaceans within French waters: Part II: The Bay of Biscay and the English Channel. *Deep-Sea Research II, 14*, 31−40.

Mirimin, L., Westgate, A., Rogan, E., Rosel, P., Read, A., Coughlan, J., & Cross, T. (2009). Population structure of short-beaked common dolphins (*Delphinus delphis*) in the North Atlantic Ocean as revealed by mitochondrial and nuclear genetic markers. *Marine Biology, 156*, 821−834.

Murphy, S., Evans, P. G. H., & Collet, A. (2008). Common dolphin *Delphinus delphis*. In S. Harris, & D. W. Yalden (Eds.), *Mammals of the British Isles* (4th ed., pp. 719−724). Southampton: The Mammal Society, 800 pp, Handbook.

Murphy, S., Pinn, E. H., & Jepson, P. D. (2013). The short-beaked common dolphin (*Delphinus delphis*) in the north-eastern Atlantic: Distribution, ecology, management and conservation status. *Oceanography and Marine Biology: An Annual Review, 51*, 193−280.

Perrin, W. F. (2018). Common dolphin *Delphinus delphis*. In B. Würsig, J. G. M. Thewissen, & K. M. Kovacs (Eds.), *Encyclopedia of Marine Mammals* (3rd edition, pp. 205−209). London: Academic Press, 1,157 pp.

Reid, J. B., Evans, P. G. H., & Northridge, S. P. (2003). *Atlas of Cetacean Distribution in North-west European Waters*. Peterborough: Joint Nature Conservation Committee, 76 pp.

Ridoux, V., van Canneyt, O., Doremus, G., & Certain, G. (2003). *Détermination des habitats préférentiels estivaux de prédateurs supérieurs pélagiques du proche Atlantique par observations aeriennes* (19 pp). Rapport contractuel CRMM-ULR/IFREMER.

Rogan, E., Breen, P., Mackey, M., Cañadas, A., Scheidat, M., Geelhoed, S., & Jessopp, M. (2018). *Aerial surveys of cetaceans and seabirds in Irish waters: Occurrence, distribution and abundance in 2015-2017*. Dublin, Ireland: Department of Communications, Climate Action & Environment and National Parks and Wildlife Service (NPWS), Department of Culture, Heritage and the Gaeltacht, 297 pp.

Santos, J., Araujo, H., Ferreira, M., Henriques, A., Miodonski, J., Monteiro, S., . . . and Vingada, J. (2012) MARPRO Project. Chapter 1: Baseline estimates of abundance and distribution of target species. Annex to the Midterm Report of Project LIFE09 NAT/PT/000038. 80 pp.

Fraser's dolphin *Lagenodelphis hosei*

Fraser's dolphin was formally described as recently as 1956. With characteristics of both common and white-beaked dolphins, it has a robust body (2.4−2.7 m length) with a short but well-defined beak and small curved dorsal fin and flippers. The body is bluish grey on the back and light pink or white on the belly with a creamy white band and parallel black band extending from the eye along the flanks towards the anus (Fig. 4.17).

Fraser's dolphin is typically a tropical oceanic species occurring in deep waters often in large schools. It is regular in the Caribbean, but occasionally reaches Florida in the western North Atlantic. In the eastern North Atlantic it occurs mainly of the coast of northwest Africa and in the Canaries, Azores and Madeira (Gomes-Pereira, Marques, Cruz, & Martins, 2013). However, there was a mass stranding of 11 animals that occurred in north Brittany on the Atlantic French coast in 1984, and a single stranding record from the Outer Hebrides, northwest Scotland in 1996 (Evans, 2008, see Fraser's dolphin).

FIGURE 4.17 Fraser's Dolphin has characteristics of both *Delphinus* and *Lagenorhynchus* with a short beak, bluish grey back, a creamy white band and parallel black band extending backwards from the eye. *Photo: Bernd Brederlau.*

References

Evans, P. G. H. (2008). Fraser's dolphin *Lagenodelphis hosei*. In S. Harris, & D. W. Yalden) (Eds.), *Mammals of the British Isles* (4th ed., pp. 731−733). Southampton: The Mammal Society, 800 pp, Handbook.

Gomes-Pereira, J. N., Marques, R., Cruz, M. J., & Martins, A. M. (2013). The little-known Fraser's dolphin *Lagenodelphis hosei* in the North Atlantic: New records and a review of distribution. *Marine Biodiversity*, *43*(4). Available from https://doi.org/10.1007/s12526-013-0159-2.

White-beaked dolphin *Lagenorhynchus albirostris*

The white-beaked dolphin is the most common dolphin inhabiting the cold temperate and low Arctic shelf waters of the North Atlantic and North Sea. In comparison with the striped or common dolphin, it is a much larger, stouter species, 2.4−2.8 m in length, with a centrally placed, taller and more recurved dorsal fin. It has a rounded snout and short thick beak generally tipped light grey or white. There can often be areas of white also on the head, and generally a black thoracic patch, surrounded by lighter areas. Most characteristic is the pale grey or white area extending along the flanks and over the otherwise dark grey or black dorsal surface behind the fin (Fig. 4.18). The white on the flanks can lead to confusion with Atlantic white-sided dolphin although the latter never has white over the back behind the dorsal fin. Juveniles have more uniform coloration, the white areas being more indistinct. The thick tail-stock gradually tapers towards the slightly notched tail flukes.

White-beaked dolphins are confined to temperate and subpolar seas (generally in sea surface temperatures of 2°C−13°C) of the North Atlantic from southwest and central east Greenland, Svalbard and the Barents Sea, south to around Cape Cod (United States) in the west and the Bay of Biscay in the east (Fig. 4.19), although they are occasionally sighted around the Iberian Peninsula.

Four principal centres of high density can be identified: (1) the Labrador Shelf including southwestern Greenland; (2) Icelandic waters; (3) the waters around Scotland and northeast England, including the central and northern North Sea and northwest coast of Scotland; and (4) the narrow shelf stretching along the Norwegian coast, extending north into the White Sea (Evans et al., 2003; Galatius & Kinze, 2016; Kinze, 2008; Reid et al., 2003). The species is rare in the Irish Sea and most of the English Channel (a small population exists in and around Lyme Bay, Devon).

The species occurs over a large part of the northern European continental shelf, mainly in waters of 50−100 m depth and almost entirely within the 200 m isobath (Evans & Smeenk, 2008; Reid et al., 2003). However, further north in the

FIGURE 4.18 The White-beaked Dolphin is characteried by a pale grey or white area extending along the flanks and over the dark back behind the tall recurved dorsal fin. *Photo: Kevin Hepworth.*

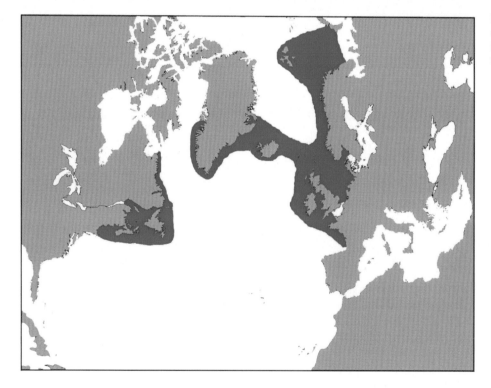

FIGURE 4.19 Overall distribution of White-beaked Dolphin (depicting those areas where the species is thought to regularly occur).

Barents Sea and west of Greenland, the species occurs at depths also of 300 to >1000 m (Fall & Skern-Mauritzen, 2014; Hansen & Heide-Jørgensen, 2013).

The SCANS survey in July 1994 arrived at a population estimate of 23,716 white-beaked dolphins (CV 0.30, 95% CI: 13,400−41,900) (revised from Hammond et al., 2002). All records were made in the North Sea and the area directly northwest of Scotland, between c. 54°−60°N, 6°W−7°E. The SCANS-II survey in July 2005, covering European continental shelf seas from SW Norway south to Atlantic Portugal, estimated 37,689 individuals (CV 0.36, 95% CI: 18,900−75,200) (revised from Hammond et al., 2013), with the majority in the North Sea and off northwest Britain. The SCANS-III survey in July 2016 over a similar but slightly larger area gave an overall abundance estimate of 36,287 (CV 0.29, 95% CI: 18,700−61,900) dolphins, with the highest densities in the northern North Sea and shelf

FIGURE 4.20   White-beaked Dolphins frequently leap clear of the water. *Photo: Jacopo Bridda.*

waters west and north of Scotland (Hammond et al., 2017). For the equivalent area (North Sea), white-beaked dolphin abundance estimates for 1994, 2005 and 2016 showed no significant trend (Hammond et al., 2017). The ObSERVE surveys of the Irish EEZ in summer and winter of 2015 and 2016 yielded an estimate in summer 2015 of 7090 (CV 0.44, 95% CI: 4100−12,200) dolphins with a much lower estimate of 3248 (CV 0.46, 95% CI: 1800−5700) in summer 2016. All of these were west of Ireland, but, interestingly, often far offshore close to the shelf edge and very few in winter (Rogan et al., 2018). Combining SCANS-III and ObSERVE survey results in summer 2016 resulted in an overall abundance estimate of 39,535 (CV 0.27, 95% CI: 23,600−66,300).

Further north, around Iceland and the Faroes (but mainly in north and southwest Iceland), the 2007 TNASS survey estimated approximately 111,000 white-beaked dolphins (CV 0.59, 95% CI: 36,346−340,114) (NAMMCO, 2018).

White-beaked dolphins do not generally move with the same speed and agility as common and striped dolphins, but like them, they will frequently breach (Fig. 4.20) and bow-ride vessels. In the ASCOBANS Agreement Area, group sizes are typically small, fewer than 10, although herds up to 30 are not uncommon, and occasionally can number 100−500 animals; they are present year-round, although in coastal waters they are most frequently observed between June and September (Evans, 1992; Evans et al., 2003; Evans & Smeenk, 2008; Kinze et al., 1997). Calves are born mainly between May and September, although births have been recorded outside this period (Kinze et al., 1997; Evans & Smeenk, 2008).

Four management units have been proposed on the basis of genetic and morphometric evidence: (1) Western North Atlantic (Canadian waters at least); (2) Icelandic waters; (3) northern Norway; and (4) the North Sea and waters around the British Isles and Ireland (Evans & Teilmann, 2009). With more widespread sampling, more differentiation may be identified. Mitochondrial DNA studies showed that genetic diversity at the nucleotide level was extremely low (Banguera-Hinestroza et al., 2010).

Threats to the species in the ASCOBANS Agreement Area are not well known but currently may be primarily from prey depletion and the possible effects of climate change. From post mortem examinations of 84 white-beaked dolphins stranded in the UK between 1991 and 2010, 32% live-stranded, 13% died from starvation, 11% from infectious disease, and 8% were diagnosed as by-catch (Deaville & Jepson, 2011).

References

Banguera-Hinestroza, E., Bjørge, A., Reid, R. J., Jepson, P., & Hoelzel, A. R. (2010). The influence of glacial epochs and habitat dependence on the diversity and phylogeography of a coastal dolphin species: *Lagenorhynchus albirostris*. *Conservation Genetics, 11*, 1823−1836.

Deaville, R. and Jepson, P.D. (compilers) (2011) *UK Cetacean Strandings Investigation Programme Final report for the period 1st January 2005 - 31st December 2010.* Zoological Society of London, UK. (http://randd.defra.gov.uk/Document.aspx?Document = FinalCSIPReport2005-2010_finalversion061211released[1].pdf).

Evans, P. G. H. (1992). *Status Review of Cetaceans in British and Irish waters.* London: UK Dept. of the Environment, 98 pp.

Evans, P. G. H., Anderwald, P., & Baines, M. E. (2003). *UK Cetacean Status Review. Report to English Nature and Countryside Council for Wales.* Oxford: Sea Watch Foundation, 160 pp.

Evans, P. G. H., & Smeenk, C. (2008). White-beaked dolphin *Lagenorhynchus albirostris.* In S. Harris, & D. W. Yalden (Eds.), *Mammals of the British Isles* (4th ed., pp. 724–727). Southampton: The Mammal Society, 800 pp, Handbook.

Evans, P.G.H. and Teilmann, J. (editors) (2009) *Report of ASCOBANS/HELCOM Small Cetacean Population Structure Workshop.* ASCOBANS/ UNEP Secretariat, Bonn, Germany. 140 pp.

Fall, J., & Skern-Mauritzen, M. (2014). White-beaked dolphin distribution and association with prey in the Barents Sea. *Marine Biology Research, 10,* 957–971.

Galatius, A., & Kinze, C. C. (2016). Lagenorhynchus albirostris (Cetacea: Delphinidae). *Mammalian Species, 48*(933), 35–47.

Hammond, P. S., Berggren, P., Benke, H., Borchers, D. L., Collet, A., Heide-Jørgensen, M. P., … Øien, N. (2002). Abundance of harbour porpoise and other cetaceans in the North Sea and adjacent waters. *Journal of Applied Ecology, 39,* 361–376.

Hammond, P.S., Lacey, C., Gilles, A., Viquerat, S., Borjesson, P., Herr, H., … and Øien, N. (2017) Estimates of cetacean abundance in European Atlantic waters in summer 2016 from the SCANS-III aerial and shipboard surveys. Available at https://synergy.standrews.ac.uk/scans3/files/2017/ 05/SCANS-III-design-based-estimates-2017-05-12-final-revised.pdf.

Hammond, P. S., Macleod, K., Berggren, P., Borchers, D. L., Burt, M. L., Cañadas, A., … Vázquez, J. A. (2013). Cetacean abundance and distribution in European Atlantic shelf waters to inform conservation and management. *Biological Conservation, 164,* 107–122.

Hansen, R. G., & Heide-Jørgensen, M. P. (2013). Spatial trends in abundance of long-finned pilot whales, white-beaked dolphins and harbour porpoises in West Greenland. *Marine Biology, 160,* 2929–2941.

Kinze, C. C. (2008). White-beaked Dolphin *Lagenorhynchus albirostris.* In W. F. Perrin, B. Würsig, & J. Thewissen (Eds.), *Encyclopedia of marine mammals* (2nd ed., pp. 1254–1258). London: Academic Press, 1352 pp.

Kinze, C. C., Addink, M., Smeenk, C., Garcia Hartmann, M., Richards, H. W., Sonntag, R. P., & Benke, H. (1997). The white-beaked dolphin (*Lagenorhynchus albirostris*) and the white-sided dolphin (*Lagenorhynchus acutus*) in the North and Baltic Seas: Review of available information. *Reports of the International Whaling Commission, 47,* 675–681.

NAMMCO (2018) *Report of the NAMMCO Scientific Working Group on Abundance Estimates.* Available at https://nammco.no/topics/sc-working-group-reports/.

Reid, J. B., Evans, P. G. H., & Northridge, S. P. (2003). *Atlas of Cetacean Distribution in North-west European Waters.* Peterborough: Joint Nature Conservation Committee, 76 pp.

Rogan, E., Breen, P., Mackey, M., Cañadas, A., Scheidat, M., Geelhoed, S., & Jessopp, M. (2018). *Aerial surveys of cetaceans and seabirds in Irish waters: Occurrence, distribution and abundance in 2015-2017.* Dublin, Ireland: Department of Communications, Climate Action & Environment and National Parks and Wildlife Service (NPWS), Department of Culture, Heritage and the Gaeltacht, 297 pp.

Atlantic white-sided dolphin *Lagenorhynchus acutus*

The Atlantic white-sided dolphin commonly inhabits the continental shelf edge of the northern North Atlantic. At sea the species superficially resembles the white-beaked dolphin in size, shape and markings. However, it is slightly slimmer and smaller at 2.1–2.6 m, and can best be identified by a distinctive white blaze on the flank and behind it a yellow or ochre band (Fig. 4.21). Neither extends over the dorsal surface, in contrast to the pale area on the flanks of the white-beaked dolphin. It has a sloping forehead with a short beak, and in the centre of its black back is a slender recurved fin. The tail-stock has a thickened keel particularly in males, which then narrows sharply close to the flukes Whilst actively swimming at speed, white-sided dolphins commonly break the surface (Fig. 4.23), sometimes in a full breach, but generally parallel to the sea surface (see also Fig. 4.21).

Atlantic white-sided dolphins are confined to temperate and subpolar seas (generally in sea surface temperatures of 3°C–12°C) of the North Atlantic, mainly found offshore from southwest Greenland, Iceland and the western Barents Sea southwards to the Georges Bank (NE United States) and southwest Ireland, although the species occurs casually south to Virginia (United States) in the west and the Bay of Biscay (47°N) in the east (Fig. 4.22). It is rare in the Irish Sea, English Channel and southern North Sea, and occurs only occasionally further east in Danish waters and the Baltic.

The following appear to represent areas of relatively high density: (1) southwest Gulf of Maine (40–42°N) and continental slope at approximately 39°N; (2) Irminger Basin between southeast Greenland and Iceland; (3) Iceland Basin south and east of Iceland; (4) Faroe Bank Channel and Faroe–Shetland Channel; (5) Halten Bank, west of Norway and (6) Rockall Trough west of Scotland and Ireland (Evans & Smeenk, 2008; Evans et al., 2003; Reeves, Smeenk, Brownell, and Kinze, 1999; Reid et al., 2003; Salo, 2004; Selzer & Payne, 1988).

No comprehensive population estimates exist. The July 2007 CODA survey in offshore northwest European waters (between approximately 44° and 61°N) saw surprisingly few Atlantic white-sided dolphins and thus was unable to

FIGURE 4.21 Atlantic white-sided Dolphin displaying its distinctive white blaze on the flank with a yellow ochre band extending back towards the tail. *Photo: Heike Vester/Ocean Sounds — Marine Mammal Research & Conservation.*

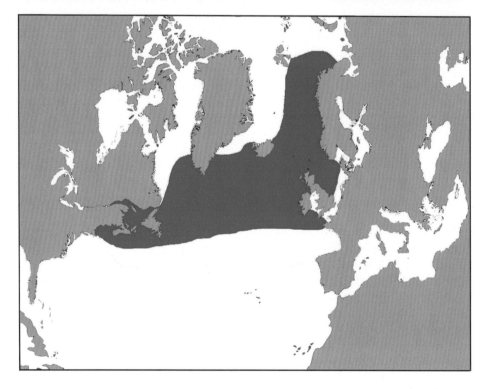

FIGURE 4.22 Overall distribution of Atlantic white-sided dolphin (depicting those areas where the species is thought to regularly occur).

derive an abundance estimate (Hammond et al., 2009). The SCANS-III survey in July 2016 over a large part of the ASCOBANS Agreement Area obtained an overall abundance estimate of 15,510 but with wide confidence limits due to the low numbers seen (CV 0.72, 95% CI: 4,389−54,800) dolphins, with greatest densities along the shelf edge northwest of Scotland (Hammond et al., 2017). The ObSERVE surveys of the Irish EEZ in 2015 and 2016 gave an abundance estimate for the summeres of the 2 years combined of 1,921 dolphins (CV 0.46, 95% CI: 1,100−3,400), with almost all sightings along the shelf edge northwest of Ireland (Rogan et al., 2018). The combined estimate from SCANS-III and ObSERVE surveys for 2016 was 17,431 (CV 0.64, 95% CI: 5,500−55,000).

Salo (2004) analysed the sightings of Atlantic white-sided dolphins from the NASS/TNASS surveys and found that the species was more commonly observed in the eastern part of Iceland and towards the Faroe Islands than elsewhere in the

FIGURE 4.23 Atlantic White-sided Dolphins are fast-moving dolphins, often creating spray when they break the surface. *Photo: Luca Lamoni.*

central and eastern North Atlantic. In that area the abundance of the species during the 2007 TNASS survey was estimated at 42,547 (CV 0.42; 95% CI: 17,537−103,225); adjustment for possible measurement bias by the primary platform decreased this estimate by 13% (NAMMCO, 2018). In British and Irish waters, there is some indication that the range of the species has shifted northwards since the 1990s, possibly linked to climate change (Evans & Bjørge, 2013).

Frequently seen in large schools that can number tens or hundreds of animals, white-sided dolphins often travel at high speeds although they do not bow-ride vessels to the same extent as white-beaked dolphins. The species favours the continental slope (mainly around 100−300 m depth) and deeper waters, particularly areas of high bottom relief and around deep submarine canyons. Over the northwest European continental shelf, it is usually recorded in groups numbering 5−50 individuals during the months of July to September (Evans, 1992; Evans et al., 2003; Evans & Smeenk, 2008; Kinze et al., 1997; Reid et al., 2003). Genetic studies using both mitochondrial DNA and microsatellites revealed no differentiation between western and eastern North Atlantic samples but some indication of differences with historical samples; the species showed relatively low genetic diversity suggesting strong changes in population size and range in the past (Banguera-Hinestroza et al., 2014; Mirimin et al., 2011).

Atlantic white-sided dolphins continue to be hunted opportunistically in drive fisheries in the Faroe Islands with 806 animals killed there between 2006 and 2010 and 430 animals between 2011 and 2015 (NAMMCO, 2010, 2015). The main threats in the ASCOBANS Agreement Area are thought to be fisheries-related bycatch and possibly prey depletion, with substantial bycatches recorded on occasions from mid-water trawl fisheries west and south of Ireland (see, e.g. Couperus, 1997; Morizur, Berrow, Tregenza, Couperus, and Pouvreau, 1999), and in the past salmon driftnets are known to have caused bycatch of the species. From postmortem examinations of 84 white-sided dolphins stranded in the United Kingdom between 1991 and 2010, 54% live-stranded, 21% died as a result of infectious disease, 14% died from starvation and 4% were diagnosed as bycatch (Deaville & Jepson, 2011).

References

Banguera-Hinestroza, E., Evans, P. G. H., Mirimin, L., Reid, R. J., Mikkelsen, B., Couperus, A. S., . . . Hoelzel, A. R. (2014). Phylogeography and population dynamics of the Atlantic white-sided dolphin (*Lagenorhynchus acutus*) in the North Atlantic. *Conservation Genetics, 15*(4), 789−802. Available from https://doi.org/10.1007/s10592-014-0578-z.

Couperus, A. S. (1997). Interactions between Dutch mid-water trawl and Atlantic White-sided dolphins SW of Ireland. *Journal of Northwest Atlantic Fisheries Science, 22*, 209−218.

Deaville, R. and Jepson, P.D. (compilers) (2011) *UK Cetacean Strandings Investigation Programme Final report for the period 1st January 2005 - 31st December 2010.* Zoological Society of London, UK. (http://randd.defra.gov.uk/Document.aspx?Document = FinalCSIPReport2005-2010_finalversion061211released[1].pdf).

Evans, P. G. H. (1992). *Status Review of Cetaceans in British and Irish waters*. London: UK Dept. of the Environment, 98 pp.

Evans, P. G. H., Anderwald, P., & Baines, M. E. (2003). *UK Cetacean Status Review. Report to English Nature and Countryside Council for Wales*. Oxford: Sea Watch Foundation, 160 pp.

Evans, P. G. H., & Bjørge, A. (2013). Impacts of climate change on marine mammals. *Marine Climate Change Impacts Partnership (MCCIP) Science Review, 2013*, 134−148. Available from https://doi.org/10.14465/2013.arc15.134-148, Published online November 28, 2013.

Evans, P. G. H., & Smeenk, C. (2008). Atlantic white-sided dolphin *Leucopleura acutus*. In S. Harris, & D. W. Yalden (Eds.), *Mammals of the British Isles* (4th ed, pp. 727−731). Southampton: The Mammal Society, 800 pp, Handbook.

Hammond, P.S., Lacey, C., Gilles, A., Viquerat, S., Borjesson, P., Herr, H., ... and Øien, N. (2017). Estimates of cetacean abundance in European Atlantic waters in summer 2016 from the SCANS-III aerial and shipboard surveys. Available at https://synergy.standrews.ac.uk/scans3/files/2017/05/SCANS-III-design-based-estimates-2017-05-12-final-revised.pdf.

Hammond, P.S., Burt, M.L., Cañadas, A., Castro, R., Certain, G., Gillespie, D., ... Winship, A. (2009). Cetacean Offshore Distribution and Abundance in the European Atlantic (CODA). Report available from SMRU, Gatty Marine Laboratory, University of St Andrews, St Andrews, Fife KY16 8LB, UK. http://biology.st-andrews.ac.uk/coda/.

Kinze, C. C., Addink, M., Smeenk, C., Garcia Hartmann, M., Richards, H. W., Sonntag, R. P., & Benke, H. (1997). The white-beaked dolphin (*Lagenorhynchus albirostris*) and the white-sided dolphin (*Lagenorhynchus acutus*) in the North and Baltic Seas: Review of available information. *Reports of the International Whaling Commission, 47*, 675−681.

Mirimin, L., Banguera-Hinestroza, E., Dillane, E., Hoelzel, A. R., Cross, T. F., & Rogan, E. (2011). Insights into genetic diversity, parentage, and group composition of Atlantic white-sided dolphins (*Lagenorhynchus acutus*) off the west of Ireland based on nuclear and mitochondrial genetic markers. *Journal of Heredity, 102*(1), 79−87.

Morizur, Y., Berrow, S. D., Tregenza, N. J. C., Couperus, A. S., & Pouvreau, S. (1999). Incidental catches of marine mammals in pelagic trawl fisheries of the north-east Atlantic. *Fisheries Research, 41*(3), 297−307.

NAMMCO (2010). *NAMMCO annual report 2009* (529 pp). Tromsø: North Atlantic Marine Mammal Commission.

NAMMCO (2015). *NAMMCO annual report 2014* (246 pp). Tromsø: North Atlantic Marine Mammal Commission.

NAMMCO (2018) *Report of the NAMMCO Scientific Working Group on Abundance Estimates*. Available at https://nammco.no/topics/sc-working-group-reports/.

Reeves, R. R., Smeenk, C., Brownell, L., & Kinze, C. C. (1999). Atlantic white-sided dolphin *Lagenorhynchus acutus* (Gray, 1828). In S. H. Ridgway, & R. J. Harrison (Eds.), *Handbook of marine mammals. Vol. 6: The second book of dolphins and porpoises* (pp. 31−56). San Diego: Academic Press.

Reid, J. B., Evans, P. G. H., & Northridge, S. P. (2003). *Atlas of Cetacean Distribution in North-west European Waters*. Peterborough: Joint Nature Conservation Committee, 76 pp.

Rogan, E., Breen, P., Mackey, M., Cañadas, A., Scheidat, M., Geelhoed, S., & Jessopp, M. (2018). *Aerial surveys of cetaceans and seabirds in Irish waters: Occurrence, distribution and abundance in 2015-2017*. Dublin, Ireland: Department of Communications, Climate Action & Environment and National Parks and Wildlife Service (NPWS), Department of Culture, Heritage and the Gaeltacht, 297 pp.

Salo, K. (2004). *Distribution of Cetaceans in Icelandic waters* (M.Sc. thesis). Odense: University of Southern Denmark.

Selzer, L. A., & Payne, P. M. (1988). The distribution of white sided (*Lagenorhynchus acutus*) and common dolphins (*Delphinus delphis*) vs. environmental features of the continental shelf of the Northeastern United States. *Marine Mammal Science, 4*, 141−153.

Risso's dolphin *Grampus griseus*

The Risso's dolphin is uncommon but widely distributed, occurring mainly in warm temperate and subtropical seas of all oceans. It is a large, stout dolphin, about 3.5 m in length. Although the young are dark grey in coloration, as they grow older the head, back and flanks become light grey, and even white in the more mature animals. Many scratches and scars typically show around the head and back, particularly in front of the dorsal fin (Fig. 4.24). The fin is relatively tall and recurved, lightening with age, particularly along the leading edge. The blunt, rounded head with slight melon and no beak and the long, pointed flippers show similarities to the pilot whales.

In the North Atlantic, the main range of the Risso's dolphin is from the tropics north to Georges Bank, northeast United States in the west and the Shetland Isles, north Scotland in the east (Fig. 4.25), favouring continental slope waters in depths of 200−1200 m, although in the British Isles it is frequently found over slopes of 50−100 m depth (Evans, 2008). Major populations exist around the oceanic archipelagos of the Azores, Madeira and Canaries. The species has been recorded occasionally north to Sable Island, Nova Scotia and Newfoundland in the west, and the Faroe Islands and west Norway in the east, apparently favouring waters mainly between 10°C and 28°C (Evans, 2008).

Although nowhere common, the major populations in northern European waters occur in the Hebrides, and the species is regular in small numbers in Shetland, Orkney and northeast Scotland, in the Irish Sea (particularly off the coasts of Co. Wexford, west Pembrokeshire, northwest Wales and the Isle of Man) and around western Ireland, particularly the

FIGURE 4.24 Risso's dolphin. As individuals mature, the head, back and flanks become light grey and even white, with scratches and scars across the body. *Photo: Peter GH Evans.*

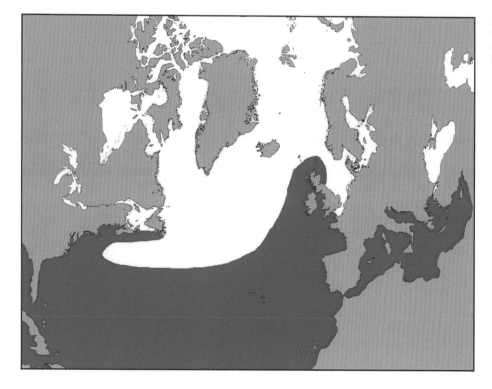

FIGURE 4.25 North Atlantic distribution of Risso's dolphin (depicting those areas where the species is thought to regularly occur).

southwest (Berrow et al., 2010; Baines & Evans, 2012; Evans et al., 2003; Reid et al., 2003; Stevens 2014; Wall et al., 2013). It is rare in the North Sea but in recent years it has been seen regularly off east Scotland; in the English Channel it occurs mainly in the western part, although also a regular visitor to the Channel Islands (Evans et al., 2003). Further south, the species is present in West France, the southern Bay of Biscay, all around the Iberian Peninsula and in the Mediterranean Sea (Azzellino et al., 2016; Kiszka et al., 2004, 2007; Notarbartolo di Sciara & Birkun, 2010; Reid et al., 2003).

The only population estimate for the species spanning most of the ASCOBANS Agreement Area has been the SCANS-III survey in July 2016 with an overall abundance of 13,584 (CV 0.44, 95% CI: 5900−31,000) dolphins, and

FIGURE 4.26 Risso's Dolphins usually form small groups, wth some evidence for segregation by age and sex. *Photo: Vittorio Fadda, Tethys Research Institute.*

the greatest densities were west of Scotland and in the Irish Sea (Hammond et al., 2017). The ObSERVE Project survey of the Irish EEZ in 2015−16 yielded a maximum overall abundance, in summer 2016, of 2630 (CV 0.41, 95% CI: 1200−5700) animals, with highest densities around the southwest and south coastal areas of Ireland (Rogan et al., 2018). Sightings during the SCANS, SCANS-II and CODA surveys were too few to derive realistic population estimates (Hammond et al., 2002, 2009, 2013). In the western North Atlantic, a population estimate of 18,250 (CV = 0.46) exists for waters off the eastern United States, and 2442 (CV 0.57) in the northern Gulf of Mexico (Palka, 2012; Waring, Josephson, Maze-Foley, & Rosel, 2016).

A photo-ID study in the North Minches, Scotland, identified at least 142 individuals (Atkinson, Gill, & Evans, 1997; Atkinson, Gill, and Evans, 1998), and another at Bardsey Island (northwest Wales) identified at least 145 individuals (De Boer, Clark, Leopold, Simmonds, & Reijnders, 2013).

In the ASCOBANS Agreement Area, Risso's dolphins are usually seen in groups of 2−20 animals, only occasionally numbering 50 or more (Evans, 2008). Calves are born mainly between March and July, and the species feeds primarily upon cephalopods − octopus, cuttlefish and small squid (Evans, 2008).

Threats to Risso's dolphin in the ASCOBANS Agreement Area are not well known but are thought to be bycatch, and possibly prey depletion and noise disturbance (Evans, 2013). From postmortem examinations of 35 Risso's dolphins stranded in the United Kingdom between 1991 and 2010, 17% were diagnosed as bycatch, 14% live-stranded, 14% died from starvation and 11% died as a result of infectious disease (Deaville & Jepson, 2011).

References

Atkinson, T., Gill, A., & Evans, P. G. H. (1997). Notes on the natural markings on Risso's Dolphins (*Grampus griseus*) photographed in the coastal waters around the Eye Peninsula, Isle of Lewis, Scotland. *European Research on Cetaceans*, *11*, 209.

Atkinson, T., Gill, A., & Evans, P. G. H. (1998). A photo-identification study of Risso's dolphins in the Outer Hebrides, Northwest Scotland. *European Research on Cetaceans*, *12*, 102.

Azzellino, A., Airoldi, S., Gaspari, S., Lanfredi, C., Moulins, A., Podestà, M., . . . Tepsich, P. (2016). Risso's dolphin, *Grampus griseus*, in the western Ligurian Sea: Trends in population size and habitat use. *Advances in Marine Biology*, *75*, 205−232.

Baines, M.E. and Evans, P.G.H. (2012) *Atlas of the Marine Mammals of Wales*. 2nd Edition. CCW Monitoring Report No. 68. 143 pp.

Berrow, S., Whooley, P., O'Connell, M., & Wall, D. (2010). *Irish Cetacean Review (2000−2009)*. Kilrush: Irish Whale and Dolphin Group, 60 pp.

Deaville, R. and Jepson, P.D. (compilers) (2011) *UK Cetacean Strandings Investigation Programme Final report for the period 1st January 2005 - 31st December 2010*. Zoological Society of London, UK. (http://randd.defra.gov.uk/Document.aspx?Document = FinalCSIPReport2005-2010_finalversion061211released[1].pdf).

De Boer, M. N., Clark, J., Leopold, M. F., Simmonds, M. P., & Reijnders, P. J. H. (2013). Photo-identification methods reveal seasonal and long-term site-fidelity of Risso's dolphins (*Grampus griseus*) in shallow waters (Cardigan Bay, Wales). *Open Journal of Marine Science*, *3*, 66−75.

Evans, P. G. H. (2008). Risso's dolphin *Grampus griseus*. In S. Harris, & D. W. Yalden) (Eds.), *Mammals of the British Isles* (4th ed, pp. 740−743). Southampton: The Mammal Society, 800 pp, Handbook.

Evans, P. G. H. (2013). The Risso's Dolphin in Europe (pp. 10−24). In: I. Chen, K. Hartman, M. Simmonds, A. Wittich, & A. J. Wright (Eds.), *Grampus griseus 200th anniversary: Risso's dolphins in the contemporary world. Report from the European Cetacean Society Conference Workshop, Galway, Ireland, 25 March 2013* (108 pp). ECS Special Publication Series No: 54.

Evans, P. G. H., Anderwald, P., & Baines, M. E. (2003). *UK Cetacean Status Review. Report to English Nature and Countryside Council for Wales.* Oxford: Sea Watch Foundation, 160 pp.

Hammond, P. S., Berggren, P., Benke, H., Borchers, D. L., Collet, A., Heide-Jørgensen, M. P., ... Øien, N. (2002). Abundance of harbour porpoise and other cetaceans in the North Sea and adjacent waters. *Journal of Applied Ecology, 39,* 361−376.

Hammond, P.S., Lacey, C., Gilles, A., Viquerat, S., Borjesson, P., Herr, H., ... and Øien, N. (2017) Estimates of cetacean abundance in European Atlantic waters in summer 2016 from the SCANS-III aerial and shipboard surveys. Available at https://synergy.standrews.ac.uk/scans3/files/2017/05/SCANS-III-design-based-estimates-2017-05-12-final-revised.pdf.

Hammond, P. S., Macleod, K., Berggren, P., Borchers, D. L., Burt, M. L., Cañadas, A., ... Vázquez, J. A. (2013). Cetacean abundance and distribution in European Atlantic shelf waters to inform conservation and management. *Biological Conservation, 164,* 107−122.

Kiszka, J., Hassani, S., & Pezeril, S. (2004). Distribution and status of small cetaceans along the French Channel coasts: using opportunistic records for a preliminary assessment. *Lutra, 47,* 33−45.

Kiszka, J., Macleod, K., Van Canneyt, O., & Ridoux, V. (2007). Distribution, relative abundance and bathymetric preferences of toothed cetaceans in the English Channel and Bay of Biscay. *ICES Journal of Marine Science,* 1033−1043.

Notarbartolo di Sciara, G. and Birkun, A. (2010) *Conserving whales, dolphins and porpoises in the Mediterranean and Black Seas.* An ACCOBAMS Status Report. ACCOBAMS, Monaco. 212 pp.

Palka, D.L. (2012) Cetacean abundance estimates in US northwestern Atlantic Ocean waters from summer 2011 line transect survey. Northeast Fisheries Science Center Reference Document 12-29. 37 pp. http://www.nefsc.noaa.gov/nefsc/publications/crd/crd1229/.

Reid, J. B., Evans, P. G. H., & Northridge, S. P. (2003). *Atlas of Cetacean Distribution in North-west European Waters.* Peterborough: Joint Nature Conservation Committee, 76 pp.

Rogan, E., Breen, P., Mackey, M., Cañadas, A., Scheidat, M., Geelhoed, S., & Jessopp, M. (2018). *Aerial surveys of cetaceans and seabirds in Irish waters: Occurrence, distribution and abundance in 2015-2017.* Dublin, Ireland: Department of Communications, Climate Action & Environment and National Parks and Wildlife Service (NPWS), Department of Culture, Heritage and the Gaeltacht, 297 pp.

Stevens, A. (2014). *A Photo-ID study of the Risso's dolphin (Grampus griseus) in Welsh coastal waters and the use of Maxent modeling to examine the environmental determinants of spatial and temporal distribution in the Irish Sea* (M.Sc. thesis) (97 pp). University of Bangor.

Wall, D., Murray, C., O'Brien, J., Kavanagh, L., Wilson, C., Ryan, C., ... Berrow, S. (2013). *Atlas of the distribution and relative abundance of marine mammals in Irish offshore waters 2005-2011.* Irish Whale and Dolphin Group, Merchants Quay, Kilrush, Co. Clare.

Waring, G. T., Josephson, E., Maze-Foley, K., & Rosel, P. E. (2016). *US Atlantic and Gulf of Mexico marine mammal stock assessments − 2015* (501 pp). NOAA Technical Memorandum NMFS-NE-238.

Melon-headed whale *Peponocephala electra*

The melon-headed whale reaches 2.3−2.7 m in length and is found predominantly in deep tropical seas (generally >1000 m) straying only very rarely into North American or European waters. Uniformly dark in colour and of slim, torpedo shape, it has a very similar appearance to the pygmy killer whale but with a more triangular relatively narrow head rather than short and round, often white lips, and flippers that are usually pointed at the tip and smooth along the rear margin rather than rounded. It has a more robust body than the rather larger all-black false killer whale, and unlike that species it tends to have a subtle grey cape pattern over the back dipping down on the flanks below the dorsal fin. The dorsal fin is also relatively tall, recurved and centrally placed on an arched back, whereas in the false killer whale it appears smaller (relative to body size), narrow but not pointed, and is further to the rear on a long, flat back.

Melon-headed whales often occur in very large tightly packed herds which may be mixed with other species, particularly Fraser's dolphin. These can travel rapidly with frequent changes in course direction. The only records of this species in the ASCOBANS Agreement Area are a stranding from Cornwall, southwest England in 1949, whilst in France, two individuals live-stranded near La Rochelle in 2003, and another one stranded in the same area in 2008 (Evans, 2008).

Reference

Evans, P. G. H. (2008). Risso's dolphin *Grampus griseus.* In S. Harris, & D. W. Yalden) (Eds.), *Mammals of the British Isles* (4th ed., pp. 740−743). Southampton: The Mammal Society, 800 pp, Handbook.

Pygmy killer whale *Feresa attenuata*

The pygmy killer whale inhabits tropical and subtropical seas of the Atlantic, Pacific and Indian Oceans, rarely occurring north of 40°N. It is one of the least known of all the delphinids, occurring largely offshore in deep waters, though sometimes seen in groups of several hundred animals. It is a relatively small whale with adults up to 2.6 m, all black or dark grey, with a slender body that tapers behind the tall, recurved, dorsal fin. The head is rounded or even bulbous and has no beak. The lips and snout are white; the flippers are long with rounded tips. The species is often confused with the false killer whale and melon-headed whale, which are both very similar in general appearance.

The main range of the species is well to the south of the ASCOBANS Agreement Area, but two sighting records were reported from the southern Bay of Biscay in April 1997 (Williams, Williams, & Brereton, 2002).

Reference

Williams, A. D., Williams, R., & Brereton, T. (2002). The sighting of pygmy killer whales (*Feresa attenuata*) in the southern Bay of Biscay and their association with cetacean calves. *Journal of the Marine Biological Association U.K.*, 82, 509–511.

False killer whale *Pseudorca crassidens*

False killer whales have a worldwide distribution mainly in deep tropical and warm temperate seas, occurring north to Maryland (United States) in the west, and Norway and the British Isles in the east. They are 5–6 m in length and have a long slender body. They are almost all-black in colour except for a blaze of grey on the chest between the flippers, and a grey area sometimes present on the sides of the head. The head is small and narrow, and tapers to overhang the lower jaw unlike the similar pygmy killer and melon-headed whales. The flippers are long, narrow and tapered with a

FIGURE 4.27   False killer whale mass stranding on the coast of North Wales in 1935. *Photo: Sea Watch Foundation photo library.*

distinctive broad hump on the front margin near the middle of the flipper. The dorsal fin, situated slightly behind the middle of the back, is tall and recurved, although its shape can vary from rounded to sharply pointed at the tip.

False killers are fast active swimmers, often forming large herds and frequently approaching vessels to bow-ride, although living in deep waters mainly far from the coast, they are only occasionally encountered.

In the eastern North Atlantic, the species occurs only occasionally north of the Bay of Biscay, though records are largely confined to a few mass strandings from the United Kingdom (in 1927, 1934 and 1935; see Fig. 4.27), a live stranding of two individuals near Ijmuiden, the Netherlands in November 1935 (presumably part of the same mass stranding at the time around the British Isles), and a handful of sightings from the Celtic and Irish Seas, northwest Ireland, northwest Scotland and the North Sea (Boran & Evans, 2008; Rogan et al., 2018).

References

Boran, J. R., & Evans, P. G. H. (2008). False killer whale *Pseudorca crassidens*. In S. Harris, & D. W. Yalden (Eds.), *Mammals of the British Isles* (4th ed., pp. 738–740). Southampton: The Mammal Society, 800 pp, Handbook.

Rogan, E., Breen, P., Mackey, M., Cañadas, A., Scheidat, M., Geelhoed, S., & Jessopp, M. (2018). *Aerial surveys of cetaceans and seabirds in Irish waters: Occurrence, distribution and abundance in 2015-2017*. Dublin, Ireland: Department of Communications, Climate Action & Environment and National Parks and Wildlife Service (NPWS), Department of Culture, Heritage and the Gaeltacht, 297 pp.

Killer whale or orca *Orcinus orca*

The killer whale or orca is a robust medium-sized whale – the largest member of the family Delphinidae. It has a global distribution that almost certainly exceeds all other cetacean species. In most regions, it is uncommon but numbers are greatest in cold temperate to polar seas. Varying in size from 5.7 to 9.5 m, adult male killer whales are about 25% larger than adult females, which range from 4.5 to 6.6 m. The males develop a very tall (up to almost 2 m), triangular and erect dorsal fin (Fig. 4.28), which is sometimes tilted forwards, and may even lie flat on the back of the animal. Immature animals and adult females have a smaller, more recurved dorsal fin and can only be distinguished by overall size and height of the dorsal fin. The coloration of killer whales is very striking – black on the back and sides, with a white belly extending as a backwards-pointing lobe up the flanks and less markedly around the throat, chin and undersides of the flippers. There is a white oval patch above and behind the eye, and a less distinct grey saddle on the back behind the dorsal fin, which shows up clearly when the animal surfaces. The species has a conical-shaped black

FIGURE 4.28 Killer whale males have a distinctive tall upright triangular fin. *Photo: Peter GH Evans.*

FIGURE 4.29 North Atlantic distribution of killer whale (depicting those areas where the species is thought to regularly occur).

head with indistinct beak, large paddle-shaped flippers, and broad tail flukes with a straight or slightly convex trailing edge and tips that particularly in adult males may curl down.

Killer whale abundance in the North Atlantic appears to be greatest in subarctic and arctic waters, although the distribution of the species extends south to the Caribbean, Azores, Madeira, Canaries and the western end of the Mediterranean (Boran, Hoelzel, & Evans, 2008; Notarbartolo di Sciara & Birkun, 2010). It is widely distributed in the North Atlantic and in coastal northern European waters, particularly around Iceland and west of Norway (Fig. 4.29). In the ASCOBANS Agreement Area, it is most commonly found off western Norway, in northern and western Scotland, and western Ireland but is rare in the Irish Sea, North Sea, the English Channel, the Bay of Biscay and off Atlantic Spain and Portugal. The species only occasionally enters the southern North Sea, inner Danish waters and the Baltic Sea.

Although recorded year-round in European seas, most sightings in near-shore waters of Scotland and Ireland occur between May and September (Berrow et al., 2010; Bolt, Harvey, Mandleberg, & Foote, 2009; Evans, 1988; Evans et al., 2003), whereas in northwest Norway they occur mainly between October and February, subsequently following the shoals of herring offshore (Similä, Holst, & Christensen, 1996), although in recent years their distribution has changed following changes in the winter distribution of herring.

No overall abundance estimates exist for the North Atlantic, but NASS sightings surveys in 1987, mainly in sea areas from Iceland to the Faroe Islands, indicated a population in that region of somewhere between 3500 and 12,500 individuals (Gunnlaugsson & Sigurjönsson, 1990). More recent TNASS surveys provide a preliminary uncorrected abundance estimate of 29,677 (CV 0.25) in 2005 (NAMMCO, 2018). The vast majority of killer whale sightings were made in the Norwegian Sea south of the Mohn ridge. They were also fairly abundant in the Icelandic–Jan Mayen survey blocks.

Photo-ID studies, supported by genetic evidence, had suggested that the population visiting coastal regions of northern Britain was quite small, possibly as few as 30 individuals, and that these were linked to the community of killer whales that follow the Icelandic summer-spawning herring, being separate from the even smaller west coast (British and Irish) community (Foote, Newton, Piertney, Willerslev, & Gilbert, 2009; Foote, Similä, Vikingsson, & Stevick, 2010; Foote et al., 2011; Samarra & Foote, 2015). Individuals range over wide distances but, nevertheless, are recorded regularly in the same localities (e.g. two easily identifiable males have been observed regularly in the Hebrides since at least 1980, but have also been recorded off eastern Scotland, western Ireland and in the case of one of these off west Wales). Further south a population of around 40 individuals occurs in and around the Strait of Gibraltar in association with bluefin tuna (Esteban et al., 2016).

Group sizes recorded in coastal waters of the British Isles and Ireland are small, usually less than 10, with a maximum of 20 recorded, but offshore north and east of Shetland in winter, larger aggregations numbering up to 300

FIGURE 4.30 Killer Whales are one of the fastest of all cetaceans attaining speeds up to 50 km per hour. *Photo: Fernando Ugarte.*

individuals have been observed, associated with pelagic trawl fisheries (Boran et al., 2008; Evans, 1988; Luque, Davis, Reid, Wang, & Pierce, 2006). Studies in the eastern North Pacific indicate that groups have a stable membership, although pods can coalesce, and individuals (particularly males) can change their association patterns. Killer whales are fast swimmers, with speeds of up to 50 km per hour, particularly when in pursuit of prey (Fig. 4.30). Breeding in northwest Europe occurs between October and March, possibly mainly between October and December (Boran et al., 2008). During winter months killer whales are found particularly offshore in the northern North Sea and west of Norway, often associated with the mackerel and herring fisheries (Couperus, 1993, 1994; Evans et al., 2003; Luque et al., 2006). In summer killer whales visit coastal waters of Scotland and Ireland where at least some have been observed taking harbour seals, as well as harbour porpoise, otter and common eider (Bolt et al., 2009; Evans et al., 2003). Whereas in the eastern North Pacific killer whales may form distinct resident, transient and offshore communities specialising in different prey, in the Hebrides individuals within the same pod of killer whales have been observed on the same day feeding upon fish and hunting marine mammals (harbour seals).

The main threats to killer whales in the eastern North Atlantic are high levels of contaminants such as PCBs, PBDEs and DDT which may be affecting reproduction (Desforges et al., 2018; Jepson et al., 2016; Wolkers, Corkeron, Van Parijs, Similå, & Van Bavel, 2007), depletion of stocks of favoured prey and, until recently, direct takes in Norwegian, Icelandic and Faroese waters (Jonsgård & Lyshoel, 1970; Øien, 1988).

References

Berrow, S., Whooley, P., O'Connell, M., & Wall, D. (2010). *Irish Cetacean Review (2000—2009)*. Kilrush: Irish Whale and Dolphin Group, 60 pp.

Bolt, H. E., Harvey, P. V., Mandleberg, L., & Foote, A. D. (2009). Occurrence of killer whales in Scottish inshore waters: Temporal and spatial patterns relative to the distribution of declining harbour seal populations. *Aquatic Conservation: Marine and Freshwater Ecosystems, 19,* 671—675.

Boran, J. R., Hoelzel, A. R., & Evans, P. G. H. (2008). Killer whale *Orcinus orca.* In S. Harris, & D. W. Yalden (Eds.), *Mammals of the British Isles* (4th ed., pp. 743—747). Southampton: The Mammal Society, 800 pp, Handbook.

Couperus, A. S. (1993). Killer whales and pilot whales near trawlers east of Shetland. *Sula, 7,* 41—52.

Couperus, A. S. (1994). Killer whales (*Orcinus orca*) scavenging on discards of freezer trawlers north-east of the Shetland islands. *Aquatic Mammals, 20,* 47—51.

Desforges, J.-P., Hall, A., McConnell, B., Rosing-Asvid, A., Barber, J. L., Brownlow, A., . . . Dietz, R. (2018). Predicting global killer whale population collapse from PCB pollution. *Science, 361,* 1373—1376.

Esteban, R., Verborgh, P., Gauffier, P., Alarcón, D., Salazar-Sierra, J. M., Giménez, J., . . . de Stephanis, R. (2016). Conservation status of killer whales, *Orcinus orca,* in the Strait of Gibraltar. *Advances in Marine Biology, 75,* 141—172.

Evans, P. G. H. (1988). Killer whales (*Orcinus orca*) in British and Irish waters. *Rit Fiskideildar, 11,* 42—54.

Evans, P. G. H., Anderwald, P., & Baines, M. E. (2003). *UK Cetacean Status Review. Report to English Nature and Countryside Council for Wales.* Oxford: Sea Watch Foundation, 160 pp.

Foote, A. D., Newton, J., Piertney, S. B., Willerslev, E., & Gilbert, M. T. (2009). Ecological morphological and genetic divergence of sympatric North Atlantic killer whale populations. *Molecular Ecology, 18*(24), 5207−5217.

Foote, A. D., Similä, T., Vikingsson, G. A., & Stevick, P. T. (2010). Movement, site fidelity and connectivity in a top marine predator, the killer whale. *Evolutionary Ecology, 24*(4), 803−814.

Foote, A. D., Vilstrup, J. T., De Stephanis, R., Verborgh, P., Abel Nielsen, S. C., Deaville, R., ... Piertney, S. B. (2011). Genetic differentiation among North Atlantic killer whale populations. *Molecular Ecology, 20*(3), 629−641.

Gunnlaugsson, T., & Sigurjönsson, J. (1990). NASS-87: Estimation of whale abundance based on observations made on board Icelandic and Faroese survey vessels. *Reports of International Whaling Commission, 40*, 571−580.

Jepson, P. D., Deaville, R., Barber, J. L., Aguilar, A., Borrell, A., Murphy, S., ... Law, R. J. (2016). PCB pollution continues to impact populations of orcas and other dolphins in European waters. *Nature Scientific Reports, 6*, 18573. Available from https://doi.org/10.1038/srep18573.

Jonsgård, A., & Lyshoel, P. B. (1970). A contribution to the biology of the killer whale, *Orcinus orca* (L.). *Norwegian Journal of Zoology, 18*, 41−48.

Luque, P. L., Davis, C. G., Reid, D. G., Wang, J., & Pierce, G. J. (2006). Opportunistic sightings of killer whales from Scottish pelagic trawlers fishing for mackerel and herring off North Scotland (UK) between 2000 and 2006. *Aquatic Living Resources, 19*, 403−410.

NAMMCO (2018) *Report of the NAMMCO Scientific Working Group on Abundance Estimates.* Available at https://nammco.no/topics/sc-working-group-reports/.

Notarbartolo di Sciara, G. and Birkun, A. (2010) *Conserving whales, dolphins and porpoises in the Mediterranean and Black Seas.* An ACCOBAMS Status Report. ACCOBAMS, Monaco. 212 pp.

Øien, N. (1988). The distribution of killer whales (*Orcinus orca*) in the North Atlantic based on Norwegian catches, 1938-1981, and incidental sightings, 1967-1987. *Rit Fiskideildar, 11*, 65−78.

Samarra, F. I. P., & Foote, A. D. (2015). Seasonal movements of killer whales between Iceland and Scotland. *Aquatic Biology, 24*, 75−79.

Similä, T., Holst, J. C., & Christensen, I. (1996). Occurrence and diet of killer whales in northern Norway: Seasonal patterns relative to the distribution and abundance of Norwegian spring-spawning herring. *Canadian Journal of Fisheries and Aquatic Science, 53*, 769−779.

Wolkers, H., Corkeron, P. J., Van Parijs, S. M., Similå, T., & Van Bavel, B. (2007). Accumulation and transfer of contaminants in killer whales (*Orcinus orca*) from Norway: Indications for contaminant metabolism. *Environmental Toxicology and Chemistry, 26*, 1582−1590.

Long-finned pilot whale *Globicephala melas*

The long-finned pilot whale is found in the temperate seas of the North Atlantic and Mediterranean, particularly in deep waters off the continental shelf, where it can occur in groups numbering hundreds of animals.

It has a robust body with long pointed sickle-shaped flippers. Adult males can reach 5.0−5.9 m and females 3.8−4.8 m. The head is distinctive in shape being rather square and bulbous, particularly in old males, and with a slightly protruding upper lip. Both the head and back are black to dark grey in colour but an anchor-shaped patch of greyish-white can be seen on the chin, which is lighter in younger individuals. The dorsal fin is fairly low, long-based and situated slightly forwards of the mid-point of the back (Fig. 4.31), varying from recurved in immature animals and adult females, to flag-shaped in adult males. There is a thick keel on the tail-stock, which is more pronounced on adult males. The tail flukes have a concave trailing edge and are deeply notched in the centre.

There is some genetic evidence for differences between pilot whales living off Greenland and those elsewhere in the Atlantic, and it has been suggested that sea surface temperature may be an isolating mechanism, perhaps through temperature-dependent distribution of their favoured squid prey (Fullard et al., 2000).

Long-finned pilot whales are widespread in temperate regions of the world, usually occurring in deep temperate and subpolar waters of 200−3000 m depth (particularly around the 1000 m isobath) seaward and along edges of the continental shelf where bottom relief is greatest. They may venture occasionally into coastal waters, entering fjords and bays. In the northeast Atlantic, the species is common and widely distributed from the Faroe Islands and Iceland south to the Bay of Biscay and Iberian Peninsula; it is also common in the Mediterranean (Boran, Evans, & Martin, 2008; Evans et al., 2003; Reid et al., 2003; Verborgh et al., 2016; see Fig. 4.32).

In the ASCOBANS Agreement Area, the main areas that appear to be favoured include the Faroe−Shetland Channel, Rockall Trough, Porcupine Bight, the southwest Approaches to the English Channel and the northern part of the Bay of Biscay (Evans et al., 2003; Hammond et al., 2009; Reid et al., 2003; Wall et al., 2013; Weir et al., 2001). The species is rare in the North Sea except in the northernmost sector, but regularly enters the English Channel (Boran et al., 2008; Camphuysen & Peet, 2006; Evans et al., 2003; Smeenk & Camphuysen, 2016).

Sightings surveys in the eastern North Atlantic in the late 1980s (Buckland et al., 1993) estimated the population at 778,000 (CV 0.295), but the difficulty of accurately estimating group size and distance of the centre of the group from

FIGURE 4.31 The dorsal fin of Long-finned Pilot Whale is low, wide at the base, and situated slightly-forwards of the mid-point of the back. *Photo: Peter GH Evans.*

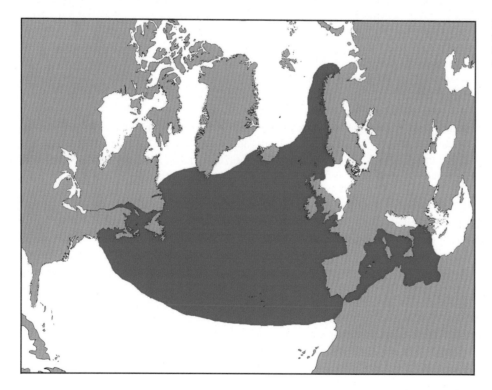

FIGURE 4.32 North Atlantic distribution of long-finned pilot whale (depicting those areas where the species is thought to regularly occur).

survey vessels imposes serious limitations to the accuracy of such estimates whilst the area of coverage was primarily to the north of the ASCOBANS Area. The TNASS survey in 2007 resulted in an abundance estimate of 128,000 (95% CI: 75,700–216,800) in the Iceland–Faroese survey area (NAMMCO, 2012, 2016). The NASS survey in 2015 resulted in a total abundance estimate of 589,691 (CV 0.38, 95% CI: 269,116–1,292,140), corrected for perception bias (NAMMCO, 2018). Density of long-finned pilot whales was highest around the Faroe Islands and to the southwest of Iceland. The above estimates are not directly comparable because they cover slightly different areas, times and survey methodology, so they cannot provide reliable trends, but they do give some idea of the approximate numbers in this region.

FIGURE 4.33 Long-finned Pilot Whale is one of the most abundant cetaceans in offshore European waters beyond the shelf edge, occurring in social groups, numbering in the tens or even hundreds of animals. *Photo: Sabina Airoldi/Tethys.*

Further south, there were too few sightings on the European continental shelf during the July 1994 and July 2005 SCANS & SCANS-II surveys to derive abundance estimates for the species (Hammond et al., 2002, 2013). Combining results from the 2005 SCANS-II and the 2007 CODA survey (in offshore waters south to NW Spain) resulted in an estimate of 123,732 (CV 0.35, 95% CI: 63,600−240,900) whales (revised from Hammond et al., 2009, 2013). The SCANS-III survey in July 2016 yielded an abundance estimate of 25,777 pilot whales (CV 0.345, 95% CI: 13,350−49,800), with the highest densities north and west of the Outer Hebrides and along the Biscay coast of Spain (Hammond et al., 2017). Note that the areas surveyed for which abundance estimates are given are not exactly equivalent to one another, excluding offshore waters beyond the shelf edge (see Figs 6.9 and 6.10). The ObSERVE surveys of the Irish EEZ in summer and winter 2015−16 yielded a maximum abundance of 7413 (CV 0.40, 95% CI: 3500−15,800) whales in summer 2016, and 9036 (CV 0.32, 95% CI: 4900−16,700) whales in winter 2016−17 (Rogan et al., 2018). Densities were greatest along the shelf edge west of Ireland. The combined estimate from the SCANS-III and ObSERVE surveys in summer 2016 was 33,190 (CV 0.28, 95% CI: 19,300−57,100).

Organised drives for pilot whales have taken place for at least 11 centuries in the Faroe Islands, with an average annual catch of 850 from 1709 to 1992 (Zachariassen, 1993) and 708 from 1992 to 2015 (Prime Minister's Office of the Faroe Islands, 2018). These drives continue to the present day. Between 1991 and 2017, annual catches have varied greatly, from zero (in 2008) to 1572 (in 1992) and 1203 (in 2017) (Prime Minister's Office of the Faroe Islands, 2018). Until the early 20th century, other drive fisheries operated opportunistically, mainly in Shetland and Orkney, but also in the Outer Hebrides and western Ireland, whilst through the 20th century small numbers have been taken at sea west of Norway, around the Faroes, Iceland and off west Greenland (Boran et al., 2008; Christensen, 1975; Fairley, 1981).

Besides direct takes of pilot whales just north of the ASCOBANS Agreement Area, the main threats are thought to be bycatch, possibly high contaminant levels and vessel strikes. From postmortem examinations of 33 pilot whales stranded in the United Kingdom between 1991 and 2010, 67% live-stranded, 6% died as a result of infectious disease, 3% were diagnosed as bycatch and 3% died from starvation (Deaville & Jepson, 2011).

References

Boran, J. R., Evans, P. G. H., & Martin, A. R. (2008). Long-finned pilot whale *Globicephala melas*. In S. Harris, & D. W. Yalden (Eds.), *Mammals of the British Isles* (4th ed., pp. 735−738). Southampton: The Mammal Society, 800 pp, Handbook.

Buckland, S. T., Bloch, D., Cattanach, K. L., Gunnlaugsson, Th, Hoydal, K., Lens, S., & Sigurjónsson, J. (1993). Distribution and abundance of long-finned pilot whales in the North Atlantic, estimated from NASS-87 and NASS-89 data. *Reports of the International Whaling Commission* (Special issue 10), 33−49.

Camphuysen, C. J., & Peet, G. (2006). *Whales and dolphins of the North Sea*. Kortenhoef: Fontaine Uitgevers, 160 pp.

Christensen, I. (1975). Preliminary report on the Norwegian fishery for small whales: Expansion of Norwegian whaling to arctic and northwest Atlantic waters, and Norwegian investigations of the biology of small whales. *Journal of the Fisheries Research Board of Canada, 32*, 1083−1094.

Deaville, R. and Jepson, P.D. (compilers) (2011) *UK Cetacean Strandings Investigation Programme Final report for the period 1st January 2005 - 31st December 2010.* Zoological Society of London, UK. (http://randd.defra.gov.uk/Document.aspx?Document = FinalCSIPReport2005-2010_finalversion061211released[1].pdf).

Evans, P. G. H., Anderwald, P., & Baines, M. E. (2003). *UK Cetacean status review report to English Nature and the Countryside Council for wales.* Oxford: Sea Watch Foundation, 160 pp.

Fairley, J. (1981). *Irish Whales and Whaling.* Belfast: Blackstaff Press, 218 pp.

Fullard, K. J., Early, G., Heide-Jorgensen, M. P., Bloch, D., Rosing-Asvid, A., & Amos, W. (2000). Population structure of long-finned pilot whales in the North Atlantic: A correlation with sea surface temperature? *Molecular Ecology, 9*(7), 949–958.

Hammond, P. S., Berggren, P., Benke, H., Borchers, D. L., Collet, A., Heide-Jørgensen, M. P., ... Øien, N. (2002). Abundance of harbour porpoise and other cetaceans in the North Sea and adjacent waters. *Journal of Applied Ecology, 39*, 361–376.

Hammond, P.S., Burt, M.L., Cañadas, A., Castro, R., Certain, G., Gillespie, D., ... Winship, A. (2009). Cetacean Offshore Distribution and Abundance in the European Atlantic (CODA). Report available from SMRU, Gatty Marine Laboratory, University of St Andrews, St Andrews, Fife KY16 8LB, UK. http://biology.st-andrews.ac.uk/coda/.

Hammond, P.S., Lacey, C., Gilles, A., Viquerat, S., Borjesson, P., Herr, H., ... and Øien, N. (2017). Estimates of cetacean abundance in European Atlantic waters in summer 2016 from the SCANS-III aerial and shipboard surveys. Available at https://synergy.standrews.ac.uk/scans3/files/2017/05/SCANS-III-design-based-estimates-2017-05-12-final-revised.pdf.

Hammond, P. S., Macleod, K., Berggren, P., Borchers, D. L., Burt, M. L., Cañadas, A., ... Vázquez, J. A. (2013). Cetacean abundance and distribution in European Atlantic shelf waters to inform conservation and management. *Biological Conservation, 164*, 107–122.

NAMMCO (2012) Report of the eighteenth meeting of the Scientific Committee. In: NAMMCO Annual Report 2011. North Atlantic Marine Mammal Commission, Tromsø, Norway. 523 pp.

NAMMCO (2016) *Report of the NAMMCO Scientific Working Group on Abundance Estimates.* NAMMCO SC/23/15. 32 pp.

NAMMCO (2018) *Report of the NAMMCO Scientific Working Group on Abundance Estimates.* Available at https://nammco.no/topics/sc-working-group-reports/.

Prime Minister's Office, Faroe Islands (2018). *Long-finned pilot whale catch statistics, 2000-2017.* Available at <https://www.whaling.fo/en/regulated/450-years-of-statistics/catches/>.

Reid, J. B., Evans, P. G. H., & Northridge, S. P. (2003). *Atlas of Cetacean Distribution in North-west European Waters.* Peterborough: Joint Nature Conservation Committee, 76 pp.

Rogan, E., Breen, P., Mackey, M., Cañadas, A., Scheidat, M., Geelhoed, S., & Jessopp, M. (2018). *Aerial surveys of cetaceans and seabirds in Irish waters: Occurrence, distribution and abundance in 2015-2017.* Dublin, Ireland: Department of Communications, Climate Action & Environment and National Parks and Wildlife Service (NPWS), Department of Culture, Heritage and the Gaeltacht, 297 pp.

Verborgh, P., Gauffier, P., Esteban, R., Gimenez, J., Cañadas, A., Salazar-Sierra, J. M., & de Stephanis, R. (2016). Conservation Status of Long-Finned Pilot Whales, *Globicephala melas*, in the Mediterranean Sea. *Advances in Marine Biology, 75*, 174–203.

Smeenk, C., & Camphuysen, C. J. (2016). Griend *Globicephala melas.* In S. Broekhuizen, K. Spoelstra, J. B. M. Thissen, K. J. Canters, & J. C. Buys (Eds.), *Atlas van de Nederlandse Zoogdieren* (pp. 362–364). Leiden: Natuur van Nederland 12, Naturalis Biodiversity Center, 432 pp., (206).

Wall, D., Murray, C., O'Brien, J., Kavanagh, L., Wilson, C., Ryan, C., ... Berrow, S. (2013). *Atlas of the distribution and relative abundance of marine mammals in Irish offshore waters 2005-2011.* Irish Whale and Dolphin Group, Merchants Quay, Kilrush, Co. Clare.

Weir, C. R., Pollock, C., Cronin, C., & Taylor, S. (2001). Cetaceans of the Atlantic Frontier, north and west of Scotland. *Continental Shelf Research, 21*, 1047–1071.

Zachariassen, P. (1993). Pilot whale catches in the Faroes, 1709-1992. *Report of the International Whaling Commission* (Special issue 14), 69–88.

Short-finned pilot whale *Globicephala macrorhynchus*

The short-finned pilot whale can be found in tropical to warm temperate seas around the world, rarely ranging north of 50°N. It is very similar to its close relative, the long-finned pilot whale, mainly distinguished by slightly shorter flippers and a more robust body, often showing a light saddle patch behind the dorsal fin (Fig. 4.34). Adult females reach 5.5 m and males 7.2 m. Like its relative, the species is fairly robust with a bulbous head, which in adult males may appear square, an indistinct beak, low and falcate dorsal fin situated one-third of the way from the head and long sickle-shaped flippers. The flippers are not as long as in long-finned pilot whales, however, being less than 20% of the body length. Coloration is black or dark brownish-grey, except for a light grey anchor-shaped chest patch, grey saddle behind the dorsal fin and a pair of parallel slanting light streaks from the top of the head towards the eye.

In the North Atlantic short-finned pilot whales are common in the West Indies, Canaries and Azores, but in the eastern North Atlantic rarely occur north of Africa, although because of its close similarity to the long-finned pilot whale, it could easily be overlooked. The most northerly confirmed records in Europe are from Atlantic France, all in the southwest: one stranded in Charente-Maritime in 1966; one stranded in the Gironde in 1988; three live-stranded at Saint-Jean-de-Luz close to the Spanish border in 2008; one stranded in the Gironde in 2011; and there were two strandings

FIGURE 4.34 The Short-finned Pilot Whale is difficult to distinguish from its close relative, the long-finned pilot whale, but often shows a light saddle patch behind the dorsal fin. *Photo: Peter GH Evans.*

reported, one at Bidart and the other at Anglet, in the French department of Pyrénées-Atlantiques in 2012 (Van Canneyt, Bouchard, Dabin, Demaret, & Doremus, 2013). In the Mediterranean, there has been one confirmed sight record of three individuals in the Adriatic Sea in 2010 (Verborgh et al., 2016). A long-term (1999–2012) photo-ID study in the Canaries has identified 1320 well-marked individuals, exhibiting a large degree of variability in site fidelity but with an apparent core-resident group of 50 individuals associated particularly with Tenerife (Servidio et al., 2019).

References

Servidio, A., Pérez-Gil, E., Pérez-Gil, M., Cañadas, A., & Hammond, P.S. (2019). Site fidelity and movement patterns of short-finned pilot whales within the Canary Islands: evidence for resident and transient populations. Aquatic Conservation.

Van Canneyt, O., Bouchard, C., Dabin, W., Demaret, F., & Doremus, G. (2013). *Les échouages de mammifères marins sur le Littoral Français en 2012* (51 pp). Rapport scientifique de l'Observatoire Pelagis, Université de La Rochelle et CNRS.

Verborgh, P., Gauffier, P., Esteban, R., Giménez, J., Cañadas, A., Salazar-Sierra, J. M., & de Stephanis, R. (2016). Conservation status of long-finned pilot whales, *Globicephala melas*, in the Mediterranean Sea. *Advances in Marine Biology, 75*, 173–203.

Family Monodontidae

Narwhal *Monodon monoceros*

The narwhal (Fig. 4.35) is a medium-sized (up to 5 m in length, excluding the tusk) cetacean living in the Arctic. It has a body shape similar to the beluga: a blunt head, no dorsal fin and flippers that turn up distally with age, especially in males. An age-related change in tail fluke shape occurs in a similar, but even more marked manner, to beluga. All but the youngest males have an unmistakable tusk; females rarely have a tusk.

The species has a stout body with a small, rounded head, bulbous forehead and very slight beak. There is just one pair of teeth in the upper jaw only, the left tooth of the male becoming greatly extended as a spiralled tusk (up to 2.7 m long) pointing forwards and erupting through the upper lip. Very rarely, both teeth may erupt and become elongated.

FIGURE 4.35 Narwhal is generally a species of the high arctic rarely entering the temperate seas of Western Europe. *Photo: Mads Peter Heide-Jørgensen.*

In females, the teeth are embedded in the skull, and rarely or never erupt. The body coloration is mottled grey-green, cream and black; older males appear lighter, partly because of the accumulation of white scar tissue. Newborns are blotchy slate grey or bluish grey; mottling increases in juveniles. The flippers are short, and have an upturned tip in adults; in very old males they may describe a complete circle, almost touching the dorsal surface.

The narwhal has a discontinuous circumpolar distribution, mainly in the high Arctic (between 60° and 85°N). The centres of distribution for the species are eastern Canada/west Greenland and east Greenland/Svalbard/Franz Josef Land; it rarely occurs outside the Arctic. There have been only seven records since the 16th century in the British Isles, and none since 1949; and elsewhere in the ASCOBANS Agreement Area the only records are from Germany (1736), the Netherlands (1912), Sweden (1992) and Belgium (2016) (Camphuysen & Peet, 2006; Evans et al., 2003; Haelters et al., 2018; Kinze, Jensen, Tougaard, & Baagøe, 2010; Martin & Evans, 2008).

References

Camphuysen, C. J., & Peet, G. (2006). *Whales and dolphins of the North Sea*. Kortenhoef: Fontaine Uitgevers, 160 pp.

Evans, P. G. H., Anderwald, P., & Baines, M. E. (2003). UK Cetacean Status Review. *Report to English Nature and the Countryside Council for Wales*. Oxford: Sea Watch Foundation, 160 pp.

Haelters, J., Kerckhof, F., Doom, M., Evans, P. G. H., Van den Neucker, T., & Jauniaux, T. (2018). New extralimital record of a narwhal (*Monodon monoceros*) in Europe. *Aquatic Mammals, 44*(1), 39–50. Available from https://doi.org/10.1578/AM.44.1.2018.39.

Kinze, C. C., Jensen, T., Tougaard, S., & Baagøe, H. J. (2010). Danske hvalfund i perioden 1998-2007 [Records of cetacean strandings on the Danish coastline during 1998-2007]. *Flora og Fauna, 116*(4), 91–99.

Martin, A. R., & Evans, P. G. H. (2008). Narwhal *Monodon monoceros*. In S. Harris, & D. W. Yalden (Eds.), *Mammals of the British Isles* (4th ed., pp. 702–704). Southampton: The Mammal Society, 800 pp, Handbook.

Beluga or white whale *Delphinapterus leucas*

The beluga is the only other medium-sized cetacean (up to 5.5 m in length) lacking a dorsal fin and with a uniform white body colour, although this is only achieved in adulthood. It is unique among cetaceans in having all, or almost all, of the neck vertebrae unfused giving the neck great flexibility.

FIGURE 4.36 The Beluga is the only white whale lacking a dorsal fin. *Photo: Caroline R Weir.*

The species has a stout body with a small head and pronounced melon. The head has a slight beak, and looks unusually small compared with the body due to the great thickness (up to 15 cm) of blubber covering the thorax and abdomen. The melon is bulbous and malleable. The skin is soft, often with small transverse ridges, and frequently scarred. Adults are pure white or (in early summer) yellowish, whereas calves are grey or grey-brown, and often blotched. Juveniles become progressively lighter with age (becoming pure white by 9 years old in males and 7 years in females), although females often are still light grey at sexual maturity. The flippers are short and rounded, progressively turned up at their tips in adulthood. The tail fluke is deeply notched and also changes shape with age, developing a lobe on each side of the trailing edge. A distinct ridge takes the place of a dorsal fin (Fig. 4.36).

Belugas have a circumpolar arctic distribution normally occurring at or near the ice edge, their seasonal distributions largely dictated by the annual sea-ice cycle. In the North Atlantic, the main populations occur off northeast Canada south to the Gulf of St Lawrence, around the coasts of Greenland (mainly in Baffin Bay), and in the Barents Sea including around Svalbard. Within the ASCOBANS Agreement Area, the species may wander southwards west of Norway, with records from southern Scandinavia, the Baltic and North Seas (Denmark, Sweden, Finland, Lithuania, Poland, Germany, the Netherlands and Belgium), and no more than a dozen, mainly sightings, from the British Isles (Camphuysen & Peet, 2006; Evans, 2010; Evans et al., 2003; Heide-Jørgensen & Wüg, 2002; Kinze Jensen, Tougaard, and Baagøe, 2010; Martin & Evans, 2008).

References

Camphuysen, C. J., & Peet, G. (2006). *Whales and dolphins of the North Sea*. Kortenhoef: Fontaine Uitgevers, 160 pp.

Evans, P.G.H. (2010) *Review of Trend Analyses in the ASCOBANS Area*. ASCOBANS AC17/Doc. 6-08 (S). 68 pp.

Evans, P. G. H., Anderwald, P., & Baines, M. E. (2003). UK Cetacean Status Review. *Report to English Nature and the Countryside Council for Wales*. Oxford: Sea Watch Foundation, 160 pp.

Heide-Jørgensen, M.-P., & Wüg, Ø. (Eds.), (2002). *Belugas in the North Atlantic and the Russian Arctic* (4). NAMMCO Scientific Publications, 270 pp.

Kinze, C. C., Jensen, T., Tougaard, S., & Baagøe, H. J. (2010). Danske hvalfund i perioden 1998-2007 [Records of cetacean strandings on the Danish coastline during 1998-2007]. *Flora og Fauna, 116*(4), 91−99.

Martin, A. R., & Evans, P. G. H. (2008). Beluga *Delphinapterus leucas*. In S. Harris, & D. W. Yalden (Eds.), *Mammals of the British Isles* (4th ed., pp. 700−702). Southampton: The Mammal Society, 800 pp, Handbook.

Family Ziphiidae

Cuvier's beaked whale *Ziphius cavirostris*

Cuvier's beaked whale has the most widespread distribution of the family Ziphiidae, occurring probably worldwide in tropical to warm temperate seas. Although reaching lengths of 5.1−6.9 m, it is, however, difficult to observe except in calm seas. Usually it provides only a brief view of its back, small, triangular or hooked fin two-thirds of the way along, and in some older adults, the light, almost white, head and back of the neck, with darker crescent-shaped areas around the eyes (Fig. 4.37). Its head is small with a sloping or slightly bulbous forehead, short indistinct beak and up-curved mouth-line. A single pair of conical teeth is situated at the tip of the lower jaw, erupting only in adult males and then protruding forward from the mouth. It has small, narrow flippers with a pointed tip, located low down on the flanks and fitting into 'flipper pockets'. Back coloration can be dark, rust brown, slate grey or fawn, with a generally paler belly and head, particularly in older males. The back and sides are usually covered with linear scars and white or cream-coloured oval blotches. Sometimes yellow diatoms may occur in patches over the body (Fig. 4.38).

Cuvier's beaked whales are record-breaking divers, with tagged animals having been recorded diving to 2992 m depth with dives lasting up to 137.5 minutes (Schorr, Falcone, Moretti, & Andrews, 2014).

Favouring deep, warm waters, Cuvier's beaked whale is the most common beaked whale in Southern Europe, off the Iberian Peninsula, in the Bay of Biscay and the Mediterranean, whereas it is relatively rare in the British Isles (although the number of records has increased a lot in recent years) and further north and east, with isolated records from Iceland, Sweden and the Netherlands (Evans, Smeenk, & Van Waerebeek, 2008; Kinze et al., 2011; see Fig. 4.39). Its presence has been recorded acoustically year-round offshore west of Ireland, with a seasonal increase in late autumn off northwest Porcupine Bank, a region of relatively high detections (Kowarski et al., 2018). It is a regular inhabitant of the waters around the Azores, Madeira and Canaries, where it is seen year-round. In the western North Atlantic, it occurs in the Caribbean, Gulf of Mexico and southeastern United States.

Around the British Isles and Ireland, there have been only a handful of actual sighting records (Berrow et al., 2010; Evans et al., 2003, 2008) in contrast to the number of sightings further south in the Bay of Biscay (Kiszka et al., 2007; Smith, 2010; Williams, Brereton, & Williams, 1999; see Fig. 4.40). On the other hand, there have been more than 100 strandings on the coasts of Britain and Ireland over the last 100 years, of which two-thirds have occurred since 1963

FIGURE 4.37 Cuvier's Beaked Whale can be almost white over much of the back and head as they become older but with a darker crescent-shaped area around the eye. *Photo: Marta Guerra/Universidad de La Laguna.*

FIGURE 4.38 Cuvier's Beaked Whale in tropical regions may have a yellow or oange film of diatoms over their body. *Photo: Natacha Aguilar.*

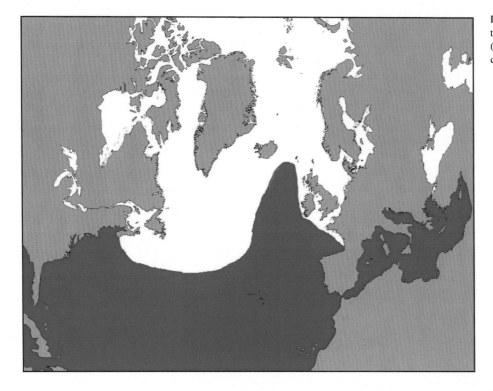

FIGURE 4.39 North Atlantic distribution of Cuvier's Beaked Whale (depicting those areas where the species is thought to regularly occur).

(Berrow & Rogan, 1997; Berrow et al., 2010; Deaville & Jepson, 2011; Evans et al., 2003; Evans et al., 2008). Almost all have come from the Atlantic coasts, mainly in late winter or spring. In early 2008 there was an unusually high number (21) of strandings of this species in the region, mainly in north and west Scotland, the cause of which could not be determined (Deaville & Jepson, 2011; Dolman et al., 2010), and between late June and September 2018, at least 76 Cuvier's beaked whales stranded in the northeastern North Atlantic, mainly on the west coasts of Scotland and Ireland but extending as far north as Iceland; as yet, the cause of death is unknown.

No abundance estimates exist for this species in the overall region, although an abundance estimate in July 2016 during the SCANS-III survey for all beaked whale species (including also northern bottlenose whale and all *Mesoplodon* species) indicated 11,394 whales (CV 0.50, 95% CI: 4500−28,900) (Hammond et al., 2017). In 2005/07,

FIGURE 4.40 Cuvier's Beaked Whale is the commonest beaked whale in the Bay of Biscay. *Photo: Maren Reichelt/Mick E Baines.*

an estimate of 12,869 beaked whales (CV 0.31, 95% CI: 7,100−23,300) has also since been calculated (see Table 6.1). The ObSERVE surveys of the Irish EEZ in the summer and winter of 2015 and 2016 gave an abundance estimate for all beaked whales in summer 2016 of 3142 (CV 0.505, 95% CI: 1200−8100) and in winter 2016−17 of 4948 (CV 0.54, 95% CI: 1800−13,500), in both seasons in deep offshore waters west of Ireland (Rogan et al., 2018). A separate estimate was made for Cuvier's beaked whale, with 237 (CV 0.71, 95% CI: 67−847) in summer 2016 and 765 (CV 0.77, 95% CI: 200−3000) in winter 2016−17 (Rogan et al., 2018). Combining estimates from SCANS-III and ObSERVE in summer 2016 yields an overall estimate for all beaked whale species of 14,536 (CV 0.41, 95% CI: 6,700−31,400).

The species appears to be particularly vulnerable to the effects of mid-frequency active sonar as used in some military exercises, and these have been shown to lead to mass strandings (see Chapter 5: Conservation threats).

References

Berrow, S., Whooley, P., O'Connell, M., & Wall, D. (2010). *Irish Cetacean Review (2000−2009)*. Kilrush: Irish Whale and Dolphin Group, 60 pp.

Deaville, R. and Jepson, P.D. (compilers) (2011) *UK Cetacean Strandings Investigation Programme Final report for the period 1st January 2005 - 31st December 2010*. Zoological Society of London, UK. (http://randd.defra.gov.uk/Document.aspx?Document = FinalCSIPReport2005-2010_finalversion061211released[1].pdf).

Dolman, S. J., Pinn, E., Reid, R. J., Barley, J. P., Deaville, R., Jepson, P. D., ... Simmonds, M. P. (2010). A note on the unprecedented stranding of 56 deep-diving odontocetes along the UK and Irish coast. *Marine Biodiversity Records, 3*, 1−8.

Evans, P. G. H., Anderwald, P., & Baines, M. E. (2003). UK Cetacean Status Review. *Report to English Nature and the Countryside Council for Wales*. Oxford: Sea Watch Foundation, 160 pp.

Evans, P. G. H., Smeenk, C., & Van Waerebeek, K. (2008). Cuvier's beaked whale *Ziphus cavirostris*. In S. Harris, & D. W. Yalden (Eds.), *Mammals of the British Isles* (4th ed., pp. 690−692). Southampton: The Mammal Society, 800 pp, Handbook.

Hammond, P.S., Lacey, C., Gilles, A., Viquerat, S., Borjesson, P., Herr, H., ... and Øien, N. (2017) Estimates of cetacean abundance in European Atlantic waters in summer 2016 from the SCANS-III aerial and shipboard surveys. Available at https://synergy.standrews.ac.uk/scans3/files/2017/05/SCANS-III-design-based-estimates-2017-05-12-final-revised.pdf.

Kinze, C.C., Schulze, G., Skora, K., & Benke, H. (2011) *Zahnwale als Gastarten in der Ostsee*. Meer und Museum 23: 53-82 (in German).

Kiszka, J., Macleod, K., Van Canneyt, O., & Ridoux, V. (2007). Distribution, relative abundance and bathymetric preferences of toothed cetaceans in the English Channel and Bay of Biscay. *ICES Journal of Marine Science*, 1033−1043.

Kowarski, K., Delarue, J., Martin, B., O'Brien, J., Made, R., Ó Cadhla, O., & Berrow, S. (2018). Signals from the deep: Spatial and temporal acoustic occurrence of beaked whales off western Ireland. *PLoS One, 13*(6), e0199431. Available from https://doi.org/10.1371/journal.pone.0199431.

Rogan, E., Breen, P., Mackey, M., Cañadas, A., Scheidat, M., Geelhoed, S., & Jessopp, M. (2018). *Aerial surveys of cetaceans and seabirds in Irish waters. Occurrence, distribution and abundance in 2015-2017*. Dublin, Ireland: Department of Communications, Climate Action & Environment and National Parks and Wildlife Service (NPWS), Department of Culture, Heritage and the Gaeltacht, 297 pp.

Schorr, G. S., Falcone, E. A., Moretti, D. J., & Andrews, R. D. (2014). First long-term behavioral records from Cuvier's beaked whales (*Ziphius cavirostris*) reveal record-breaking dives. *PLoS One, 9*(3), e92633. Available from https://doi.org/10.1371/journal.pone.0092633.

Smith, J. (2010) The Ecology of Cuvier's beaked whale, *Ziphius cavirostris* (Cetacea Ziphiidae), in the Bay of Biscay (Ph.D. thesis). University of Southampton.

Williams, A. D., Brereton, T. M., & Williams, R. (1999). Seasonal variation in the occurrence of beaked whales in the southern Bay of Biscay. *European Research on Cetaceans, 13*, 275–280.

Northern bottlenose whale *Hyperoodon ampullatus*

The northern bottlenose whale is one of two members of the little-known family of beaked whales most likely to be encountered in the ASCOBANS Agreement Area. However, it is a deep diver (regularly to depths greater than 800 m) and can remain underwater for 70 minutes, possibly more (Hooker & Baird, 1999), and as a result is infrequently observed. Its most distinctive feature is the bulbous forehead and short dolphin-like beak that can be seen if the animal spy-hops (Fig. 4.41) or breaches as it occasionally does (Fig. 4.42). The species frequently approaches vessels and has also been observed to lob-tail. The body is long (7.0–8.5 m in females; 8.0–9.5 m in males), relatively robust and cylindrical in shape. Coloration varies from chocolate-brown to greenish-brown above, often lighter on the flanks, lightening to buff or cream all over with age. The blow is bushy, rising to a height of about 2 m, and slightly forward pointing.

Northern bottlenose whales have a strongly recurved dorsal fin two-thirds along the back, which may lead to confusion with minke and sei whales unless other more distinctive features such as head shape are seen. The fin of the bottlenose whale is generally more erect and hooked than that of the minke whale whereas an adult sei whale is almost twice the length of a bottlenose whale.

The species is confined to the North Atlantic from warm temperate to Arctic seas, particularly occurring in the vicinity of deep ocean abysses of >500 m depth. Its main range extends from Baffin Island and west Greenland south to New England in the west, and from Svalbard to the southern tip of the Iberian Peninsula in the east, including around the oceanic archipelago of the Azores, although it occurs casually further south (to approximately 15°N), having been recorded in the Caribbean in the west and the Canaries in the east (Fig. 4.43).

Main regions of concentration, identified from former whaling activities, appear to be west of Norway, west of Svalbard, north of Iceland, in the Davis Strait off Labrador, off the Faroes and in The Gully off eastern Canada

FIGURE 4.41 Northern Bottlenose Whale commonly spy-hop, displaying its bulbous forehead and distinct beak. *Photo: Paul H Ensor/SMRU.*

FIGURE 4.42 Northern Bottlenose Whales occasionally breach showing a robust cylindrical body wth a recurved dorsal fin located two-thirds along the back. *Photo: Saana Isojunno/SMRU.*

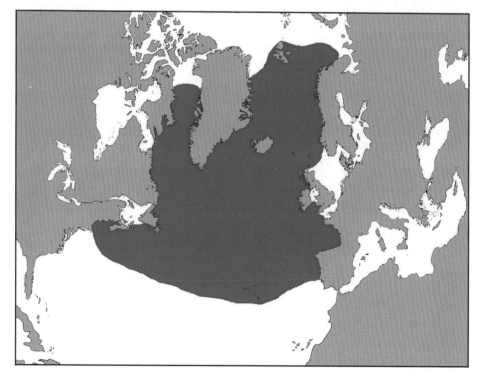

FIGURE 4.43 Overall distribution of Northern Bottlenose Whale (depicting those areas where the species is thought to regularly occur).

(Benjaminsen, 1972; Benjaminsen and Christensen, 1979; Compton, 2005; Reeves, Mitchell, & Whitehead, 1993; see Figs. 4.44 and 4.45). Within the ASCOBANS Agreement Area, northern bottlenose whales have been sighted primarily in waters exceeding 1000 m, such as the Faroe−Shetland Channel, Rockall Trough and southern Bay of Biscay (Evans et al., 2003; Hooker et al., 2008; Reid et al., 2003; Walker, Telfer, & Cresswell, 2001; Weir, 2001; Williams Brereton, and Williams, 1999). Acoustic detections confirmed the occurrence of northern bottlenose whales west of Ireland in small numbers in late summer and autumn, particularly on the southern edge of the Rockall Trough (Kowarski et al., 2018).

FIGURE 4.44 Northern Bottlenose Whale in northern Iceland. *Photo: Marianne Rasmussen.*

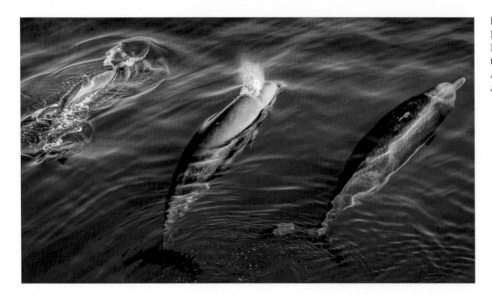

FIGURE 4.45 A pod of Northern Bottlenose Whales off Jan Mayen Island, one of the hotspot areas for the species in the NE Atlantic. *Photo: Christian Harboe-Hansen/ SMRU.*

In summer northern bottlenose whales may move towards northwest European shelf waters, where most records occur between April and September (Reid et al., 2003). Peak sightings in the Bay of Biscay occur between June and August (Carlisle et al., 2001, Coles et al., 2001; Smith, 2010), off northern Scotland in August (Evans et al., 2003), and in the Faroes in August and September, when catches are also highest (Bloch, Desportes, Zachariassen and Christensen, 1996). The seasonal pattern of acoustic detections west of Ireland, confined to the months of July to September (Kowarski et al., 2018), support this. The species occasionally enters the North Sea, which may result in strandings.

During the July 2005 SCANS-II, July 2007 CODA survey and SCANS-III/ObSERVE surveys in summer 2016 of the shelf seas and slopes of the ASCOBANS Agreement Area, there were too few sightings of northern bottlenose whale to arrive at any robust population estimate.

In adjacent seas to the north, NASS in 1987 and 1989 yielded an overall population estimate of about 40,000 animals (NAMMCO, 1993; Vikingsson, 1993). From these surveys, Gunnlaugsson and Sigurjonsson (1990) estimated regional population sizes of 4925 (CV 0.16) off Iceland and 902 (CV 0.45) off the Faroe Islands. Subsequently, Pike, Gunnlaugsson, Víkingsson, Desportes and Mikkelson (2003) estimated 27,900 (CV 0.67) for a survey in 1995 and 28,000

(CV 0.22) in 2001. However, major challenges for obtaining accurate population estimates for this species are the fact that these can be negatively biased by the whales' long dives that result in them being undetected, whilst their tendency to be attracted to vessels may create a positive bias. These estimates should therefore be viewed with some caution, and no attempt was made for a population estimate from the TNASS survey in 2007 or the NASS survey in 2015.

The hunting of northern bottlenose whales goes back to at least the 16th century, although the largest numbers were taken between the middle of the 19th century and first half of the 20th century, with an estimated 65,000 killed between about 1872 and 1972 (Mitchell, 1977). In the eastern North Atlantic, most catches were off western (Møre) and northern (Andenes) Norway but also in the deep canyon between Iceland and the Faroes, north of Iceland and around Svalbard (Whitehead & Hooker, 2011). As with other beaked whales, the species may be vulnerable to exposure to mid-frequency active sonar as well as other sources of underwater noise (Miller et al., 2015).

References

Benjaminsen, T. (1972). On the biology of the bottlenose whale, *Hyperoodon ampullatus* (Forster). Norweg. *Journal of Zoology, 20*, 233–241.

Benjaminsen, T., & Christensen, I. (1979). The natural history of the bottlenose whale, *Hyperoodon ampullatus* (Forster). In H. E. Winn, & B. L. Olla (Eds.), *Behavior of marine animals* (pp. 143–164). New York: Plenum.

Bloch, D., Desportes, G., Zachariassen, M., & Christensen, I. (1996). The northern bottlenose whale in the Faroe Islands, 1584-1993. *Journal of Zoology, 239*, 123–140.

Carlisle, A., Coles, P., Cresswell, G., Diamond, J., Gorman, L., Kinley, C., ... Walker, D. (2001). A report on the whales, dolphins and porpoises of the Bay of Biscay and English Channel, 1999. *Orca Report, 1*, 5–53.

Coles, P., Diamond, J., Harrop, H., Macleod, K., & Mitchell, J. (2001). A report on the whales, dolphins and porpoises of the Bay of Biscay and English Channel, 2000. *Orca Report, 2*, 9–61.

Compton, R. (2005) Predicting key habitat and potential distribution of northern bottlenose whales (Hyperoodon ampullatus) in the Northwest Atlantic Ocean (M.Sc. thesis). University of Plymouth.

Evans, P. G. H., Anderwald, P., & Baines, M. E. (2003). UK Cetacean Status Review. *Report to English Nature and the Countryside Council for Wales*. Oxford: Sea Watch Foundation, 160 pp.

Gunnlaugsson, T., & Sigurjönsson, J. (1990). NASS-87: Estimation of whale abundance based on observations made on board Icelandic and Faroese survey vessels. *Reports of International Whaling Commission, 40*, 571–580.

Hooker, S., & Baird, R. W. (1999). Deep-diving behaviour of the northern bottlenose whale, *Hyperoodon ampullatus* (Cetacea: Ziphiidae). *Proceedings of the Royal Society, London, B, 266*, 671–676.

Hooker, S. K., Gowans, S., & Evans, P. G. H. (2008). Northern bottlenose whale *Hyperoodon ampullatus*. In S. Harris, & D. W. Yalden (Eds.), *Mammals of the British Isles* (4th ed., pp. 685–690). Southampton: The Mammal Society, 800 pp. Handbook.

Kowarski, K., Delarue, J., Martin, B., O'Brien, J., Made, R., Ó Cadhla, O., & Berrow, S. (2018). Signals from the deep: Spatial and temporal acoustic occurrence of beaked whales off western Ireland. *PLoS ONE, 13*(6), e0199431. Available from https://doi.org/10.1371/journal.pone.0199431.

Miller, P. J. O., Kvadsheim, P. H., Lam, F. P. A., Tyack, P. L., Curé, C., DeRuiter, S. L., ... Hooker, S. K. (2015). First indications that northern bottlenose whales are sensitive to behavioural disturbance from anthropogenic noise. *Royal Society Open Science, 2*, 140484. Available from https://doi.org/10.1098/rsos.140484.

Mitchell, E. (1977). Evidence that the northern bottlenose whale is depleted. *Reports of the International Whaling Commission, 27*, 195–203.

NAMMCO (1993). *Report of the scientific committee working group on northern bottlenose and killer whales*. Tromsø: North Atlantic Marine Mammal Commission.

Pike, D. G., Gunnlaugsson, T., Víkingsson, G. A., Desportes, G., & Mikkelson, B. (2003). *Surface abundance of northern bottlenose whales (Hyperoodon ampullatus) from NASS-1995 and 2001 shipboard surveys*. Tromsø: North Atlantic Marine Mammal Commission.

Reeves, R. R., Mitchell, E., & Whitehead, H. (1993). Status of the northern bottlenose whale, *Hyperoodon ampullatus*. *Canadian Field-Naturalist, 107*, 490–508.

Reid, J. B., Evans, P. G. H., & Northridge, S. P. (2003). *Atlas of Cetacean Distribution in North-west European Waters*. Peterborough: Joint Nature Conservation Committee, 76 pp.

Smith, J. (2010) The Ecology of Cuvier's beaked whale, Ziphius cavirostris (Cetacea Ziphiidae), in the Bay of Biscay. PhD thesis, University of Southampton.

Vikingsson, G. (1993). *Northern bottlenose whale (Hyperoodon ampullatus)*. Availability of data and status of research in Iceland. NAMMCO SC-WG/NBK1/7 1-4.

Walker, D., Telfer, M., & Cresswell, G. (2001). The status and distribution of beaked whales (Ziphiidae) in the Bay of Biscay. *European Research on Cetaceans, 15*, 278–282.

Weir, C. R. (2001). Sightings of beaked whale species (Cetacea: Ziphiidae) in the waters to the north and west of Scotland and the Faroe Islands. *European Research on Cetaceans, 14*, 239–243.

Whitehead, H., & Hooker, S. K. (2011). *Status of northern bottlenose whales* (11 pp). IWC SC/63/SM4.

Williams, A. D., Brereton, T. M., & Williams, R. (1999). Seasonal variation in the occurrence of beaked whales in the southern Bay of Biscay. *European Research on Cetaceans, 13*, 275–280.

True's beaked whale *Mesoplodon mirus*

True's beaked whale (Fig. 4.46) is one of the least known species of this genus in the North Atlantic. Because of the great difficulty in distinguishing it from other *Mesoplodon* species, it has rarely been identified with certainty at sea. The most diagnostic feature is a short but clearly defined beak sloping into a slightly bulbous forehead, with a single pair of slightly laterally compressed teeth, oval in cross-section, directed forward and upward at the extreme tip of the lower jaw, exposed above the mouth-line only in adult males. In females and young these are smaller and concealed below the gum.

The species has a long spindle-shaped body (approximately 4−5.5 m length). It is more robust than Sowerby's beaked whale, the body shape being closer to Cuvier's beaked whale. It has a small head with a bulge in front of the crescent-shaped blowhole, a short but pronounced beak and a pair of throat grooves forming a V-shape below. The flippers are relatively small and narrow, often tucked into 'flipper pockets' and there is a triangular or slightly recurved dorsal fin situated almost two-thirds along the back. The tail flukes have a slightly concave trailing edge and are usually unnotched although some have a slight median notch. The back is slate grey in colour, often with pale spots and linear scars, which may be in closely spaced pairs. The belly is lighter, and pale patches may be present in the anal and genital regions. In some individuals a clearly delimited diagonal white blaze extends dorsally backwards from the beak across the blowhole to the top of the melon, and ventrally to the eye and start of the mouth-line (Fig. 4.47); a dark eye patch may also be present.

The range of this species is poorly known but it appears to be widespread and antitropical, occurring in deep waters of the temperate Atlantic extending in a disjunct manner to the South Atlantic, southern Indian and Pacific Oceans, with records from eastern North America, northwest Europe, Brazil, South Africa, Mozambique, Madagascar, southern Australia and the Tasman coast of New Zealand (Aguilar de Soto et al., 2017; Evans, Herman, and Kitchener, 2008; Jefferson et al., 2015; Mead, 1989).

There have been 11 stranding records from Europe since 1899, 10 of which are from western Ireland (Berrow et al., 2010; Evans et al., 2008). The other European record is from France (Barriety, 1962), although the species has been reported also from the Canaries and Azores, confirmed by genetics (Aguilar de Soto et al., 2017; Vonk & Martin, 1988). Sightings identified as this species on the basis of uniform bluish grey coloration, a short but prominent beak and, in one case, indication of a tooth at the extreme front of the jaw, have also been made in the Canaries, the Azores and the Bay of Biscay (Aguilar de Soto et al., 2017; Evans et al., 2008; Steiner, Gordon, & Beer, 1998; Weir, Stokes,

FIGURE 4.46 True's Beaked Whale is one of the least known and rarely seen species. This image shows two individuals thought to be of this species breaching in the Canaries. *Photo: Antonio Portales.*

FIGURE 4.47 Mother and calf True's Beaked Whale in the Azores, showing pale blaze over the melon. *Photo: Petra Sziama.*

Martin, & Cermeno, 2004). These suggest a distribution possibly influenced by a southern branch of the North Atlantic Current. From these few records its favoured habitat is likely to be adjacent to deep ocean basins and trenches at approximately 500−2500 m depth, such as the Porcupine Seabight, around the Azores and Canaries.

References

Aguilar de Soto, N., Martin, V., Silva, M., Edler, R., Reyes, C., Carrillo, M., . . . Carroll, E. (2017). True's beaked whale (*Mesoplodon mirus*) in Macaronesia. *PeerJ, 5*, e3059. Available from https://doi.org/10.7717/peerj.3059.

Barriety, L. (1962). Echouage a Bidart d'un *Mesoplodon mirus. Centre de Recherches scientifiques, Biarritz, Bulletin, 4*, 93−94. (in French).

Berrow, S., Whooley, P., O'Connell, M., & Wall, D. (2010). *Irish Cetacean Review (2000−2009)*. Kilrush: Irish Whale and Dolphin Group, 60 pp.

Evans, P. G. H., Herman, J. S., & Kitchener, A. C. (2008). True's beaked whale *Mesoplodon mirus*. In S. Harris, & D. W. Yalden (Eds.), *Mammals of the British Isles* (4th ed., pp. 694−696). Southampton: The Mammal Society, 800 pp, Handbook.

Jefferson, T. A., Pitman, R. L., & Webber, M. A. (2015). *Marine Mammals of the World*. Amsterdam: Elsevier.

Mead, J. G. (1989). Beaked whales of the genus *Mesoplodon*. In S. H. Ridgway, & R. Harrison (Eds.), *Handbook of Marine Mammals. Vol. 4: River dolphins and the larger toothed whales* (pp. 349−430). London: Academic Press.

Steiner, L., Gordon, J., & Beer, C. (1998). Marine mammals of the Azores. *European Research on Cetaceans, 12*, 79.

Vonk, R., & Martin, V. (1988). First list of odontocetes from the Canaries. *European Research on Cetaceans, 2*, 31−35.

Weir, C. R., Stokes, J., Martin, C., & Cermeno, P. (2004). Three sightings of *Mesoplodon* species in the Bay of Biscay: First confirmed True's beaked whales (*M. mirus*) for the north-east Atlantic? *Journal of the Marine Biological Association UK, 84*(5), 1095−1100.

Gervais' beaked whale *Mesoplodon europaeus*

Gervais' beaked whale (Fig. 4.48) is another beaked whale of warm waters, rarely seen at sea when it is difficult to distinguish from other *Mesoplodon* species, particularly True's beaked whale. A diagnostic feature is the short but clearly defined, slender beak, and relatively straight mouth-line, with one pair of laterally compressed, triangular teeth in the lower jaw (exposed above the gum only in adult males) about one-third along the gape from the tip of the snout. In transverse section the rostrum is ventrally flattened whereas in Blainville's beaked whale for example it is dorsally flattened. In females and young the pair of teeth is usually smaller and concealed below the gum.

The species has a long spindle shaped body (approximately 4.5−5 m in length), proportionately small head with a prominent bulge on the forehead in front of the crescent-shaped blowhole, and a pronounced, slender beak, with a pair

FIGURE 4.48 Gervais' Beaked Whale off Tenerife, Canary Islands. *Photo: Sergio Hanquet.*

of deep grooves forming a V-shape below. It has relatively small narrow flippers often tucked into 'flipper pockets' and a triangular or slightly recurved dorsal fin situated almost two-thirds along the back. Unnotched tail flukes may have a small (approximately 3 cm) median projection on the slightly concave trailing edge. Coloration on the back is dark grey or indigo, in some animals becoming medium or light grey on the lower flanks and belly, with a single pair of scars, especially in adult males. Juveniles have a white belly. A distinctive pattern of pale and dark stripes often occurs over the back. Juveniles have a white belly. Similar to True's beaked whale, the species may exhibit a dark eye patch.

In some adult females a patch of white, about 15 cm in diameter, extends from just anterior of the genital slit to a point just posterior of the anus.

Gervais' beaked whale is known only from the Atlantic where it apparently favours warm temperate and subtropical waters. Although the type specimen was found floating in the English Channel in 1848 and taken to France where it was described by Gervais (1855), most records come from the western Atlantic, from Long Island, New York in the north to the eastern Caribbean in the south (but with most records south of North Carolina). It is the most common species of *Mesoplodon* recorded from the eastern United States and the Gulf of Mexico (Jefferson & Schiro, 1997; Mead, 1989).

In recent years several strandings of the species have been recorded in the eastern North Atlantic. These include records from western Ireland, Portugal, southern Spain, the Azores, Mauritania and Guinea-Bissau (Evans, Herman, & Kitchener, 2008; Reiner, 1980), although most records have come from the Canaries (Aguilar de Soto et al., 2017; Carrillo & Martin, 1999). Extralimital records from the Mediterranean Sea include a dead animal stranded near Castiglionello, Livorno, in Italy and a live stranding thought to be of this species from Turkey (Notarbartolo di Sciara & Birkun, 2010).

References

Aguilar de Soto, N., Martin, V., Silva, M., Edler, R., Reyes, C., Carrillo, M., . . . Carroll, E. (2017). True's beaked whale (*Mesoplodon mirus*) in Macaronesia. *Peer J, 5*, e3059. Available from https://doi.org/10.7717/peerj.3059.

Carrillo, M., & Martin, V. (1999). First sighting of Gervais' beaked whale (*Mesoplodon europaeus* Gervais, 1855) (Cetacea; Ziphiidae) from the North Oriental Atlantic coast. *European Research on Cetaceans, 13*, 53.

Evans, P. G. H., Herman, J. S., & Kitchener, A. C. (2008). Gervais' beaked whale *Mesoplodon europaeus*. In S. Harris, & D. W. Yalden (Eds.), *Mammals of the British Isles* (4th ed., pp. 696−697). Southampton: The Mammal Society, 800 pp, Handbook.

Jefferson, J. A., & Schiro, A. J. (1997). Distribution of cetaceans in the offshore Gulf of Mexico. *Mammal Review, 27*, 27−50.

Mead, J. G. (1989). Beaked whales of the genus *Mesoplodon*. In S. H. Ridgway, & R. Harrison (Eds.), *Handbook of marine mammals. Vol. 4: River dolphins and the larger toothed whales* (pp. 349−430). London: Academic Press.

Notarbartolo di Sciara, G. and Birkun, A. (2010) *Conserving whales, dolphins and porpoises in the Mediterranean and Black Seas*. An ACCOBAMS Status Report. ACCOBAMS, Monaco. 212 pp.

Reiner, F. (1980). First record of an Antillean beaked whale, *Mesoplodon europaeus* Gervais 1855, from Republica Popular da Guine-Bissau. *Museo Marinos, Cascias, Memorias, Serie Zoologie, 1*, 1−8.

Sowerby's beaked whale *Mesoplodon bidens*

The most frequently recorded species of the genus *Mesoplodon* in northern European seas is the Sowerby's beaked whale. However, like others of this genus, it is rarely seen alive, occurring typically in very deep waters, and usually coming to the surface only briefly. It has a long (up to 5.5 m in males; 5.1 m in females), slender tapering body, usually dark grey in colour, relatively small narrow flippers often tucked into 'flipper pockets', and a small and slender recurved dorsal fin situated almost two-thirds along the back. Light spots and scratches scattered over the back may also be seen. The head is small with a slightly concave forehead, well-defined relatively long beak, protruding, moderately arched lower jaw (Fig. 4.49) and, in adult males, a single pair of conspicuous teeth extruding from the middle of the beak, which distinguishes it from other related species. A prominent bulge often shows in front of the crescent-shaped blowhole.

Known only from the temperate North Atlantic, Sowerby's beaked whale seems to occur more commonly in European seas than in eastern North America, with a distribution centred upon deep waters of the mid- and eastern North Atlantic somewhat north of other *Mesoplodon* species. In northern European waters, there have been confirmed sightings south of Iceland, in the Norwegian Sea, west of Norway, around the Faroe Islands, north and west of the British Isles and Ireland, in the Channel Approaches and in the Bay of Biscay (Evans et al., 2008; MacLeod, 2000, 2005; see Fig. 4.50). Further south and west it has been observed regularly around the Azores and Madeira. Most strandings in Europe come from the British Isles. These have occurred mainly in the Northern Isles of Scotland and along the coast of eastern Britain although there are a number of records also from the English Channel, western coasts of Britain and Ireland and those European countries bordering the North Sea. Strandings within the relatively shallow North Sea, however, may reflect passive drift from further north. West of Ireland acoustic detections indicated a widespread distribution but with the highest occurrence off northwest Ireland, and particularly in the month of May (Kowarski et al., 2018).

There are no population estimates for the species but it appears to be comparatively uncommon, forming small groups rarely exceeding 10 individuals. Like other beaked whales, its preferred habitat is deep ocean basins and trenches of 700 m depth or more, rarely being observed away from these areas. Threats are unknown but, similar to other beaked whales, probably are mainly related to disturbance from underwater noise such as mid-frequency active sonar.

FIGURE 4.49 Sowerby's beaked whale has a small head and well-defined relatively long beak. *Photo: Justin Hart.*

FIGURE 4.50 Overall distribution of Sowerby's beaked whale (depicting those areas where the species is thought to regularly occur).

References

Evans, P. G. H., Herman, J. S., & Kitchener, A. C. (2008). Sowerby's beaked whale *Mesoplodon bidens*. In S. Harris, & D. W. Yalden (Eds.), *Mammals of the British Isles* (4th ed., pp. 692–694). Southampton: The Mammal Society, 800 pp, Handbook.

Kowarski, K., Delarue, J., Martin, B., O'Brien, J., Made, R., Ó Cadhla, O., & Berrow, S. (2018). Signals from the deep: Spatial and temporal acoustic occurrence of beaked whales off western Ireland. *PLoS ONE, 13*(6), e0199431. Available from https://doi.org/10.1371/journal.pone.0199431.

MacLeod, C. D. (2000). Review of the distribution of *Mesoplodon* species (order Cetacea, family Ziphiidae) in the North Atlantic. *Mammal Review, 30*, 1–8.

MacLeod, C. D. (2005). Niche partitioning, distribution and competition in North Atlantic beaked whales (Ph.D. thesis). University of Aberdeen.

Gray's beaked whale *Mesoplodon grayi*

Gray's beaked whale is primarily a cool temperate species of the Southern Hemisphere, most records coming from south of 30°S. However, there is one extraordinary, extralimital record from the Northern Hemisphere — a stranding at Loosduinen near The Hague, the Netherlands, in December 1927, confirmed in 2001 by DNA analysis of a tooth (Smeenk & Camphuysen, 2016).

The species has the typical slender spindle-shaped body (approximately 5–5.5 m length) and small head of other beaked whales of the genus *Mesoplodon*. It has a small triangular to recurved dorsal fin about two-thirds of the way along the back, small and narrow flippers and usually unnotched tail flukes. Diagnostic features include an extremely long narrow beak that becomes white in adults, the white coloration extending over much of the front of the forehead. Most of the rest of the body is grey, sometimes tinged tan, although there is a white patch around the genital region. Young animals have much darker beaks and lighter bellies than adults. The mouth-line is straight, there is a single pair of shallow throat grooves, and the melon bulges slightly between the beak and the blowhole. There are two triangular teeth set in the middle of the lower jaw, which erupt to form tusks in adult males, the tips of which show above the mouth-line.

Reference

Smeenk, C., & Camphuysen, C. J. (2016). Spitssnuitdolfijn van Gray *Mesoplodon grayi*. In S. Broekhuizen, K. Spoelstra, J. B. M. Thissen, K. J. Canters, & J. C. Buys (Eds.), *Atlas van de Nederlandse Zoogdieren* (p. 330). Leiden: Natuur van Nederland 12. Naturalis Biodiversity Center, 432 pp.

Blainville's beaked whale *Mesoplodon densirostris*

Blainville's beaked whale is one of the most widely distributed species of the genus *Mesoplodon*, recorded from tropical and warm temperate seas of all oceans. It is more commonly seen at sea than other North Atlantic *Mesoplodon* species, suggesting that it may be quite common in some areas, although it is also easier to identify. A high arching prominence near the corner of the mouth gives the species a prominent contour, at the root of which, in adult males, a massive pair of flattened, triangular teeth is exposed, tilting slightly forward and often encrusted with barnacles (Fig. 4.51). It is usually dark bluish grey or black on the back, lighter grey on the abdomen, and commonly with large grey or pale blotches (sometimes pinkish) over the upper surface of the body, which may disappear after death. The head is often flattened directly in front of the blowhole, from which protrudes a moderately long, slender beak (Fig. 4.52) (in Gervais' beaked whale, the rostrum is ventrally flattened). The body has the spindle shape typical of the genus, although it is slightly smaller than Sowerby's beaked whale, adults reaching lengths of between 4 and 5 m.

FIGURE 4.51 Adult Blainville's Beaked Whales have a high arched jawline, at the root of which, in adult males, a massive pair of flattened, triangular teeth is exposed, often encrusted with barnacles. *Photo: Bernd Brederlau.*

FIGURE 4.52 Juvenile Blainville's beaked whale. *Photo: Natacha Aguilar.*

In the Atlantic it is mostly recorded from the southeastern United States, the Gulf of Mexico and the Caribbean, but with records in the west ranging from Nova Scotia to southern Brazil. In the eastern North Atlantic the species has been recorded from Spain, Portugal, Madeira and the Canaries (Evans et al., 2008), most records coming from the Canaries where sightings of the species frequently occur (Ritter & Brederlau, 1999). There are also single extralimital stranding records from Iceland (1910), west Wales (1993), the Netherlands (2005) and southwest England (2013) (Camphuysen & Peet, 2006; Evans, Aguilar de Soto, Herman and Kitchener, 2008; Herman, Kitchener, Baker, & Lockyer, 1994; Petersen & Ólafsson, 1986), and three strandings in France (1998, 2002 and 2008), including a live stranding of two individuals on the south Biscay coast in 2008 (Van Canneyt, Dars, Gonzalez, & Dorémus, 2009). The main threats appear to be related to disturbance from underwater noise such as mid-frequency active sonar

References

Camphuysen, C. J., & Peet, G. (2006). *Whales and dolphins of the North Sea*. Kortenhoef: Fontaine Uitgevers, 160 pp.

Evans, P. G. H., Aguilar de Soto, N., Herman, J. S., & Kitchener, A. C. (2008). Blainville's beaked whale *Mesoplodon densirostris*. In S. Harris, & D. W. Yalden (Eds.), *Mammals of the British Isles* (4th ed., pp. 697−699). Southampton: The Mammal Society, 800 pp, Handbook.

Herman, J. S., Kitchener, A. C., Baker, J. R., & Lockyer, C. (1994). The most northerly record of Blainville's beaked whale, *Mesoplodon densirostris*, from the eastern Atlantic. *Mammalia, 58*, 657−661.

Petersen, A. and Ólafsson, E. (1986) Dýralíf Suðurnesja [The animals of Suðurnes.] Pp. 31−48. In: Suðurnes (Náttúrufar, minjar og landnýting) [Suðurnes (Natural history, monuments and land use).] Staðarvalsnefnd. 82 pp. (In Icelandic).

Ritter, F., & Brederlau, B. (1999). Abundance and distribution of cetaceans off La Gomera (Canary Islands). *European Research on Cetaceans, 13*, 98.

Van Canneyt, O., Dars, C., Gonzalez, L., & Dorémus, G. (2009). *Les échouages de mammifères marins sur le littoral français en 2008* (31 pp). Rapport CRMM pour le Ministère de l'Ecologie, de l'Energie, du Développement Durable et de la Mer, Direction de l'eau et de la biodiversité, Programme Observatoire du Patrimoine Naturel.

Family Kogiidae

Pygmy sperm whale *Kogia breviceps*

The genus *Kogia* contains two closely related species, pygmy sperm whale (*Kogia breviceps*) and dwarf sperm whale (*Kogia simus*). Both species are among the least known of cetaceans and were only recognised as separate species around 50 years ago. At sea they are difficult to detect except in extremely calm conditions, let alone to distinguish from one another. The size and position of the dorsal fin are amongst the more diagnostic of features.

Pygmy sperm whales have small stout bodies reaching lengths of only 2.7−3.8 m. They surface relatively slowly, and in calm seas may lie still for a period, showing their blunt squarish head, dark slightly humped back and small recurved dorsal fin, the latter of which is proportionately smaller than dwarf sperm whale, typically less than 8% of the snout to fin length (Fig. 4.53A,B). The conical-shaped head, one-sixth of body length (cf. one-third in sperm whale) becomes squarer with age. There is no beak but an underslung jaw. The blowhole is situated on the top of the head as in most other toothed whales, but generally more than 10% along from the snout tip. The right nostril passes through a valve structure (termed the 'museau de singe'), which is found only in the families *Kogiidae* and *Physeteridae*, and is believed to be used for sound production. Also unique to these families is the oil-filled spermaceti organ, lying above the skull behind a large melon of fatty tissue. There are short ill-defined grooves in the throat region, which give the superficial appearance of gills. The body colour is dark blue-grey on the back, outer margins of flippers and upper surface of tail flukes, lightening to pale grey on the flanks and dull white belly (sometimes with a pinkish tinge). The skin may have a wrinkled appearance. The flippers are relatively long (up to 14% of body length), wide at the base tapering to a rounded point, and are set far forward on the sides near the head. The low, hooked dorsal fin is placed a little way behind the centre of the back. The tail has a concave trailing edge with a distinct median notch.

The pygmy sperm whale has an apparently worldwide distribution, occurring in tropical, subtropical and temperate seas of both hemispheres. It is rarely sighted, most information coming from strandings, the majority of which have

FIGURE 4.53   (A) Pygmy sperm whale stranding at Dinas Dinlle, North Wales. (B) Pygmy Sperm Whale stranding, showing clearly its blunt squarish head and small dorsal fin a little way behind the centre of the back. *(A) Photo: CSIP-ZSL. (B) Photo: Caroline R Weir.*

occurred on the coasts of North America. Records in the North Atlantic range from equatorial waters north to Nova Scotia in the west and to Britain and Ireland in the east, the species generally appearing to favour more temperate waters than its cousin, the dwarf sperm whale.

In the ASCOBANS Agreement Area, most strandings have been from Atlantic European coasts, with (up to 2017) 32 from France, 17 from the United Kingdom, 10 from Ireland and, from southern North Sea countries, just a single record from the Netherlands in 1925 (Berrow & Rogan, 1997; Berrow et al., 2010; Deaville & Jepson, 2011; Evans et al., 2003; Leaper & Evans, 2008; Smeenk & Camphuysen, 2016; Van Canneyt et al., 2016, 2017). There has been an increase in strandings in northern Europe since the year 2000, which possibly reflects increased sea temperatures in the region. There have also been strandings recorded in Spain (Nores & Pérez, 1982; Santos et al., 2006) and Portugal (Reiner, 1981; Sequeira, Inácio, & Reiner, 1992; Sequeira, Inácio, Silva, & Reiner, 1996) despite the fact that they do not have the same level of strandings recording effort as the countries to the north. Although not easy to see and identify at sea, there has also been a handful of sightings in the Bay of Biscay and in British and Irish waters, the furthest north being at 56°N off northwest Ireland (Evans et al., 2003; Leaper & Evans, 2008).

References

Berrow, S., Whooley, P., O'Connell, M., & Wall, D. (2010). *Irish Cetacean Review (2000–2009)*. Kilrush: Irish Whale and Dolphin Group, 60 pp.

Berrow, S. D., & Rogan, E. (1997). Cetaceans stranded on the Irish coast. *Mammal Review, 27*, 51–75.

Deaville, R. and Jepson, P.D. (compilers) (2011) *UK Cetacean Strandings Investigation Programme Final report for the period 1st January 2005 - 31st December 2010*. Zoological Society of London, UK. (http://randd.defra.gov.uk/Document.aspx?Document = FinalCSIPReport2005-2010_finalversion061211released[1].pdf)

Evans, P. G. H., Anderwald, P., & Baines, M. E. (2003). UK Cetacean Status Review. *Report to English Nature and the Countryside Council for Wales*. Oxford: Sea Watch Foundation, 160 pp.

Leaper, R., & Evans, P. G. H. (2008). Pygmy sperm whale *Kogia breviceps*. In S. Harris, & D. W. Yalden) (Eds.), *Mammals of the British Isles* (4th ed., pp. 683–685). Southampton: The Mammal Society, 800 pp, Handbook.

Nores, C., & Pérez, M. C. (1982). *Primera cita de un cachalote pigmeo, Kogia breviceps (Blainville, 1838) (Mammalia, Cetacea) para las costas peninsulares españolas*. Bol. Cien. Nat. I.D.E.A., No. 29.

Reiner, F. (1981). Nota sobre a ocorréncia de um cachaloteanão *Kogia breviceps* Blainville na Praia de Salgueiros – Vila Nova de Gaia. *Memórias do Museo do Mar, série zoológica, 2*(15), 1–12. (in Portuguese).

Santos, M. B., Pierce, G. J., López, A., Reid, R. J., Ridoux, V., & Mente, E. (2006). Pygmy sperm whales *Kogia breviceps* in the Northeast Atlantic: New information on stomach contents and strandings. *Marine Mammal Science, 22*(3), 600–616.

Sequeira, M., Inácio, A., & Reiner, F. (1992). *Arrojamentos de Mamíferos Marinhos da Costa Portuguese entre 1978 e 1988 [Marine mammal strandings along the Portuguese coast between 1978 and 1988], . Estudos de Biologia e Conservação da Natureza* (Vol. 7). Lisboa: SNPRCN, 48 pp.

Sequeira, M., Inácio, A., Silva, M. A., & Reiner, F. (1996). *Arrojamentos de Mamíferos Marinhos da Costa Continental Portuguese entre 1989 e 1994 [Marine mammal strandings along the continental Portuguese coast between 1989 and 1994]*, . *Estudos de Biologia e Consercação da Natureza* (Vol. 19). Lisboa: Instituto da Conservaçao da Natureza, 52 pp.

Smeenk, C., & Camphuysen, C. J. (2016). Dwergpotvis *Kogia breviceps*. In S. Broekhuizen, K. Spoelstra, J. B. M. Thissen, K. J. Canters, & J. C. Buys (Eds.), *Atlas van de Nederlandse Zoogdieren* (p. 324). Leiden: Natuur van Nederland 12. Naturalis Biodiversity Center, 432 pp.

Van Canneyt, O., Dars, C., Matthieu, A., Dabin, W., Demaret, F., Dorémus, G., . . . Spitz, J. (2016). *Les échouages de mammifères marins sur le littoral françcais en 2015* (48 pp). Réseau National Echouages (RNE). Rapport annuel, Agences des airesmarines protégées.

Dwarf sperm whale *Kogia sima*

The dwarf sperm whale (Fig. 4.54A,B) was not recognised as a distinct species until the mid-1960s and our knowledge of it is rudimentary. It is very similar in appearance to the pygmy sperm whale, having a triangular or squarish head,

FIGURE 4.54 (A) Dwarf sperm whale is scarcely larger than a porpoise, with a triangular dorsal fin, generally larger and more erect than pygmy sperm whale. (B) Dwarf sperm whale in Cornwall. *(A) Photo: Marine Discovery Penzance. (B) Photo: Glenn Overington.*

narrow underslung lower jaw and a small robust body that tapers rapidly behind the dorsal fin. As with the pygmy sperm whale, the head shape and light-coloured false gill slit give it a rather shark-like appearance. It is smaller than the pygmy sperm whale, adults generally reaching lengths of just 2.5 m. The dorsal fin is proportionately larger and more erect, typically 9%−16% of the snout to fin length, and generally located nearer the middle of the back giving an overall appearance more like a dolphin. The blowhole is also positioned further forward than in its close relative, usually <10% from the tip of the snout. There is usually a pair of short throat grooves similar to those in beaked whales. The flippers are small with bluntish tips and situated close to the head. It is brownish-grey in colour over the back and flanks, with a white or pinkish belly.

Dwarf sperm whales have a worldwide distribution but appear to be more confined to tropical and subtropical seas than the pygmy sperm whale. As with that species its range is known mainly from strandings only. In the North Atlantic, it has been recorded from equatorial waters to Virginia in the west and to the British Isles (Cornwall) in the east. In the ASCOBANS Agreement Area the species has been confirmed off the coast of Portugal, where also several unidentified *Kogia* sightings have been made (Santos et al., 2012). Elsewhere there have been just seven records (up to 2017) − five from France (1986, 1991, 1999, 2010 (sighting), 2015), one from Spain (1987) and one from the United Kingdom (2011) (Sea Watch Foundation, 2011; Valverde & Camiñas, 1997; Van Canneyt et al., 2016; Van Canneyt et al., (2016)). There are also two extralimital records from the Mediterranean both from Italy − a dead stranding in 1988 and a live stranding in 2002 (Notarbartolo di Sciara, 2016; Notarbartolo di Sciara & Birkun, 2010).

References

Notarbartolo di Sciara, G. (2016). Marine mammals in the Mediterranean Sea: An overview. *Advances in Marine Biology, 75*, 1−36.

Notarbartolo di Sciara, G. and Birkun, A. (2010) *Conserving whales, dolphins and porpoises in the Mediterranean and Black Seas.* An ACCOBAMS Status Report. ACCOBAMS, Monaco. 212 pp.

Santos, J., Araújo, H., Ferreira, M., Henriques, A., Miodonski, J., Monteiro, S., . . . Vingada, J. (2012). Chapter I: Baseline estimates of abundance and distribution of target species. *MARPRO Conservação de especies marinhas protegidas em Portugal continental* (80 pp). Project LIFE09/NAT/PT/000038.

Valverde, J. A., & Camiñas, J. A. (1997). On dwarf sperm whale *Kogia simus* (Owen 1866), Cetacea, Physeteroidea, in South Spain: A correction. *European Research on Cetaceans, 10*, 168−171.

Van Canneyt, O., Dars, C., Matthieu, A., Dabin, W., Demaret, F., Dorémus, G., . . . Spitz, J. (2016). *Les échouages de mammifères marins sur le littoral français en 2015. Réseau National Echouages (RNE)* (48 pp). Rapport annuel, Agences des airesmarines protégées.

Family Physeteridae

Sperm whale *Physeter macrocephalus*

The sperm whale is the largest of all the toothed whales, adult males reaching lengths of 11−18 m and adult females 8−12 m. It has a very distinctive shape, its large square-shaped head occupying at least one-third of the body and being particularly pronounced in adult males. This readily distinguishes it from the more streamlined torpedo-shaped body of rorquals such as blue, fin, sei or minke whale. Sperm whales are also distinctive in lacking a dorsal fin, having instead a triangular or dorsal hump two-thirds along the body, followed by a spinal ridge (Fig. 4.55). There are also corrugations to the skin giving it a shrivelled appearance. Another diagnostic feature is the single bushy blow directed forwards at a sharp angle to the left, rising to a height of about 1.5 m. The lower jaw is narrow and underslung, and there are 2−10 short deep throat grooves. The flippers are short and spatulate in shape. Coloration is brownish-grey to black, with white areas around the mouth and frequently on the belly, and often there may be scratches and scars over the head and back. When sperm whales make a steep dive, they characteristically throw their broad, triangular and deeply notched tail flukes into the air.

Sperm whales have a worldwide distribution, occurring in deep waters of all seas. Females and juvenile males have a more limited range than adolescent and mature males, being confined more or less to warmer waters (generally with

FIGURE 4.55 From above, the Sperm Whale appears cigar shaped, with a dorsal hump in place of a fin, two-thirds along the body and a blow directed forwards and to the left. *Photo: Sabina Airoldi/Tethys.*

sea surface temperatures above 15°C) between approximately 45°N and approximately 45°S. Young males accompany females in tropical and subtropical waters but from the ages of 14−21 years, they move increasingly to higher latitudes. Only large males are found at the highest latitudes, sometimes occurring even close to ice edge, but generally in the most productive deep waters.

In the North Atlantic, the species is widely distributed mainly off the continental shelf, along the mid-Atlantic Ridge and around oceanic archipelagos (Azores, Madeira, Canaries, Cape Verde). It can be found from Davis Strait and Newfoundland in the west, and Norway and Iceland in the east, south to breeding grounds in the Caribbean in the west, and the Iberian Peninsula, Macaronesia and the Mediterranean Sea in the east (Fig. 4.56). It favours waters exceeding 200 m, and usually 500−2000 m depth.

In the ASCOBANS Agreement Area the known preferred feeding areas for males include deeper waters along the continental slope west of Portugal and north of Spain within the Bay of Biscay, west and north of the British Isles and Ireland including the edge of the Porcupine Bank, the Rockall Trough and the Faroe−Shetland Channel, and west of Norway (Hammond et al., 2009; Evans et al., 2003; Øien, 2009; Øien (2009); Reid et al., 2003; Wall et al., 2013; Weir et al., 2001). Abundance estimates from NASS surveys indicated 4319 whales (CV 0.20) in 1995 and 6207 (CV 0.22) in 1996−2001 in Norwegian and adjacent waters (Øien, 2009) and 11,185 (CV 0.34) in 2001 in Icelandic waters (Gunnlaugsson, Vikingsson, & Pike, 2009). During the July 2007 CODA survey sperm whale predicted density was highest in the northwestern waters of the Iberian Peninsula (656 animals, CV 0.34) and in the inner part of the Bay of Biscay (477 animals, CV 0.33). The overall estimate of sperm whale abundance in those offshore areas covered by the CODA survey was 2569 (CV 0.26, 95% CI: 1600−4200) (revised from Hammond et al., 2009). In July 2016, the overall estimate of abundance for the wider region covered by the SCANS-III survey was 13,518 (CV 0.405, 95% CI: 6200−29,600), mainly off the shelf north-west of Scotland but also in the Bay of Biscay (Hammond et al., 2017). Too few sightings of sperm whales were made during the ObSERVE surveys of the Irish EEZ in summer and winter 2015 and 2016 for an abundance estimate (Rogan et al., 2018).

The TNASS survey around Iceland and the Faroes in 2007 provided an abundance estimate (corrected for perception bias but not availability bias[1]) of 12,220 whales (CV 0.38, 95% CI: 5808−25,717). Sperm whales were found throughout the survey area but were seen in greatest numbers around the Faroes and to the south and west of Iceland. In 2015 the abundance for these survey areas was estimated to be 23,311 (CV 0.59, 95% CI: 7789−69,771), with 91% of this total around the Faroes. The Norwegian surveys in 2002−07 yielded an abundance estimate of 9402 whales (CV 0.22),

---

1. The terms availability and perception bias are explained in Chapter 6, Conservation research.

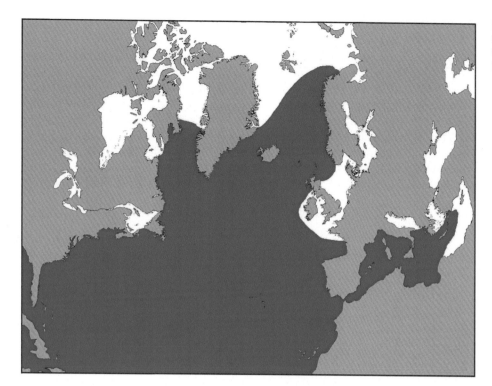

FIGURE 4.56 North Atlantic distribution of sperm whale (depicting those areas where the species is thought to regularly occur).

FIGURE 4.57 Sperm Whales generally inhabit deep waters beyond the shelf edge. *Photo: Peter GH Evans.*

uncorrected for both perception and availability bias, with the great majority of sightings made over the deep waters of the Norwegian Sea, south of the Mohn Ridge between Jan Mayen and Bear Island. In 2015 the equivalent estimate of abundance in the Norwegian Sea was 4896 (CV 0.26) (NAMMCO, 2018). Corrections for availability and/or perception bias may well increase these estimates.

Formerly, sightings around the British Isles were almost exclusively of larger males, singly or in small groups; more recently more records of smaller individuals in larger groups have been noted, including some mass strandings of immature males (Evans, 1997; Evans et al., 2003; Smeenk & Evans, 2018). The number of strandings here has increased markedly since the 1980s (Berrow, Evans, & Sheldrick, 1993; Evans, 1997). Between 1913 and 1986, an average of 0.6 whales stranded per year, whereas between 1987 and 2010, this had risen to an average of 4.8 per year. It is unclear whether this is due to changes in whale distribution, or to increased mortality from unknown causes.

Sperm whales often stray into the North Sea and even the Baltic, well outside their normal habitat, resulting in strandings of single animals or loose groups of up to 29 (Jacques & Lambertson, 1997; Kinze et al., 2010; see Fig. 4.58A,B). Such strandings have been recorded as long ago as 1577 when 13 live-stranded on the Dutch coast. The largest mass stranding in the North Sea was recorded in January to February 2016 when 29 animals came ashore on the coasts of Germany, the Netherlands, France and eastern Britain. It has been suggested that these 'navigational errors' are because animals become disorientated and eventually strand, due to poor navigation abilities in shallow water, and/ or that they follow squid prey and then run out of food (Evans, 1997; Smeenk, 1997). Peaks of strandings in the North Sea have occurred in the late 18th century and in the 1990s to early 21st century (Smeenk & Evans, 2018), possibly

(A)

(B)

FIGURE 4.58 (A) Stranding of a male Sperm Whale in Germany. (B) Sperm whale stranding in Lincolnshire, UK. *(A) Photo: ITAW-TiHo, courtesy of Ursula Siebert. (B) Photo: CSIP-ZSL.*

reflecting population recovery after cessation of hunting (Evans, 1997), although there may be multiple causes including effects of climate change (Pierce, Ward, Brownlow, & Santos, 2018).

Most strandings both in the North Sea and on Atlantic coasts occur during winter (November–March) (Berrow et al., 2010; Deaville & Jepson, 2011; Jauniaux, Brosens, Jaquinet, & Coignoul, 1998; Jepson, 2005; Smeenk, 1997; Smeenk & Evans, 2018; Tougaard & Kinze, 1999). During Scottish and Irish whaling in the early 20th century, most sperm whales were caught between June and August (Fairley, 1981; Thompson, 1928). Sightings in British and Irish waters have been mainly between July and December, with increasing evidence of small groups remaining at high latitudes into winter, when most coastal sightings and strandings take place (Berrow et al., 1993; Evans, 1997; Evans et al., 2003; Smeenk, 1997; Tougaard & Kinze, 1999). This lends some support to the hypothesis that winter strandings are the result of sperm whales remaining at these latitudes after the summer, potentially to exploit the squid *Gonatus fabricii* in the Norwegian Sea when the squid lose their ability to actively swim before spawning, but then becoming short of food.

Current threats to sperm whales include ship strikes, sound disturbance, plastic ingestion, entanglement in fishing gear and changes in prey availability.

References

Berrow, S., Evans, P. G. H., & Sheldrick, M. (1993). An analysis of sperm whale *Physeter macrocephalus* stranding and sighting records from Britain and Ireland. *Journal of Zoology, London, 230*, 333–337.

Berrow, S., Whooley, P., O'Connell, M., & Wall, D. (2010). *Irish Cetacean Review (2000–2009)*. Kilrush: Irish Whale and Dolphin Group, 60 pp.

Deaville, R. and Jepson, P.D. (compilers) (2011) *UK Cetacean Strandings Investigation Programme Final report for the period 1st January 2005 - 31st December 2010.* Zoological Society of London, UK. (http://randd.defra.gov.uk/Document.aspx?Document = FinalCSIPReport2005-2010_finalversion061211released[1].pdf).

Evans, P. G. H. (1997). Ecology of Sperm Whales (*Physeter macrocephalus*) in the Eastern North Atlantic, with special reference to sightings and strandings records from the British Isles (pp. 37–46). In T. G. Jacques & R. H. Lambertsen (Eds.), *Sperm whale deaths in the north sea science and management* (Vol. 67, 133 pp). Bulletin de L'Institut Royal des Sciences Naturelles de Belgique. Biologie.

Evans, P. G. H., Anderwald, P., & Baines, M. E. (2003). UK Cetacean Status Review. *Report to English Nature and the Countryside Council for Wales.* Oxford: Sea Watch Foundation, 160 pp.

Fairley, J. S. (1981). *Irish whales and whaling.* Dublin: Blackstaff Press, 218 pp.

Gunnlaugsson, T., Vikingsson, G. A., & Pike, D. G. (2009). Combined line-transect and cue-count estimate of sperm whale abundance in the North Atlantic, from Icelandic NASS-2001 shipboard survey. In C. Lockyer, & D. Pike (Eds.), *North Atlantic sightings surveys. Counting whales in the North Atlantic 1987-2001* (pp. 73–80). NAMMCO Scientific Publications 7, 244 pp.

Hammond, P.S., Burt, M.L., Cañadas, A., Castro, R., Certain, G., Gillespie, D., ... Winship, A. (2009). Cetacean Offshore Distribution and Abundance in the European Atlantic (CODA). Report available from SMRU, Gatty Marine Laboratory, University of St Andrews, St Andrews, Fife KY16 8LB, UK. http://biology.st-andrews.ac.uk/coda/.

Hammond, P.S., Lacey, C., Gilles, A., Viquerat, S., Borjesson, P., Herr, H., ... and Øien, N. (2017). Estimates of cetacean abundance in European Atlantic waters in summer 2016 from the SCANS-III aerial and shipboard surveys. Available at https://synergy.standrews.ac.uk/scans3/files/2017/05/SCANS-III-design-based-estimates-2017-05-12-final-revised.pdf.

T. G. Jacques, & R. H. Lambertson (Eds.) (1997). *Sperm whale deaths in the North Sea science and management* (Vol. 67, 133 pp). Bulletin de L'Institut Royal des Sciences Naturelles de Belgique, Biologie.

Jauniaux, T., Brosens, L., Jaquinet, E., & Coignoul, F. (1998). Postmortem investigations on winter stranded sperm whales from the coasts of Belgium and the Netherlands. *Journal of Wildlife Diseases, 34*, 99–109.

P. D. Jepson (Ed.) (2005). *Trends in cetacean strandings around the UK coastline and cetacean and marine turtle post-mortem investigations, 2000 to 2004 inclusive (contract CRO 238)* (85 pp). Report to Defra, UK Cetacean Strandings Investigation Programme. London: Institute of Zoology.

Kinze, C. C., Jensen, T., Tougaard, S., & Baagøe, H. J. (2010). Danske hvalfund i perioden 1998-2007 [Records of cetacean strandings on the Danish coastline during 1998-2007]. *Flora og Fauna, 16*(4), 91–99.

NAMMCO (2018) *Report of the NAMMCO Scientific Working Group on Abundance Estimates.* Available at https://nammco.no/topics/sc-working-group-reports/.

Øien, N. (2009). Distribution and abundance of large whales in Norwegian and adjacent waters based on ship surveys 1995-2001. In C. Lockyer, & D. Pike (Eds.), North Atlantic sightings surveys. Counting whales in the North Atlantic 1987-2001 (7, pp. 31–47). NAMMCO Scientific Publications, 244 pp.

Øien, N. (2009). Distribution and abundance of large whales in Norwegian and adjacent waters based on ship surveys 1995–2001. In C. Lockyer, & D. Pike (Eds.), *North Atlantic Sightings Surveys. Counting whales in the North Atlantic 1987-2001* (Volume 7, pp. 31–47). NAMMCO Scientific Publications, 244 pp.

Pierce, G. J., Ward, N., Brownlow, A., & Santos, M. B. (2018). Analysis of historical and recent diet and strandings of sperm whales (*Physeter macrocephalus*) in the North Sea. *Lutra, 61*, 71–86.

Reid, J. B., Evans, P. G. H., & Northridge, S. P. (2003). *Atlas of Cetacean Distribution in North-west European Waters.* Peterborough: Joint Nature Conservation Committee, 76 pp.

Rogan, E., Breen, P., Mackey, M., Cañadas, A., Scheidat, M., Geelhoed, S., & Jessopp, M. (2018). *Aerial surveys of cetaceans and seabirds in Irish waters: Occurrence, distribution and abundance in 2015-2017.* Dublin, Ireland: Department of Communications, Climate Action & Environment and National Parks and Wildlife Service (NPWS), Department of Culture, Heritage and the Gaeltacht, 297 pp.

Smeenk, C. (1997). Strandings of sperm whales *Physeter macrocephalus* in the North Sea: History and patterns (pp. 15–28). In T. G. Jacques & R. H. Lambertsen (Eds.), *Sperm whale deaths in the North Sea: Science and management* (Vol. 67). Bulletin de L'Institut Royal des Sciences Naturelles de Belgique, Biologie.

Smeenk, C., & Evans, P. G. H. (2018). Review of sperm whale *Physeter macrocephalus* L., 1758 strandings around the North Sea. *Lutra, 61,* 29–70.

Thompson, D. 'A. W. (1928). On Whales Landed at the Scottish Whaling Stations during the Years 1908-1914 and 1920-1927. *Scientific Investigations Fishery Board of Scotland, 3,* 1–40.

Tougaard, J., and Kinze, C.C. (Eds.) (1999). *Proceedings from the workshop on sperm whale strandings in the North Sea: The event – the action – the aftermath. Romø, Denmark, 26–27 May 1998, Biological Papers 1.* Esbjerg: Fisheries and Maritime Museum.

Wall, D., Murray, C., O'Brien, J., Kavanagh, L., Wilson, C., Ryan, C., ... Berrow, S. (2013). *Atlas of the distribution and relative abundance of marine mammals in Irish offshore waters 2005-2011.* Irish Whale and Dolphin Group, Merchants Quay, Kilrush, Co. Clare.

Weir, C. R., Pollock, C., Cronin, C., & Taylor, S. (2001). Cetaceans of the Atlantic Frontier, north and west of Scotland. *Continental Shelf Research, 21,* 1047–1071.

INFRAORDER Mysticeti, the baleen whales

Family Balaenidae (right whales)

North Atlantic right whale *Eubalaena glacialis*

The North Atlantic right whale (Fig. 4.59), along with its recently separated close relative, the North Pacific right whale *Eubalaena japonica*, are the rarest of all the great whales. The species has a very stout body, 13.5–17 m in length, and a massive head that forms 25% of the total body length in adults, and up to 35% in juveniles, and a narrow rostrum with

FIGURE 4.59 The North Atlantic Right Whale is black in colour and, like other right whales, lacks a dorsal fin. It has a large head, and both the head and jaws have several large white, grey or yellowish callosities, infested with parasites. *Photo: New England Aquarium, courtesy of Scott Kraus.*

highly arched jawline. It has very long (2–2.8 m), narrow dark brown, grey or black baleen plates in each side of the upper jaw, and both the head and jaws have several large white, grey or yellowish callosities, which are infested with parasites. Numerous hairs also occur on the upper jaw and the chin. Of the large baleen whales, only the right whales and their relative the bowhead lack a dorsal fin altogether. Instead, the dorsal surface is smooth and ridge-less. On diving, the broad (up to 6 m tip to tip) black tail flukes are raised above the water; they are deeply notched but with a distinct concave smooth trailing edge. The body colour is generally black, occasionally with a white belly and chin patches.

The blow of the right whale (and its high Arctic relative, the bowhead whale) is diagnostic, being V-shaped as a result of the widely separated blowholes. It may reach 7 m in height. Feeding involves skimming with the mouth wide-open, the narrow rostrum raised into the air and baleen plates partially exposed above the water.

The North Atlantic right whale is normally restricted to the temperate zone between 20° and 70°N (Kraus & Rolland, 2007; Monsarrat et al., 2015). Its population of around 500 individuals occurs almost entirely along the eastern coast of the United States from the Gulf of Mexico and Florida north to the Gulf of St Lawrence and coasts of Nova Scotia, Newfoundland, Labrador and south Greenland. In the eastern North Atlantic, it once ranged from northwest Africa, the Azores and the Mediterranean, north to the Bay of Biscay, west Ireland, the Hebrides, Shetland, Faroe Islands, Iceland and Svalbard. Since the 1920s, however, sightings have been sporadic, with isolated records from the Canaries, Madeira, Spain, Portugal, United Kingdom, Ireland and Iceland (Brown, 1976, 1986; Evans et al., 2003; Kraus & Evans, 2008; Kraus & Rolland, 2007), with a photographic record of a possible live sighting off Schouwen, Zeeland (the Netherlands) in July 2005 (Camphuysen & Smeenk, 2016) and another off Penmarc'h, French Biscay in June 2019.

Historically there is good evidence that eastern North Atlantic right whales calved in Cintra Bay, Mauritania during winter (Reeves, 2001). Whaling records indicate that the species occurred off the Hebrides and Ireland in early summer (mainly June), possibly having spent the winter in the Bay of Biscay since the former Basque fishery operated mainly between October and February; they may then have subsequently moved on to feeding areas west of Norway (Aguilar, 1986; Collett, 1909; Fairley, 1981; Smith, Barthelmess, & Reeves, 2006; Thompson, 1928). The few recent sightings in British and Irish waters have all occurred between May and September (Evans et al., 2003; Kraus & Evans, 2008).

The current North Atlantic population is estimated to be less than 6% of historical carrying capacity which is considered to be between c. 9,100 and 21,300 individuals before human exploitation (Monsarrat et al., 2016).

Individually recognisable right whales photographed off Iceland and Norway have been matched with animals from the western North Atlantic (Knowlton et al., 1992), and preliminary genetic data indicate that separation between eastern and western Atlantic stocks is unlikely (Kraus & Evans, 2008).

North Atlantic right whales suffer significant mortality from ship strikes and entanglement in fishing gear.

References

Aguilar, A. (1986). A review of old Basque whaling and its effect on the right whales (*Eubalaena glacialis*) of the North Atlantic. *Report of the International Whaling Commission* (Special issue 10), 191–200.

Brown, S. G. (1976). Modern whaling in Britain and the north-east Atlantic Ocean. *Mammal Review, 6*, 25–36.

Brown, S. G. (1986). Twentieth century records of right whales (*Eubalaena glacialis*) in the northeast Atlantic Ocean. *Reports of International Whaling Commission* (Special Issue, 10), 121–127.

Camphuysen, C. J., & Smeenk, C. (2016). Noordkaper *Eubalaena glacialis*. In S. Broekhuizen, K. Spoelstra, J. B. M. Thissen, K. J. Canters, & J. C. Buys (Eds.), *Atlas van de Nedrlandse Zoogdieren* (pp. 302–303). Leiden: Natuur van Nederland 12. Naturalis Biodiversity Center, 432 pp.

Collett, R. (1909). A few notes on the whale *Balaena glacialis* and its capture in recent years in the North Atlantic by Norwegian whalers. *Proceedings of the Zoological Society of London, 7*, 91–97.

Evans, P. G. H., Anderwald, P., & Baines, M. E. (2003). UK Cetacean Status Review. *Report to English Nature and the Countryside Council for Wales*. Oxford: Sea Watch Foundation, 160 pp.

Fairley, J. (1981). *Irish Whales and Whaling*. Belfast: Blackstaff Press, 218 pp.

Knowlton, A. R., Sigurjonsson, J., Ciano, J. N., & Kraus, S. D. (1992). Long-distance movements of North Atlantic right whales, (Eubalaena glacialis). *Marine Mammal Science, 8*(4), 397–405.

Kraus, S. D., & Evans, P. G. H. (2008). Northern right whale *Eubalaena glacialis*. In S. Harris, & D. W. Yalden) (Eds.), *Mammals of the British Isles* (4th ed., pp. 658–662). Southampton: The Mammal Society, 800 pp, Handbook.

Kraus, S. D., & Rolland (Eds.), (2007). *The urban whale. North Atlantic right whales at the crossroads*. Cambridge, MA: Harvard University Press, 543 pp.

Monsarrat, S., Pennino, M. G., Smith, T. D., Reeves, R. R., Meynard, C. N., Kaplan, D. M., & Rodrigues, A. S. L. (2015). Historical summer distribution of the endangered North Atlantic right whale (*Eubalaena glacialis*): A hypothesis based on environmental preferences of a congeneric species. *Diversity and Distributions, 21*, 925–937.

Monsarrat, S., Pennino, M. G., Smith, T. D., Reeves, R. R., Meynard, C. N., Kaplan, D. M., & Rodrigues, A. S. L. (2016). A spatially explicit estimate of the prewhaling abundance of the endangered North Atlantic right whale. *Conservation Biology, 30*, 783−791.

Reeves, R. R. (2001). Overview of catch history, historic abundance and distribution of right whales in the western North Atlantic and in Cintra Bay, West Africa. *Journal of Cetacean Research and Management* (Special Issue 2), 187−192.

Smith, T. D., Barthelmess, K., & Reeves, R. R. (2006). Using historical records to relocate a long-forgotten summer feeding ground of North Atlantic right whales. *Marine Mammal Science, 22*(3), 723−734.

Thompson, D. 'A. W. (1928). On Whales Landed at the Scottish Whaling Stations during the Years 1908−1914 and 1920−1927. *Scientific Investigations Fishery Board of Scotland*, 1928, No. 3: 1-40.

Bowhead whale *Balaena mysticetus*

The bowhead whale (Fig. 4.60) only superficially resembles its relative, the North Atlantic right whale. It has a very rotund body, 18−20 m in length, a large (up to approximately 40% of its body) though relatively narrow head and a strongly arched lower jaw. Like other right whales it is black in colour and has no dorsal fin but a distinctive feature is the prominent muscular bulge in the area of the blowhole with an obvious depression behind. Bowheads have a very thick skin (up to 2.5 cm) and blubber layer (up to 28 cm), and the longest baleen plates (up to 5.2 m) of any whale. The plates are dark grey to brown or black, generally with lighter fringes, which from the front may show as a white patch in front of the lower jaw. The flippers are large, fan-shaped with blunt tips. The tail flukes are wide and tapered at the tips with no central notch, and there is often a light grey or white band across the tail-stock.

Bowheads occupy arctic and subarctic regions of the Northern Hemisphere, generally between about 60° and 85°N. They favour the ice edge around the Arctic Ocean and the waters surrounding Greenland and Svalbard across to Novaya Zemlya, migrating northwards in summer into the high Arctic as the ice retreats. In recent years, however, the species has been recorded much further south − in Cape Cod Bay, New England (eastern United States) in 2012 and 2014, as well as off the coast of Cornwall (United Kingdom) (February 2015, May 2016), Northern Ireland (May 2016), France (May 2016) and Belgium (March−April 2017) (possibly all European sightings are of the same individual). This may reflect the break-up and southwards drift of ice that the Arctic is experiencing.

Heavily exploited by whalers in the Arctic Ocean, in Baffin Bay off Greenland and the Barents Sea north of Norway, the population seriously declined during the early 20th century, reaching a low in the 1920s of approximately

FIGURE 4.60 The Bowhead Whale rarely occurs in the North Atlantic south of the Arctic Circle; it is a long-lived species with a life span believed to exceed 200 years. *Photo: Caroline R Weir.*

3000. In the eastern Canadian Arctic it has been estimated that 25,000—40,000 bowhead whales were taken by Basque whalers during the period 1530—1610, followed by approximately 28,000 during the pelagic whaling phase during 1719—1915 (Ross, 1993). The prewhaling size of the east Canada—west Greenland population was thus probably at least 25,000, well above the likely current level. Based on aerial surveys and genetic capture—recapture, Rekdal et al., (2015) concluded that the number of bowhead whales using west Greenland waters in the winter/spring period increased from 1998 to 2006 but has levelled off since then. Numbers in the east Canada—west Greenland subpopulation were estimated at 4000—10,500 in 2013 from aerial surveys (Doniol-Valcroze et al., 2015), or 4500—11,000 in 2008—12 from genetic mark—recapture (Frasier et al., 2015).

The prewhaling size of the east Greenland—Svalbard—Barents Sea population has been estimated to be 33,000—65,000 (Allen & Keay, 2006). This subpopulation was originally the largest of the bowhead whale subpopulations, but was heavily depleted by premodern commercial whaling from 1611 to the last recorded capture in 1911 (Ross, 1993). The only record of catches by modern whaling is of four whales taken near Svalbard in 1932 (Ruud, 1937). The population size for the east Greenland—Svalbard—Barents Sea subpopulation may be increasing slightly, and is now known to be greater than 50 mature animals, but there are still probably fewer than 250 mature individuals (Cooke & Reeves, 2018).

The greatest threats currently are likely to be climate change, ship strikes, disturbance from underwater noise and entanglement in fishing gear.

References

Allen, R. C., & Keay, I. (2006). Bowhead whales in the eastern Arctic, 1611-1911: Population reconstruction with historical whaling records. *Environment and History, 12*(1), 89—113.

Cooke, J. G., & Reeves, R. (2018). *Balaena mysticetus. East Greenland-Svalbard-Barents Sea subpopulation.* The IUCN Red List of Threatened Species 2018, e. T2472A50348144.

Doniol-Valcroze, T., Gosselin, J.-F., Pike, D., Lawson, J., Asselin, N., Hedges, K., & Ferguson, S. (2015). *Abundance estimate of the Eastern Canada — West Greenland bowhead whale population based on the 2013 high arctic cetacean survey.* DFO Canadian Science Advisory Secretariat Research Document, 2015/058.

Frasier, T. R., Petersen, S. D., Postma, L., Johnson, L., Heide-Jørgensen, M. P., & Ferguson, S. H. (2015). *Abundance estimates of the Eastern Canada-West Greenland bowhead whale (Balaena mysticetus) population based on genetic capture-mark-recapture analyses.* DFO Canadian Science Advisory Secretariat Research Document 2015/008.

Rekdal, S. L., Hansen, R. G., Borchers, D., Bachmann, L., Laidre, K. L., Wiig, Ø., ... Heide-Jørgensen, M. P. (2015). Trends in bowhead whales in West Greenland: Aerial surveys vs. genetic capture-recapture analyses. *Marine Mammal Science, 31*(1), 133—154.

Ross, W. G. (1993). Commercial whaling in the North Atlantic Sector. In J. J. Burns, J. J. Montague, & C. J. Cowles (Eds.), *The bowhead whale* (pp. 511—561). Society for Marine Mammalogy.

Ruud, J. T. (1937). Grønlandshvalen. *Norsk Hvalfangsttid, 26*(7), 254.

Family Balaenopteridae (rorquals)

Humpback whale *Megaptera novaeangliae*

The humpback whale (Fig. 4.61) is one of the most distinctive of all rorquals. It has a more robust body than other rorquals, 13—15 m in length, and a relatively short head, the top of which is flattened and covered by a number of fleshy knobs or tubercles (Fig. 4.62). These extend over the lower jaw, which also has a rounded protuberance near the tip. The ridge along the midline of the top of the head is indistinct unlike the other rorquals. The dorsal fin is very variable both in size and shape, ranging from a small triangular knob to a larger distinctly recurved fin, placed nearly two-thirds along the back. The species has a broad, bushy blow, which reaches a height of 2.5—3 m, although in larger individuals, particularly after a long dive, it may be taller and more slender.

The flippers are very long — nearly one-third the total body length. They are scalloped with knobs or bumps along the leading edge, and are white underneath and sometimes also on the upper surface. Humpbacks often raise a flipper

FIGURE 4.61 Humpback Whale are relatively acrobatic, often displaying their short head, robust body, and long white flippers. *Photo: Peter GH Evans.*

FIGURE 4.62 The head of the Humpback Whale is flattened and covered by a number of fleshy knobs or tubercles. *Photo: Peter GH Evans.*

and slap it against the water or lie on their side or back with one or both flippers in the air. The tail flukes are very broad and distinctly notched, but are commonly scalloped with knobs along the trailing edge giving an irregular appearance. The undersurface is partially or completely white, often forming a unique pattern that can be used for individual identification. Humpbacks frequently breach right out of the water, particularly in their tropical breeding grounds.

The humpback whale has a worldwide distribution in all seas, occurring even occasionally to the ice edge. It is a highly migratory species, feeding in summer in high latitudes, and mating and calving in winter in tropical waters, although a few overwinter on the feeding grounds. The species shows strong individual fidelity to feeding areas; in the North Atlantic these include the Gulf of Maine, Gulf of St Lawrence, Newfoundland/Labrador, Greenland, Iceland and

FIGURE 4.63 North Atlantic Distribution of Humpback Whale (depicting those areas where the species is thought to regularly occur).

Norway. Matching of photographically and genetically identified individuals indicates that the eastern North Atlantic population migrates primarily to the West Indies (Martin, Katona, Mattila, Hembree, & Waters, 1984; Stevick et al., 2003, Stevick, Øien, & Mattila, 1998; Stevick et al., 2006), although some animals winter in the Cape Verde Islands (Hazevoet & Wenzel, 2000; Jann et al., 2003; Reiner, dos Santos, & Wenzel, 1996); genetic analysis suggests a third, unknown, breeding area. Despite fidelity to specific feeding grounds, however, whales from all North Atlantic areas appear to mix spatially and genetically in the West Indies in winter.

The species is more common on the west side of the North Atlantic, spending the summer in Baffin Bay and along the New England coast. In the eastern North Atlantic it occurs around Iceland, Norway, the British Isles and Ireland (Fig. 4.63). North Atlantic humpbacks winter in the West Indies, with small numbers wintering around the Cape Verde Islands off Northwest Africa (a minimum estimate of 260 individuals of part of the archipelago was obtained by mark−recapture photo-ID; Ryan, Wenzel, Súarez, & Berrow, 2014).

Sightings from around the British Isles and Ireland have increased markedly since the early 1980s; occurring in three main areas: (1) the Northern Isles south to eastern England; (2) the northern Irish Sea north to West Scotland and (3) the Celtic Sea from southwest Ireland to southwest Wales and southwest England (Camphuysen & Peet, 2006; Clapham & Evans, 2008; Evans et al., 2003; Ryan et al., 2016), but with a few sightings and strandings elsewhere including the southern North Sea (Leopold, Rotshuizen, & Evans, 2018).

In shelf waters of the ASCOBANS Agreement Area, humpbacks occur mainly between May and September, and this was the period when catches from whaling activities were highest (Brown, 1976; Clapham & Evans, 2008; Thompson, 1928). However, particularly from the British Isles and Ireland south to the Iberian Peninsula, sightings may occur at anytime of year, including November to March, as indicated also from acoustic detections with SOSUS hydrophone arrays offshore in the North Atlantic at these latitudes (Clark & Charif, 1998; Charif, Clapham, & Clark, 2001). Sightings in Ireland, occurring mainly along the south coast, increase through the summer to peak in September to December, rapidly declining between January and May (Berrow et al., 2010).

Overall the North Atlantic population has recovered well from exploitation, estimated at somewhere between 9400 and 16,400 in 1992, with the great majority occurring in the west central part (Smith et al., 1999; Stevick et al., 2003). NASS surveys around Iceland in the central North Atlantic gave an abundance estimate of 10,521 (95% CI: 3700−24,600) in 1995 and 14,662 (95% CI: 9400−29,900) in 2001, mainly to the east and north of Iceland but also to the west (Paxton et al., 2009). In 2007 the TNASS survey in the region of Iceland and the Faroes estimated total abundance, including all species certainty categories, uncorrected for availability bias, at 10,031 (95% CI: 4962−20,278).

FIGURE 4.64 Humpback Whale in northern Iceland. The species occurs in greatest numbers around Iceland, Bear Island, and in the Barents Sea. *Photo: Mike J Tetley.*

This is 46% lower than the equivalent corrected estimate for 2007. As in previous surveys, humpbacks were most commonly sighted to the north and northwest of Iceland (NAMMCO, 2018).

In the Barents and Norwegian Seas, the Norwegian survey estimate was 1059 (CV 0.25) in 1995, and 1450 (CV 0.29) in 1996–2001 (Øien, 2009). For the period 2002 to 2007, and 2015, a total estimate from Norwegian surveys was 7352 (CV 0.31), uncorrected for perception and availability bias (NAMMCO, 2018). Humpbacks during those surveys were found mainly around Bear Island, with two additional concentrations in the northern Barents Sea and north and east of Iceland. The NASS/Norwegian surveys, as yet, have not been corrected for availability bias nor some for perception bias so they cannot provide reliable trends, but they do give some idea of the approximate numbers in this region.

During SCANS and CODA surveys the numbers observed in the ASCOBANS Agreement Area were too low to derive abundance estimates, as was also the case for the ObSERVE surveys of the Irish EEZ.

Threats to humpbacks include entanglement in fishing gear, ship strikes and disturbance from underwater noise.

References

Berrow, S., Whooley, P., O'Connell, M., & Wall, D. (2010). *Irish Cetacean Review (2000–2009)*. Kilrush: Irish Whale and Dolphin Group, 60 pp.

Brown, S. G. (1976). Modern whaling in Britain and the north-east Atlantic Ocean. *Mammal Review, 6*(1), 25–36.

Camphuysen, C. J., & Peet, G. (2006). *Whales and dolphins of the North Sea*. Kortenhoef: Fontaine Uitgevers, 160 pp.

Charif, R. A., Clapham, P. J., & Clark, C. W. (2001). Acoustic detections of singing humpback whales in deep waters off the British Isles. *Marine Mammal Science, 17*, 751–768.

Clapham, P., & Evans, P. G. H. (2008). Humpback whale *Megaptera novaeangliae*. In S. Harris, & D. W. Yalden (Eds.), *Mammals of the British Isles* (4th ed., pp. 663–665). Southampton: The Mammal Society, 800 pp, Handbook.

Clark, C. W., & Charif, R. A. (1998). *Acoustic monitoring of large whales to the west of Britain and Ireland using bottom-mounted hydrophone arrays, October 1996 - September 1997* (pp. 1–25). Joint Nature Conservation Committee Report No. 281.

Evans, P. G. H., Anderwald, P., & Baines, M. E. (2003). UK Cetacean Status Review. *Report to English Nature and the Countryside Council for Wales*. Oxford: Sea Watch Foundation, 160 pp.

Hazevoet, C. J., & Wenzel, F. W. (2000). Whales and dolphins (Mammalia, Cetacea) of the Cape Verde Islands, with special reference to the Humpback Whale *Megaptera novaeangliae* (Borowski, 1781). *Contributions to Zoology, 69*, 197–211.

Jann, B., Allen, J., Carrillo, M., Hanquet, S., Katona, S. K., Martin, R. R., ... Wenzel, F. W. (2003). Migration of a humpback whale (*Megaptera novaeangliae*) between the Cape Verde Islands and Iceland. *Journal of Cetacean Research and Management, 5*(2), 125–129.

Leopold, M. F., Rotshuizen, E., & Evans, P. G. H. (2018). From nought to 100 in no time: How humpback whales (*Megaptera novaeangliae*) came into the southern North Sea. *Lutra, 61*, 165–188.

Martin, A. R., Katona, S. K., Mattila, D., Hembree, D., & Waters, T. D. (1984). Migration of humpback whales between the Caribbean and Iceland. *Journal of Mammalogy, 65*, 330–333.

NAMMCO (2018) *Report of the NAMMCO Scientific Working Group on Abundance Estimates.* Available at https://nammco.no/topics/sc-working-group-reports/.

Øien, N. (2009). Distribution and abundance of large whales in Norwegian and adjacent waters based on ship surveys 1995-2001 (pp. 31–47). In C. Lockyer, & D. Pike (Eds.), *North Atlantic sightings surveys. Counting whales in the North Atlantic 1987-2001* (7). NAMMCO Scientific Publications, 244 pp.

Paxton, C. G. M., Burt, M. L., Hedley, S. L., Vikingsson, G., Gunnlaugsson, Th, & Desportes, G. (2009). Density surface fitting to estimate the abundance of humpback whales based on the NASS-95 and NASS-2001 aerial and shipboard surveys (pp. 143–159). In C. Lockyer, & D. Pike (Eds.), *North Atlantic sightings surveys. Counting whales in the North Atlantic 1987-2001* (7). NAMMCO Scientific Publications, 244 pp.

Reiner, F., dos Santos, M. E., & Wenzel, F. W. (1996). Cetaceans of the Cape Verde archipelago. *Marine Mammal Science, 12,* 434–443.

Ryan, C., Wenzel, F. W., Súarez, P. L., & Berrow, S. D. (2014). An abundance estimate for humpback whales, *Megaptera novaeangliae* breeding around Boa Vista, Cape Verde Islands. *Zoologia Caboverdiana, 5*(1), 20–28.

Ryan, C., Whooley, P., Berrow, S. D., Barnes, C., Massett, N., Strietman, W. J., ... Schmidt, C. (2016). A longitudinal study of humpback whales in Irish waters. *Journal of the Marine Biological Association of the United Kingdom, 96*(4), 877–883.

Smith, T. D., Allen, J., Clapham, P. J., Hammond, P. S., Katona, S., Larsen, F., ... Øien, N. (1999). An ocean-basin-wide mark-recapture study of the North Atlantic humpback whale *(Megaptera novaeangliae). Marine Mammal Science, 15,* 1–32.

Stevick, P. T., Allen, J., Clapham, P. J., Friday, N., Katona, S. K., Larsen, F., ... Hammond, P. S. (2003). North Atlantic humpback whale abundance and rate of increase four decades after protection from whaling. *Marine Ecology Progress Series, 258,* 263–273.

Stevick, P. T., Allen, J., Clapham, P. J., Friday, N., Katona, S. K., Larsen, F., ... Hammond, P. S. (2006). Population spatial structuring on the feeding grounds in North Atlantic humpback whales (*Megaptera novaeangliae*). *Journal of Zoology, London, 270,* 244–255.

Stevick, P. T., Øien, N., & Mattila, D. K. (1998). Migration of a humpback whale (*Megaptera novaeangliae*) between Norway and the West Indies. *Marine Mammal Science, 14,* 162–165.

Thompson, D. A. W. (1928). On Whales Landed at the Scottish Whaling Stations during the Years 1908-1914 and 1920-1927. *Scientific Investigations Fishery Board of Scotland, 3,* 1–40.

Common minke whale *Balaenoptera acutorostrata*

The minke whale (Fig. 4.65), at 6.5–8.5 m length, is the smallest and commonest of the baleen whales inhabiting the cold temperate shelf seas of the North Atlantic. The head is more pointed and triangular than other rorquals (Fig. 4.66), the body slimmer, with a relatively tall, recurved dorsal fin, nearly two-thirds along the back. Only under particular atmospheric conditions is the blow seen clearly, occurring almost simultaneously with the appearance of the fin. There is a single prominent ridge along the middle of the top of the head forward of the blowhole. Occasionally there is a light chevron on the back behind the head, as in the fin whale. Otherwise the head, back and upper flanks are dark grey to

FIGURE 4.65 Minke Whales can enlarge their throat grooves to take in gulpfuls of small fish or plankton. *Photo: Pia Anderwald.*

black, but with areas of light grey often on the flanks, one just above and behind the flippers and the other in front of and below the dorsal fin. The slim, pointed flippers each have a distinctive transverse white band, which sometimes extends over almost the entire upper surface (Fig. 4.66).

The species has a cosmopolitan distribution from the tropics to the ice edge in both hemispheres, though it is more uncommon in equatorial waters. It is widespread along the Atlantic seaboard of Europe from Norway south to the southern tip of Portugal, as well as in the North Sea, although abundance is greatest in the north (Hammond et al., 2009, 2013; Reid et al., 2003 see Fig. 4.67). Around the British Isles and Ireland the highest numbers occur off the north and west coasts of Scotland and the Hebrides, the west and south coasts of Ireland, the central part of the Irish

FIGURE 4.66 The Minke Whale has a pointed, triangular head and slim, pointed flippers each with a distinctive transverse white band. *Photo: SWF Photo Library.*

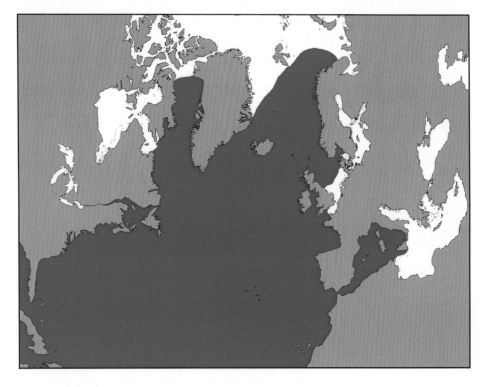

FIGURE 4.67 North Atlantic distribution of minke whale (depicting those areas where the species is thought to regularly occur).

Sea including the Celtic Deep, and in the northern and central North Sea including around the Dogger Bank; it is rare in the southernmost North Sea and eastern half of the English Channel (Anderwald et al., 2011; Baines & Evans, 2012; Berrow et al., 2010; Camphuysen & Peet, 2006; De Boer, 2010; Evans et al., 2003; Hammond et al., 2013, 2017; Reid et al., 2003; Robinson, Tetley, & Mitchelson-Jacob, 2009). In the western English Channel south to southwest Portugal, it is present but uncommon out to the edge of the continental shelf, but is largely absent from the deeper parts of the Bay of Biscay (Evans et al., 2003; Hammond et al., 2009, 2013, 2017; Reid et al., 2003). The species is a casual visitor to inner Danish waters, the Baltic and Mediterranean (mainly the western part) (Kinze et al., 2010; Notarbartolo di Sciara & Birkun, 2010).

Minke whales from high latitudes are thought to migrate southwards to winter in lower latitudes (Risch et al., 2014); at midlatitudes, however, such as around the British Isles and Ireland, at least some have been recorded in every month of the year (Anderwald & Evans, 2007).

There is some genetic evidence for two sympatric stocks existing in the North Atlantic, with overlapping ranges, but otherwise no evidence of population structure (Anderwald et al., 2011). The implication is that minke whales range extensively across the North Atlantic seasonally, but segregate to some extent on at least two breeding grounds (as yet unidentified). Interestingly an animal taken off Jan Mayen in 1996 has been genetically identified as an Antarctic minke whale (*Balaenoptera bonaerensis*), while in 2007 a female taken off the coast of Svalbard was determined to be a result of mating between a *Balaenoptera acutorostrata* father and a *B. bonaerensis* mother (Glover et al., 2010). Until then Antarctic minke whales, distinguishable externally by having no white on its flippers, had only been observed in the Southern Hemisphere.

The only published population estimates for minke whales covering much of the ASCOBANS Agreement Area are from the SCANS, SCANS-II, SCANS-III and CODA surveys, although the Norwegian surveys for minke whales have covered the northern part of the North Sea. In July 1994, the SCANS survey of the North Sea, English Channel and Celtic Sea estimated 9685 individuals (CV 0.23, 95% CI: 6200−15,100) (revised from Hammond et al., 2002). A more extensive line transect survey (SCANS-II) over the Northwest European continental shelf in July 2005 gave an overall estimate of 15,249 (CV 0.31, 95% CI: 8400−27,600) (revised from Hammond et al., 2013). And the offshore CODA survey in 2007 yielded a population estimate of 6765 (CV 0.99, 95% CI: 1300−34,200) (Hammond et al., 2009). This latter estimate has very wide confidence intervals and was uncorrected for animals missed along the track-line, and is therefore negatively biased. In July 2016 over a wider area, an overall estimate of 14,759 (CV 0.33, 95% CI: 7900−27,500) was obtained, with greatest numbers in shelf waters of the central/northern North Sea and west of Britain (Hammond et al., 2017). For the equivalent area (North Sea) surveyed in 1994, 2005 and 2016, there was no evidence for a significant change in numbers (Hammond et al., 2013, 2017). The ObSERVE surveys of the Irish EEZ in summer and winter 2015 and 2016 gave highest abundance estimates in summer: 6680 (CV 0.50, 95% CI:

FIGURE 4.68 The Minke Whale is the commonest baleen whale in the northern North Atlantic. *Photo: Mike J Tetley.*

3600–12,300) in 2015 and 6579 (CV 0.50, 95% CI: 3600–12,100) in 2016, and those were the seasons when densities nearer the coast were greater (Rogan et al., 2018). Densities were highest southwest of Ireland, in the Irish Sea and in an area at the edge of the continental shelf in the Porcupine Seabight. Combined abundance estimates from the SCANS-III & ObSERVE surveys in summer 2016 give 21,158 minke whales (CV 0.27, 95% CI: 12,500–35,800). Note that the survey areas between years are not equivalent.

A population estimate for the entire northeastern North Atlantic (based upon data from 2008 to 2013) gave 90,000 individuals (95% CI: 60,000–130,000) (IWC website: www.iwc.int), with an additional 50,000 (95% CI: 30,000–85,000) in the central North Atlantic (2005–07) (see also Lockyer & Pike, 2009). The TNASS surveys in 2015 estimated 42,515 animals (CV 0.31, 95% CI: 22,896–78,942) in the area around Iceland and the Faroes.

The stock seasonally inhabiting the Norwegian and Barents Seas was estimated at 63,700 individuals (95% CI: 44,000–92,200) during the late 1980s (Schweder et al., 1993), 112,100 (95% CI: 92,200–136,300) in 1995 (Schweder et al., 1997) and c. 80,500 (95% CI: 60,100–107,800) over the years 1996–2001 (Skaug et al., 2004), whilst preliminary estimates from the Norwegian surveys between 2002 and 2016 indicate between 80,000 and 90,000 but with some re-distribution of animals (NAMMCO, 2018). Abundance estimates around Jan Mayen were about 50% higher in 2016 compared with the estimates from the cycles 1996–2001 and 2002–07, and four times that of 2010, whereas in the Norwegian Sea abundance has decreased steadily during the recent survey cycles. In the Svalbard area the abundance in 2014 had decreased to 45% of the level observed in 2008, having peaked that year for the period from the late 1980s (NAMMCO, 2018). Assessing minke whale numbers is particularly challenging. Not only is the species inconspicuous at sea, and may react to survey vessels, but it also seems to respond relatively quickly to changes in prey availability resulting in population range shifts (Anderwald et al., 2012).

Threats to minke whales, besides direct takes, include entanglement in fishing gear, prey depletion, ship strikes and disturbance from underwater noise.

References

Anderwald, P., Daníelsdottir, A. K., Haug, T., Larsen, F., Lesage, V., Reid, R. J., . . . Hoelzel, A. R. (2011). Possible cryptic stock structure for minke whales in the North Atlantic: Implications for conservation and management. *Biological Conservation, 144*, 2479–2489.

Anderwald, P., & Evans, P. G. H. (2007). Minke whale populations in the North-Atlantic – an overview with special reference to UK Waters. In K. P. Robinson, P. T. Stevick, & C. D. MacLeod (Eds.), *An integrated approach to non-lethal research on minke whales in European waters* (Vol. 4, pp. 8–13). European Cetacean Society Special Publication Series.

Anderwald, P., Evans, P. G. H., Dyer, R., Dale, A., Wright, P. J., & Hoelzel, A. R. (2012). Spatial scale and environmental determinants in minke whale habitat use and foraging. *Marine Ecology Progress Series, 450*, 259–274.

Baines, M.E. and Evans, P.G.H. (2012) *Atlas of the Marine Mammals of Wales*. 2nd Edition. CCW Monitoring Report No. 68. 143 pp.

Berrow, S., Whooley, P., O'Connell, M., & Wall, D. (2010). *Irish Cetacean Review (2000–2009)*. Kilrush: Irish Whale and Dolphin Group, 60 pp.

Camphuysen, C. J., & Peet, G. (2006). *Whales and dolphins of the North Sea*. Kortenhoef: Fontaine Uitgevers, 160 pp.

De Boer, M. N. (2010). Spring distribution and density of minke whale *Balaenoptera acutorostrata* and other marine mammals in the Central North Sea. *Marine Ecology Progress Series, 408*, 265–274.

Evans, P. G. H., Anderwald, P., & Baines, M. E. (2003). UK Cetacean Status Review. *Report to English Nature and the Countryside Council for Wales*. Oxford: Sea Watch Foundation, 160 pp.

Glover, K. A., Kanda, N., Haug, T., Pastene, L. A., Øien, N., Goto, M., . . . Skaug, H. J. (2010). Migration of Antarctic minke whales to the Arctic. *PLoS One, 5*(12), e15197. Available from https://doi.org/10.1371/journal.pone.0015197.

Hammond, P. S., Berggren, P., Benke, H., Borchers, D. L., Collet, A., Heide-Jørgensen, M. P., . . . Øien, N. (2002). Abundance of harbour porpoise and other cetaceans in the North Sea and adjacent waters. *Journal of Applied Ecology, 39*, 361–376.

Hammond, P.S., Burt, M.L., Cañadas, A., Castro, R., Certain, G., Gillespie, D., . . . Winship, A. (2009). Cetacean Offshore Distribution and Abundance in the European Atlantic (CODA). Report available from SMRU, Gatty Marine Laboratory, University of St Andrews, St Andrews, Fife KY16 8LB, UK. http://biology.st-andrews.ac.uk/coda/.

Hammond, P.S., Lacey, C., Gilles, A., Viquerat, S., Borjesson, P., Herr, H., . . . and Øien, N. (2017). Estimates of cetacean abundance in European Atlantic waters in summer 2016 from the SCANS-III aerial and shipboard surveys. Available at https://synergy.standrews.ac.uk/scans3/files/2017/05/SCANS-III-design-based-estimates-2017-05-12-final-revised.pdf.

Hammond, P. S., Macleod, K., Berggren, P., Borchers, D. L., Burt, M. L., Cañadas, A., . . . Vázquez, J. A. (2013). Cetacean abundance and distribution in European Atlantic shelf waters to inform conservation and management. *Biological Conservation, 164*, 107–122.

Kinze, C. C., Jensen, T., Tougaard, S., & Baagøe, H. J. (2010). Danske hvalfund i perioden 1998–2007 [Records of cetacean strandings on the Danish coastline during 1998-2007]. *Flora og Fauna, 16*(4), 91–99.

Lockyer, C., & Pike, D. (Eds.), (2009). *North Atlantic sightings surveys. Counting whales in the North Atlantic 1987-2001*. NAMMCO Scientific Publications 7, 244 pp.

NAMMCO (2018) *Report of the NAMMCO Scientific Working Group on Abundance Estimates.* Available at https://nammco.no/topics/sc-working-group-reports/.

Notarbartolo di Sciara, G. and Birkun, A. (2010) *Conserving whales, dolphins and porpoises in the Mediterranean and Black Seas.* An ACCOBAMS Status Report. ACCOBAMS, Monaco. 212 pp.

Reid, J. B., Evans, P. G. H., & Northridge, S. P. (2003). *Atlas of Cetacean Distribution in North-west European Waters.* Peterborough: Joint Nature Conservation Committee, 76 pp.

Risch, D., Manuel Castellote, C., Clark, C. W., Davis, G. E., Dugan, P. J., Hodge, L. E. W., ... Van Parijs, S. M. (2014). Seasonal migrations of North Atlantic minke whales: Novel insights from large-scale passive acoustic monitoring networks. *Movement Ecology*, 2(24). Available from https://doi.org/10.1186/s40462-014-0024-3.

Robinson, K. P., Tetley, M. J., & Mitchelson-Jacob, E. G. (2009). The distribution and habitat preference of coastally occurring minke whales (*Balaenoptera acutorostrata*) in north-east Scotland. *Journal of Coastal Conservation*, 13(1), 39−48.

Rogan, E., Breen, P., Mackey, M., Cañadas, A., Scheidat, M., Geelhoed, S., & Jessopp, M. (2018). *Aerial surveys of cetaceans and seabirds in Irish waters: Occurrence, distribution and abundance in 2015-2017.* Dublin, Ireland: Department of Communications, Climate Action & Environment and National Parks and Wildlife Service (NPWS), Department of Culture, Heritage and the Gaeltacht, 297 pp.

Schweder, T., Øien, N., & Host, G. (1993). Estimates of abundance of the Northeastern Atlantic minke whales in 1989. *Report of the International Whaling Commission*, 43, 323−331.

Schweder, T., Skaug, H. J., Dimakos, X. K., Langas, M., & Øien, N. (1997). Abundance of northeastern Atlantic minke whales, estimates for 1989 and 1995. *Reports of the International Whaling Commission*, 47, 453−483.

Skaug, H. J., Øien, N., Schweder, T., & Bøthun, G. (2004). Abundance of minke whales (*Balaenoptera acutorostrata*) in the Northeast Atlantic: variability in time and space. *Canadian Journal of Fisheries and Aquatic Science*, 61, 870−886.

Sei whale *Balaenoptera borealis*

The sei whale is a large slender rorqual typically 12−17 m in length with a relatively slender head and slightly arched forehead, similar to that of a fin whale but rounder than in the blue whale. A single prominent ridge occurs along the middle of the top of the head, and the tip of the upper jaw tends to be downturned. The head, back and flanks are usually a dark steely black or dark grey colour. The blow of the sei whale resembles that of the fin whale but is lower and less dense, typically rising to a height of 3 m. The dorsal fin tends to show almost simultaneously with the blow, and both remain in view for relatively long periods before a normally shallow dive. The dorsal fin is almost erect and strongly recurved (Fig. 4.69). It is taller than those of other large rorquals, and is placed slightly less than two-thirds along the back and thus further forward than that of the blue or fin whale.

FIGURE 4.69 The Sei Whale is similar in external appearance to the fin whale but has a taller more erect dorsal fin, strongly recurved at the top. *Photo: Caroline R Weir.*

The species has a worldwide distribution, occurring mainly offshore in deep waters from the tropics to the polar seas of both hemispheres, with seasonal latitudinal migrations. In the central North Atlantic summering populations are concentrated in deep waters north of Iceland, with a concentration of animals observed in June just north and southwest of the Charlie Gibbs Fracture Zone (53°N) over the mid-Atlantic ridge (Skov et al., 2008; Waring et al., 2008). It is also seen regularly in small numbers in the Azores and Madeira (Prieto, Janiger, Silva, Waring, & Gonçalves, 2012). In the western North Atlantic the species is reported in the summer from two main locations, the Nova Scotia shelf and Labrador, and in the winter, from Florida, the Gulf of Mexico and the Caribbean, although some in the latter area may actually be misidentified Bryde's whales (Horwood, 1987; Jonsgård & Darling, 1977; Mitchell & Chapman, 1977). In the eastern North Atlantic it is thought to winter off northwest Africa, Spain, Portugal and in the Bay of Biscay, migrating north to summering grounds off Shetland, the Faroes, Norway and Svalbard (Fig. 4.70) (Evans et al., 2003; Horwood, 1987; Prieto et al., 2012).

In the ASCOBANS Agreement Area, sei whales are rarely seen, most records coming from waters >200 m deep (mainly in depths of 500−3000 m), such as west of Norway, between the Faroes and Northern Isles of Scotland and in the Rockall Trough, although the species occurs occasionally in the shelf seas off Shetland, the Hebrides, western and southwest Ireland (Evans et al., 2003; Reid et al., 2003; Weir et al., 2001). A few strandings have occurred in Europe, mainly around the North Sea along the coasts of the United Kingdom (17, from 1913 to 2015), Belgium (1, in 1984) and the Netherlands (4, in 1811, 1972, 1986 and 2005) (Deaville, 2013, 2014; Evans, 2008; Smeenk & Camphuysen, 2016). There are two stranding records from Danish waters (in 1955 and 1980), whilst the type specimen comes from a stranding in 1819 from the German Baltic (Kinze, Thøstesen and Olsen, 2018).

During the operation of whale fisheries in Scotland and western Ireland between 1903 and 1928, sei whales were caught west of Co. Mayo, Ireland mainly in May and June, along the shelf edge near St Kilda mainly in June, and off Shetland mainly in July to August, although catches could occur anytime between April and October; numbers caught were highest north of Shetland (Fairley, 1981; Thompson, 1928). Recent sightings around the United Kingdom and Ireland have been primarily between July and November, particularly August (Berrow et al., 2010; Evans, 1992; Evans et al., 2003). The species is seen regularly in the Bay of Biscay in autumn and winter (Coles, 2001). Whaling activities in the early 20th century caught numbers of sei whales off northwest Spain as well as on the Atlantic side of the Strait of Gibraltar (Sanpera & Aguilar, 1992); there is some uncertainty whether the species was also ever taken in any numbers off central Portugal (Prieto et al., 2012).

FIGURE 4.70 North Atlantic distribution of sei whale (depicting those areas where the species is thought to regularly occur).

FIGURE 4.71 The Sei Whale is generally a pelagic species occurring well beyond the continental shelf. *Photo: Maren Reichelt/Mick E Baines.*

There have been no abundance estimates covering the entire North Atlantic. The most comprehensive survey result published so far has been from the NASS survey in July and August 1989 when 10,300 sei whales (CV 0.27) were counted, mainly in the waters of the central North Atlantic south of Iceland (Cattanach et al., 1993). In July 2007 the CODA survey along the edge of the continental shelf of the ASCOBANS Agreement Area detected sei whales in only one of the four survey blocks, off northwestern Spain, where 366 animals (CV 0.33) were estimated (Hammond et al., 2009).

References

Berrow, S., Whooley, P., O'Connell, M., & Wall, D. (2010). *Irish Cetacean Review (2000–2009)*. Kilrush: Irish Whale and Dolphin Group, 60 pp.

Carlisle, A., Coles, P., Cresswell, G., Diamond, J., Gorman, L., Kinley, C., ... Walker, D. (2001). A report on the whales, dolphins and porpoises of the Bay of Biscay and English Channel, 1999. *Orca Report, 1*, 5–53.

Cattanach, K. L., Sigurjönsson, J., Buckland, S. T., & Gunnlaugsson, Th (1993). Sei whale abundance, estimated from Icelandic and Faroese NASS-87 and NASS-89 data. *Reports of the International Whaling Commission, 43*, 315–321.

Coles, P., Diamond, J., Harrop, H., Macleod, K., & Mitchell, J. (2001). A report on the whales, dolphins and porpoises of the Bay of Biscay and English Channel, 2000. *Orca Report, 2*, 9–61.

Deaville, R. (2013). *UK Cetacean Strandings Investigation Programme* (69 pp). Annual Report to Defra for the period 1st January–31st December 2012 (Contract number MB0111). London: Institute of Zoology.

Deaville, R. (2014). *UK Cetacean Strandings Investigation Programme* (74 pp). Annual Report to Defra for the period 1st January–31st December 2013 (Contract number MB0111). London: Institute of Zoology.

Evans, P. G. H. (1992). *Status Review of Cetaceans in British and Irish waters*. London: UK Dept. of the Environment, 98 pp.

Evans, P. G. H. (2008). Sei whale *Balaenoptera borealis* (pp. 672–675. In S. Harris, & D. W. Yalden (Eds.), *Mammals of the British Isles* (4th ed.). Southampton: The Mammal Society, 800 pp, Handbook.

Evans, P. G. H., Anderwald, P., & Baines, M. E. (2003). UK Cetacean Status Review. *Report to English Nature and the Countryside Council for Wales*. Oxford: Sea Watch Foundation, 160 pp.

Fairley, J. S. (1981). *Irish Whales and Whaling*. Dublin: Blackstaff Press, 218 pp.

Horwood, J. (1987). *The sei whale*. London: Academic Press, 375 pp.

Jonsgård, Å., & Darling, K. (1977). On the biology of the eastern North Atlantic sei whale, *Balaenoptera borealis* Lesson. *Reports of the International Whaling Commission, 1*(Special issue), 121–123.

Kinze, C. C., Thøstesen, C. B., & Olsen, M. T. (2018). Cetacean stranding records along the Danish coastline: Records for the period 2008-2017 and a comparative review. *Lutra, 61*(1), 87–105.

Mitchell, E. D., & Chapman, D. G. (1977). Preliminary assessment of stocks of North Atlantic sei whales (*Balaenoptera borealis*). *Reports of the International Whaling Commission* (Special issue 1), 117–120

Prieto, R., Janiger, D., Silva, M. A., Waring, G. T., & Gonçalves, J. M. (2012). The forgotten whale: A bibliometric analysis and literature review of the North Atlantic sei whale *Balaenoptera borealis*. *Mammal Review, 42*(3), 235–272.

Reid, J. B., Evans, P. G. H., & Northridge, S. P. (2003). *Atlas of Cetacean Distribution in North-west European Waters*. Peterborough: Joint Nature Conservation Committee, 76 pp.

Sanpera, C., & Aguilar, A. (1992). Modern whaling off the Iberian Peninsula during the 20th century. *Reports of the International Whaling Commission, 42*, 723–730.

Skov, H., Gunnlaugsson, T., Budgell, W. P., Horne, J., Nottestad, L., Olsen, E., ... Waring, G. (2008). Small-scale spatial variability of sperm and sei whales in relation to oceanographic and topographic features along the Mid-Atlantic Ridge. *Deep-Sea Research Part II, 55*, 254–268.

Smeenk, C., & Camphuysen, C. J. (2016). Noordse vinvis *Balaenoptera borealis*. In S. Broekhuizen, K. Spoelstra, J. B. M. Thissen, K. J. Canters, & J. C. Buys (Eds.), *Atlas van de Nederlandse Zoogdieren* (pp. 308–309). Leiden: Natuur van Nederland 12. Naturalis Biodiversity Center, 432 pp.

Thompson, D. 'A. W. (1928). On Whales Landed at the Scottish Whaling Stations during the Years 1908-1914 and 1920-1927. *Scientific Investigations Fishery Board of Scotland, 3*, 1–40, 1928.

Waring, G. T., Nøttestad, L., Olsen, E., Skov, E., & Vikingsson, G. (2008). Distribution and density estimates of cetaceans along the mid-Atlantic Ridge during summer 2004. *Journal of Cetacean Research and Management, 10*, 137–146.

Weir, C. R., Pollock, C., Cronin, C., & Taylor, S. (2001). Cetaceans of the Atlantic Frontier, north and west of Scotland. *Continental Shelf Research, 21*, 1047–1071.

Bryde's whale *Balaenoptera brydei*

The close similarity between Bryde's whale and the sei whale has long made it difficult to separate the two species at sea. The main distinguishing feature is the fact that most Bryde's whales have three prominent ridges on the top of the head (Fig. 4.72), whereas the sei whale, and other rorquals, usually only have one. However, these ridges are not always well pronounced so that sightings of the species outside its apparent normal range could readily be overlooked.

The body length of Bryde's whale overlaps that of the sei whale, with adult males up to 15 m and females up to 16.5 m in length. Like that species Bryde's whale has a pointed head, slender streamlined body and a tall fairly erect recurved dorsal fin two-thirds along the back. The blow is often poorly visible above the surface unlike for the sei whale. The body is dark grey above and light grey, sometimes with a pinkish tinge, on the belly and lower flanks.

Bryde's whale has a worldwide distribution mainly in tropical and subtropical seas, and is rarely reported at latitudes higher than 40 degrees. In the North Atlantic, the species occurs in the Caribbean, the Cape Verde Islands and around the Canaries. There is a single extralimital record from the ASCOBANS Agreement Area, with a stranded animal at Kyndyby, Denmark in September 2000 (Kinze, Jensen, Tougaard and Baagøe, 2010). However, as noted above, the species may occasionally occur in the region and be overlooked.

FIGURE 4.72 The Bryde's Whale may be distinguished from the closely similar sei whale by having three prominent ridges on the top of the head. *Photo: Mark Somerville.*

Reference

Kinze, C. C., Jensen, T., Tougaard, S., & Baagøe, H. J. (2010). Danske hvalfund i perioden 1998-2007 [Records of cetacean strandings on the Danish coastline during 1998-2007]. *Flora og Fauna*, *16*(4), 91–99.

Fin whale *Balaenoptera physalus*

The fin whale is the second largest of all mammals with adults reaching a length of 17.5−24 m. Like all rorquals it has a slender streamlined body. The head resembles that of the blue whale, but is narrower and more V-shaped. Like the blue whale there is a single prominent ridge along the middle of the top of the head, but it is not quite so flat. There is a characteristic extension of white, which is sometimes yellowish-tinged on the right side to include the front baleen plates, mouth cavity and lower lip on that side (Fig. 4.73). Sometimes a light grey streak or chevron extends from here onto the top of the neck. Otherwise the head, back and flanks are uniformly dark grey or brown. When a fin whale surfaces its blow resembles an elongated inverted cone and rises to a height of 4−6 m, then followed by a long shallow roll. The dorsal fin is variable in shape. Though small, it is taller than the blue whale's fin, much more obvious and backward-pointing and placed two-thirds along the back (Fig. 4.74). There is a prominent ridge along the tail-stock between the dorsal fin and the flukes. The flippers are long and tapered.

The fin whale has a worldwide distribution, mainly in temperate and polar seas of both hemispheres. It is uncommon but widely distributed in deep waters of the North Atlantic from Baffin Bay, Iceland and Norway south to the Gulf of Mexico and Greater Antilles in the west and the Iberian Peninsula and Mediterranean (where a separate resident population exists) in the east (Fig. 4.75). The species sometimes enters the North Sea and occasionally inner Danish waters, with three strandings even in the German and Polish Baltic (Camphuysen & Peet, 2006; Camphuysen & Smeenk, 2016; Evans et al., 2003; Jensen & Kinze, 2010; Kinze, Thøstesen, & Olsen, 2018).

Although fin whales may show seasonal latitudinal migration, remaining in polar seas only during summer, those which are further south at the latitude of the British Isles and Ireland appear to be present year-round (Berrow et al., 2010; Clark & Charif, 1998; Evans et al., 2003; Wall et al., 2013), whilst acoustic detections indicate that at least male North Atlantic fin whales are present year-round throughout most of the range (Morano et al., 2012).

The species is most commonly recorded in deep waters (400−2000 m depth) off the edge of the continental shelf. However, it is much more likely to occur over the continental slope than blue or sei whale, and in some localities (e.g.

FIGURE 4.73 The Fin Whale has a characteristic white area over the right lower jaw including the lower lip, mouth cavity and front baleen plates. *Photo: Mark Somerville.*

FIGURE 4.74 Although variable in size and shape, the dorsal fin of the Fin Whale is taller than the blue whale's fin, much more obvious and backward-pointing. *Photo: Peter GH Evans.*

FIGURE 4.75 North Atlantic distribution of fin whale (depicting those areas where the species is thought to regularly occur).

southwest Ireland and St George's Channel) it occurs in areas of less than 200 m depth. It appears to favour localities with high topographic variation — underwater sills or ledges, upwellings and frontal zones between mixed and stratified waters with high zooplankton concentrations (Evans, 1990; Ingram, Walshe, Johnston, & Rogan, 2007; Panigada et al., 2008; Skern-Mauritzen, Skaug, & Øien, 2009).

The fin whale is the commonest large whale in the eastern North Atlantic, Bay of Biscay and Mediterranean (Fig. 4.76). The July 2007 CODA survey along the shelf edge of the ASCOBANS Agreement Area obtained a model-based abundance estimate of 9019 fin whales (CV 0.11, 95% CI: 7300–11,200), with the greatest numbers in the

FIGURE 4.76 The Fin Whale is the commonest large whale in Europe, occurring particularly along the continental slope. *Photo: Kirsten Thompson.*

Bay of Biscay (Hammond et al., 2009). The 2016 SCANS-III survey, covering a wider area, yielded an overall abundance estimate of 18,142 fin whales (CV 0.32, 95% CI: 9800−33,600), with the main concentration again in the Bay of Biscay (Hammond et al., 2017). The ObSERVE surveys of the Irish EEZ in the summer and winter of 2015 and 2016 gave a maximum overall abundance estimate of 781 (CV 0.39, 95% CI: 374−1628) fin whales in winter 2016−17, concentrated along the shelf edge (Rogan et al., 2018). These estimates were not corrected for g(0). The estimate for summer 2016 was 95 fin whales (CV 0.5, 95% CI: 26−342). Combining SCANS-III and ObSERVE survey results for summer 2016 yielded an overall estimate of 18,240 fin whales (CV 0.32, 95% CI: 9900−33,700). Further south high densities were predicted in front of the Galician coasts of northwest Spain (3206 animals) in 2007, in depths of between 1000 and 3000 m (Hammond et al., 2009; ICES, 2014). The MARPRO Project also found high densities in the Algarve region between Sagres and Albufera (Santos et al., 2012). Areas of local importance around the British Isles and Ireland include the Faroe−Shetland Channel, Rockall Trough, Porcupine Bank and the south coast of Ireland east into the St George's Channel (Berrow et al., 2010; Evans et al., 2003; Wall et al., 2013).

Further north NASS surveys gave abundance estimates of 19,672 (95% CI: 12,100−29,000) in 1995 and 24,887 (95% CI: 18,200−30,200) in 2001 in the central North Atlantic, with the highest densities in the Irminger Sea between Iceland and Greenland (Vikingsson et al., 2009). The NASS survey in 2015 for the region around Iceland and the Faroes yielded an abundance estimate, uncorrected for availability bias, of 38,931 (CV 0.18, 95% CI: 27,097−55,933) (NAMMCO, 2018)

The equivalent estimates for Norwegian waters were 5034 (CV 0.21) in 1995 and 6409 (CV 0.18) in 1996−2001 (Øien, 2009). Between 2002 and 2007 and in 2015, fin whales in Norwegian waters were most abundant west of Svalbard, in the Barents Sea, and in the Iceland-Jan Mayen survey blocks, yielding an overall abundance estimate, uncorrected for perception and availability bias, of 10,102 (CV 0.24) over both time periods (NAMMCO, 2018). The NASS/Norwegian surveys, as yet, have not been corrected for availability bias nor some for perception bias so they cannot provide reliable trends, but they do give some idea of the approximate numbers in this region.

There is some indication of a seasonal movement northwards in the spring and summer, and then southwards in the autumn and winter (Evans et al., 2003). Most sightings in northern Britain occur between June and August, whereas in southern Britain and Ireland, they are between September and February (Berrow et al., 2010; Evans et al., 2003). In the Bay of Biscay 70% of sightings from regular ferry surveys were made north of 45°N in August, whereas in October less than 40% of sightings occurred north of this latitude (Brereton & Williams, 2001). Some seasonal movement through the Strait of Gibraltar has been noted between the Atlantic and Mediterranean, eastwards during November to April and westwards during May to October (Gauffier et al., 2018), although there is also a population in the Mediterranean that appears to be resident (Notarbartolo di Sciara, Castellote, Druon, & Panigada, 2016).

North Atlantic fin whales were hunted extensively during the latter half of the 19th century and first half of the 20th century, leading to marked declines in the population (Clapham, Young, & Brownell, 1999). In the last half century or more, however, there is indication that the species is recovering; the annual increase in the northern North Atlantic is estimated in recent years to be around 4%, and possibly as much as 10% between Iceland and Greenland (Vikingsson et al., 2009).

Besides direct takes in Icelandic waters, the main threats currently facing fin whales appear to be ship strikes and entanglement in fishing gear, although disturbance from underwater noise may also impact individuals.

References

Berrow, S., Whooley, P., O'Connell, M., & Wall, D. (2010). *Irish Cetacean Review (2000−2009)*. Kilrush: Irish Whale and Dolphin Group, 60 pp.

Brereton, T., & Williams, A. D. (2001). A low-cost method to determine bathypelagic and seasonal occupancy of fin whales *Balaenoptera physalus in the Bay of Biscay. European Research on Cetaceans*, 14, 14−18.

Camphuysen, C. J., & Peet, G. (2006). *Whales and dolphins of the North Sea*. Kortenhoef: Fontaine Uitgevers, 160 pp.

Camphuysen, C. J., & Smeenk, C. (2016). Gewone vinvis *Balaenoptera physalus*. In S. Broekhuizen, K. Spoelstra, J. B. M. Thissen, K. J. Canters, & J. C. Buys (Eds.), *Atlas van de Nederlandse Zoogdieren* (pp. 312−314). Leiden: Natuur van Nederland 12. Naturalis Biodiversity Center, 432 pp.

Clapham, P. J., Young, S. B., & Brownell, J. R. L. (1999). Baleen whales: Conservation issues and the status of the most endangered populations. *Mammal Review*, 29, 35−60.

Clark, C.W. and Charif, R.A. (1998) *Acoustic monitoring of large whales to the west of Britain and Ireland using bottom-mounted hydrophone arrays, October 1996 - September 1997*. Joint Nature Conservation Committee Report No. 281: 1−25.

Evans, P. G. H. (1990). European cetaceans and seabirds in an oceanographic context. *Lutra*, 33, 95−125.

Evans, P. G. H., Anderwald, P., & Baines, M. E. (2003). UK Cetacean Status Review. *Report to English Nature and the Countryside Council for Wales*. Oxford: Sea Watch Foundation, 160 pp.

Gauffier, P., Verborgh, P., Giménez, J., Esteban, R., Salazar Sierra, J. M., & de Stephanis, R. (2018). Contemporary migration of fin whales through the Strait of Gibraltar. *Marine Ecology Progress Series*, 588, 215−228. Available from https://doi.org/10.3354/meps12449.

Hammond, P.S., Burt, M.L., Cañadas, A., Castro, R., Certain, G., Gillespie, D., ... Winship, A. (2009). Cetacean Offshore Distribution and Abundance in the European Atlantic (CODA). Report available from SMRU, Gatty Marine Laboratory, University of St Andrews, St Andrews, Fife KY16 8LB, UK. http://biology.st-andrews.ac.uk/coda/.

Hammond, P.S., Lacey, C., Gilles, A., Viquerat, S., Borjesson, P., Herr, H., ... and Øien, N. (2017). Estimates of cetacean abundance in European Atlantic waters in summer 2016 from the SCANS-III aerial and shipboard surveys. Available at https://synergy.standrews.ac.uk/scans3/files/2017/05/SCANS-III-design-based-estimates-2017-05-12-final-revised.pdf.

ICES (2010). Report of the Working Group on Marine Mammal Ecology (WGMME), 12−15 April 2010, Horta, The Azores. ICES CM 2010/ACOM:24. 212 pp.

Ingram, S. N., Walshe, L., Johnston, D., & Rogan, E. (2007). Habitat partitioning and the influence of benthic topography and oceanography on the distribution of fin and minke whales in the Bay of Fundy, Canada. *Journal of the Marine Biological Association, U.K.*, 87, 149−156.

Jensen, T., & Kinze, C. C. (2010). Finnwal- und Buckelwalsichtungen in der Ostsee 2003-2010. *Verhalten in einem "fremden" Gew. sser. Meer und Museum*, 23, 185−198.

Kinze, C. C., Thøstesen, C. B., & Olsen, M. T. (2018). Cetacean stranding records along the Danish coastline: Records for the period 2008-2017 and a comparative review. *Lutra*, 61(1), 87−105.

Morano, J. L., Salisbury, D. P., Rice, A. N., Conklin, K. L., Falk, K. L., & Clark, C. W. (2012). Seasonal and geographical patterns of fin whale song in the western North Atlantic Ocean. *Journal of the Acoustical Society of America*, 132(2), 1207−1212. Available from https://doi.org/10.1121/1.4730890.

NAMMCO (2018) *Report of the NAMMCO Scientific Working Group on Abundance Estimates*. Available at https://nammco.no/topics/sc-working-group-reports/.

Notarbartolo di Sciara, G., Castellote, M., Druon, J.-N., & Panigada, S. (2016). Fin Whales, *Balaenoptera physalus*: At Home in a Changing Mediterranean Sea? *Advances in Marine Biology*, 75, 75−101. Available from https://doi.org/10.1016/bs.amb.2016.08.002.

Øien, N. (2009). Distribution and abundance of large whales in Norwegian and adjacent waters based on ship surveys 1995-2001. In C. Lockyer, & D. Pike (Eds.), *North Atlantic sightings surveys. Counting whales in the North Atlantic 1987-2001* (7, pp. 31−47). NAMMCO Scientific Publications, 244 pp.

Panigada, S., Zanardelli, M., NacKenzie, M., Donovan, C., Mélin, F., & Hammond, P. S. (2008). Modelling habitat preferences for fin whales and striped dolphins in the Pelagos Sanctuary (Western Mediterranean Sea) with physiographic and remote sensing variables. *Remote Sensing of the Environment*, 112, 3400−3412.

Rogan, E., Breen, P., Mackey, M., Cañadas, A., Scheidat, M., Geelhoed, S., & Jessopp, M. (2018). *Aerial surveys of cetaceans and seabirds in Irish waters: Occurrence, distribution and abundance in 2015-2017*. Dublin, Ireland: Department of Communications, Climate Action & Environment and National Parks and Wildlife Service (NPWS), Department of Culture, Heritage and the Gaeltacht, 297 pp.

Skern-Mauritzen, M., Skaug, H. J., & Øien, N. (2009). Line transects, environmental data and GIS: Cetacean distribution, habitat and prey selection along the Barents Sea shelf edge. In C. Lockyer, & D. Pike (Eds.), *North Atlantic sightings surveys. Counting whales in the North Atlantic 1987-2001* (7, pp. 179−199). NAMMCO Scientific Publications, 244 pp.

Vikingsson, G. A., Pike, D. G., Dsportes, G., Øien, N., Gunnlaugsson, Th, & Bloch, D. (2009). Distribution and abundance of fin whales (*Balaenoptera physalus*) in the Northeast and Central Atlantic as inferred from the North Atlantic Sightings Surveys 1987-2001. In C. Lockyer, & D. Pike (Eds.), *North Atlantic sightings surveys. Counting whales in the North Atlantic 1987-2001* (pp. 49−72). NAMMCO Scientific Publications 7, 244 pp.

Wall, D., Murray, C., O'Brien, J., Kavanagh, L., Wilson, C., Ryan, C., . . . Berrow, S. (2013). *Atlas of the distribution and relative abundance of marine mammals in Irish offshore waters 2005-2011*. Irish Whale and Dolphin Group, Merchants Quay, Kilrush, Co. Clare.

Blue whale *Balaenoptera musculus*

At 23−27 m in length the blue whale is the largest mammal to have ever lived. The head is broad, flat and U-shaped, with a single ridge extending from a raised area forward of the blowhole towards the tip of the snout. The head and most of the body is characteristically pale bluish grey, mottled with grey or greyish white. It has a very small dorsal fin that varies in shape from nearly triangular to moderately recurved and is situated distinctly more than two-thirds along the back so that it is seen only just prior to a dive, and sometimes after the blow (Fig. 4.77). The blow itself is tall and slender, rising vertically to a height of 9−12 m. On diving the species lifts its tail at a slight angle, whereas in fin, sei and minke whale the tail is rarely visible above the surface. The tail is broad and triangular in shape, with slender, pointed tips to the flukes and only a slight central notch (Fig. 4.79).

Blue whales are found worldwide in all seas. Although rare after a century of exploitation, the species occurs regularly in deep waters of the North Atlantic, from the Caribbean to the Davis Strait/southern Greenland in the west, and from the Canaries, Cape Verdes and West Africa to Jan Mayen, Svalbard and the Barents Sea in the east (Fig. 4.78). Blue whales are thought to spend the winter in tropical and subtropical seas where they breed, and then migrate to feed during summer months in cold temperate and polar waters. Sightings of the species in the Azores and Canaries during winter and spring (Silva, Prieto, Jonsen, Baumgartner, & Santos, 2013), and at high latitudes off Norway and Iceland between May and September support this, although analyses of sounds recorded from bottom-mounted SOSUS arrays in the mid-Atlantic suggest that some individuals at least remain at high latitudes throughout the winter (Clark & Charif, 1998; Charif & Clark, 2000).

During the NASS surveys conducted in 1987, 1989, 1995 and 2001, blue whales were most commonly sighted in the central North Atlantic off western and southern Iceland, and to a lesser extent northeast of Iceland, whereas they

FIGURE 4.77 The Blue Whale is usually found in deep waters of the North Atlantic beyond the continental shelf. The individual in this image is about to make a deep dive, revealing its broad triangular tail and slender, pointed tips with a slight central notch. *Photo: Peter GH Evans.*

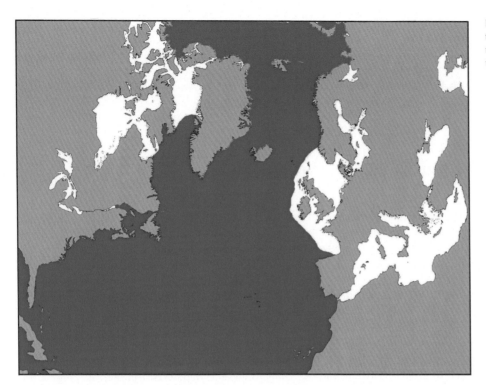

FIGURE 4.78 North Atlantic distribution of blue whale (depicting those areas where the species is thought to regularly occur).

FIGURE 4.79 The Blue Whale has a very small dorsal fin situated distinctly more than two-thirds along the bluishgrey back. *Photo: Peter GH Evans.*

were very rare or absent in the northeast Atlantic, with only 13 sightings in total in Norwegian waters during the four surveys (Pike, Vikingsson, Gunnlaugsson, & Øien, 2009). Total abundance was lowest in 1987 when the estimate was 222 (95% CI: 115−440) and highest in 1995 when 979 whales (95% CI: 137−2542) were estimated. Since then the population has shown signs of increase. The NASS survey around Iceland and the Faroes in 2015 yielded an abundance estimate, uncorrected for availability bias, of 2993 whales (CV 0.40, 95% CI: 1386−6463), concentrated mainly north of Iceland and off the east coast of Greenland (NAMMCO, 2018).

Blue whales are usually found in deep waters of 400−1000 m depth. Whaling in the early 20th century west of the British Isles and Ireland revealed small numbers during summer off the edge of the continental shelf (Brown, 1976; Thompson, 1928). More recently, sightings (Baines, Reichelt, & Griffin, 2017; Evans, 1992; Evans et al., 2003; Reid et al., 2003; Weir et al., 2001) and acoustic monitoring (Clark & Charif, 1998; Charif & Clark, 2000) have revealed small numbers in the deep waters of the Faroe—Shetland Channel and Rockall Trough, south to the Bay of Biscay and the Iberian Peninsula (Aguilar, Grau, Sanpera, & Donovan, 1983; Covelo, Ocio, & Hidalgo, 2017; Hammond et al., 2017; Sanpera, Aguilar, Grau, Jover, & Mizroch, 1984). There have been no strandings within the ASCOBANS Agreement Area since 1923.

The main threat currently facing blue whales appears to be ship strikes although changes in prey availability through climate change might also impact the species.

References

Aguilar, A., Grau, E., Sanpera, C., & Donovan, G. (1983). Report of the 'Ballena 1' whale marking and sighting cruise in the waters off western Spain. *Reports of the International Whaling Commission, 31*, 457–459.

Baines, M., Reichelt, M., & Griffin, D. (2017). An autumn aggregation of fin (*Balaenoptera physalus*) and blue whales (*B. musculus*) in the Porcupine Seabight, southwest of Ireland. *Deep Sea Research Part II: Topical Studies in Oceanography, 141*, 168–177.

Brown, S. G. (1976). Modern whaling in Britain and the north-east Atlantic Ocean. *Mammal Review, 6*, 25–36.

Charif, R. A., & Clark, C. W. (2000). *Acoustic monitoring of large whales off north west Britain and Ireland: A two-year study, October 1996-September 1998.* Aberdeen: JNCC.

Clark, C. W., & Charif, R. A. (1998). *Acoustic monitoring of large whales to the west of Britain and Ireland using bottom-mounted hydrophone arrays, October 1996 - September 1997* (pp. 1–25). Joint Nature Conservation Committee Report No. 281.

Covelo, P., Ocio, G., & Hidalgo, J. (2017). First record of a live blue whale (*Balaenoptera musculus*) in the Iberian Peninsula after three decades. *Galemys, 29*, 34–37.

Evans, P. G. H. (1992). *Status Review of Cetaceans in British and Irish waters.* London: UK Dept. of the Environment, 98 pp.

Evans, P. G. H., Anderwald, P., & Baines, M. E. (2003). UK Cetacean Status Review. *Report to English Nature and the Countryside Council for Wales.* Oxford: Sea Watch Foundation, 160 pp.

Hammond, P.S., Lacey, C., Gilles, A., Viquerat, S., Borjesson, P., Herr, H., . . . and Øien, N. (2017). Estimates of cetacean abundance in European Atlantic waters in summer 2016 from the SCANS-III aerial and shipboard surveys. Available at https://synergy.standrews.ac.uk/scans3/files/2017/05/SCANS-III-design-based-estimates-2017-05-12-final-revised.pdf.

NAMMCO (2018) *Report of the NAMMCO Scientific Working Group on Abundance Estimates.* Available at https://nammco.no/topics/sc-working-group-reports/.

Pike, D. G., Vikingsson, G. A., Gunnlaugsson, Th, & Øien, N. (2009). A note on the distribution and abundance of blue whales (*Balanoptera musculus*) in the Central and Northeast North Atlantic. In C. Lockyer, & D. Pike (Eds.), *North Atlantic sightings surveys. Counting whales in the North Atlantic 1987-2001* (pp. 19–29). NAMMCO Scientific Publications 7, 244 pp.

Reid, J. B., Evans, P. G. H., & Northridge, S. P. (2003). *Atlas of Cetacean Distribution in North-west European Waters.* Peterborough: Joint Nature Conservation Committee, 76 pp.

Sanpera, C., Aguilar, A., Grau, E., Jover, L., & Mizroch, S. A. (1984). Report of the 'Ballena 2' whale marking and sighting cruise in the Atlantic waters off Spain. *Reports of the International Whaling Commission, 34*, 663–666.

Silva, M. A., Prieto, R., Jonsen, I., Baumgartner, M. F., & Santos, R. S. (2013). North Atlantic blue and fin whales suspend their spring migration to forage in middle latitudes: Building up energy reserves for the journey? *PLoS One, 8*(10), e76507.

Thompson, D. 'A. W. (1928). On Whales Landed at the Scottish Whaling Stations during the Years 1908-1914 and 1920-1927. *Scientific Investigations Fishery Board of Scotland, 3*, 1–40, 1928.

Weir, C. R., Pollock, C., Cronin, C., & Taylor, S. (2001). Cetaceans of the Atlantic Frontier, north and west of Scotland. *Continental Shelf Research, 21*, 1047–1071.

Chapter 5

Conservation threats

Cetaceans in Europe face a wide variety of threats as a result of human activities. Some of these, such as hunting, have taken place for thousands of years, whereas others, such as developments in offshore renewable energy, have existed for no more than a couple of decades. I will briefly review each of these below, with a focus upon the ASCOBANS Agreement Area.

Hunting

Rock carvings in western Norway (e.g. at Alta and Rødoy— see Fig. 5.1) and whale bones in the kitchen middens of Neolithic settlements at sites such as Jarlshof in the Shetland Isles and Skara Brae in Orkney are testament to the fact that humans were exploiting whales at least 6000 years ago (and probably for a lot longer). Many of these were likely to have been live strandings or freshly dead carcasses taken advantage of by local coastal communities, but rock carvings showing people pursuing whales indicate that there was active hunting in those times. Artefacts such as axe heads suggest that hunting was by hand harpoon from open wooden boats.

The coastal slow-moving whale species such as the North Atlantic right whale (Fig. 5.2) and the Atlantic gray whale were probably the main targets, and centuries of hunting eventually wiped out gray whale populations in the 17th century, and reduced right whale populations to around 300 animals (now almost entirely confined to the western North Atlantic). Whaling was concentrated in the southern North Sea, English Channel and Bay of Biscay (the Basque fishery operating definitely from the 11th century onwards, and possibly as early as the 7th century).

Smaller cetaceans were also hunted: in medieval times; there was an international trade in harbour porpoises along the North Sea coasts of Europe eastwards into the inner Baltic Sea (where porpoises were hunted possibly as far back as the Stone Age and with written records going back at least to 1357); and from Atlantic islands such as the Faroes, Shetland, Orkney and the Outer Hebrides, there were organised drive fisheries particularly targeting long-finned pilot whales. In the Faroe Islands the latter have taken place for at least 11 centuries and continue to this day. The average yearly catch averaged 850 from 1709 to 1992 and 708 from 1993 to 2015 (Zachariassen, 1993, Faroese Government statistics; Figs. 5.3 and 5.4).

The northern bottlenose whale has also been the subject of exploitation in the eastern North Atlantic, mainly by Norway, with two main periods: between 1882 and 2014 when approximately 50,000 were taken, and between 1955 and 1972 when 5000 were taken (Holt, 1977; Jonsgård, 1977).

Commercial whaling has tended to concentrate upon the larger whales. During the late 18th and early 19th century, the increased demand for lamp oil resulted in the spread of whaling from sailing vessels for sperm whales. These ranged across oceans although local fisheries started in the Atlantic islands such as Madeira and the Azores. These continued right up until 1989.

The hunting of baleen whales, such as blue, fin, sei and humpback whale, only started in the mid-19th century when steam power started to replace sail, vessels could travel faster, and whales could be killed using deck-mounted harpoon guns (Tønnesen & Johnsen 1982). Most such whaling in the eastern North Atlantic occurred around Iceland, Norway, the Faroe Islands), north and west Scotland, western Ireland and in the Bay of Biscay. Whale stocks were clearly becoming depleted by the middle of the 20th century, leading to widespread public concern and an eventual moratorium on commercial whaling in 1986. Besides the direct mortality from these whaling activities, it is likely that being the target of hunts imposed great stress upon individual whales. In this context, of particular interest is a recent worldwide study of cortisol levels in ear plugs of baleen whales collected during the 20th century, indicating that levels of this stress hormone are correlated both spatially and temporally at least in part with whaling activity (Trumble et al., 2018).

With the decline in the stocks of great whales, from the 1940s onwards hunting turned increasingly to the smaller minke whale (Figs. 5.5 and 5.6). Despite the moratorium, Norway has continued taking minkes, with approximately

European Whales, Dolphins, and Porpoises. DOI: https://doi.org/10.1016/B978-0-12-819053-1.00005-3

FIGURE 5.1 Rock carvings at Alta, arctic Norway, indicate that marine mammals were exploited by humans several thousand years ago: (A) harbour porpoise; (B) brown bear and narwhal. *Photos: Peter GH Evans.*

FIGURE 5.2 The North Atlantic right whale is now an endangered species, the European population being all but wiped out after centuries of hunting involving Basque whalers during the Middle Ages. *Photo: New England Aquarium, courtesy of Scott Kraus.*

FIGURE 5.3 For more than a thousand years, the Faroese have operated a drive fishery (referred to locally as the grind) targeting mainly long-finned pilot whales: (A) boats herding pilot whales; (B) whales cut up for local consumption by the island communities. *Photos: Faroese Museum of Natural History, courtesy of Bjarni Mikkelsen.*

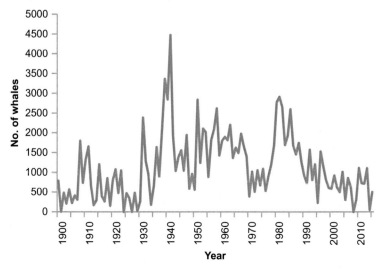

FIGURE 5.4 Graph of pilot whale catches in the Faroe Islands.

24,300 animals killed since 1978 (between 600 and 800 annually since 1998) (IWC statistics). However, it is not thought that these catch levels are causing minke whale populations to decline in Europe.

Reduction of prey from fishing

Over the last half-century, there have been major changes in the sizes of the stocks of many commercial fish species, due to a combination of fishing pressure (Fig. 5.7) and changes in the environment. Many fish stocks are being exploited at levels that are unsustainable. In 2010 the Oslo and Paris Convention (OSPAR), taking advice from the International Council for Exploration of the Sea (ICES), estimated that more than 80% of all 130 commercial fish

FIGURE 5.5 Minke whale with harpoon embedded, North Sea. *Photo: Sea Watch Foundation Photo Library.*

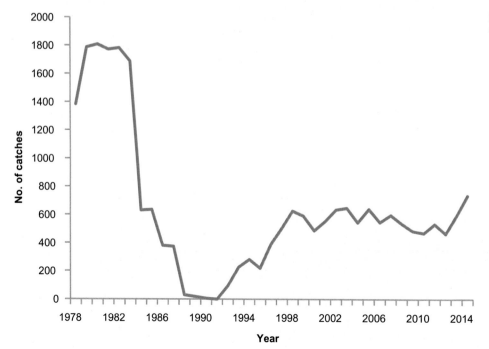

FIGURE 5.6 Graph of minke whale catches, Norway.

FIGURE 5.7 Fishing vessels packed Stornoway harbour at the height of the pelagic fisheries for species such as herring and mackerel. *Photo: Peter GH Evans.*

stocks assessed were overfished. Several of these are important prey for a variety of cetacean species. Notable examples include herring, sandeel, sprat, mackerel, cod and whiting.

In the region, particularly the North Sea, herring stocks collapsed in the mid-1960s, leading to widespread protection (Fig. 5.8). Since then, they have increased from the mid-1980s, particularly from around 2000 onwards. Whiting abundance has declined since the 1980s and sandeel stocks collapsed throughout much of the North Sea from 2000. Sprat, cod and mackerel stocks have also been much reduced since around 1970, although there has been some recovery since approximately 2010, particularly in sprat and mackerel stocks. Elsewhere, there have been long-term declines of sole and cod in the Irish Sea, and whiting in the Celtic Seas (OSPAR, 2010, 2017; ICES data, 2018). In the Kattegat, cod stocks have declined markedly over the last fifty years (HELCOM, 2013, 2018), whilst between 2001−16, stocks of herring in the Gulf of Riga (Baltic Sea Region) and sprat in the western Baltic have been rated as at unfavourable status (HELCOM, 2018).

We still have little idea what effect overfishing may have upon cetaceans. Although many species appear to have quite catholic diets and therefore may be able to switch from a fish species that has become scarce to another that is more abundant, the fact that so many stocks of several fish species have declined would suggest there is likely to be an overall impact for some. The changes we observe in distribution patterns of particular cetacean species may well reflect changes in the availability of prey resources (Evans, 1990). That could for example account for the southward shift of harbour porpoise in the western North Sea between the 1990s and 2000s, following the decline in species such as sandeel. Although we do not know if there is lower overall energy intake, which then might affect survival or reproduction of the animals, in the United Kingdom there was a general rise in incidence of starvation as a cause of death amongst porpoises examined postmortem between the 1990s and 2000s, although it has remained fairly constant since then (Deaville, 2016; Deaville & Jepson, 2011).

Incidental capture in fishing gear

The most obvious cause of death in cetaceans from a human activity is that of accidental entanglement in fishing gear. It has probably occurred ever since humans started fishing, but became a major concern from the early 1970s. By-caught dolphins, whales and porpoises commonly suffer cuts and abrasions to the skin, often showing on the beak and flippers (Fig. 5.9). This results from the scraping or incision of rope, netting or twine into the skin as the animal tries to escape. Animals that get caught in very fine and loosely set gillnets, twist or writhe, causing the netting to become more tangled and to tighten around them. By-caught dolphins are commonly recorded with broken teeth, beaks

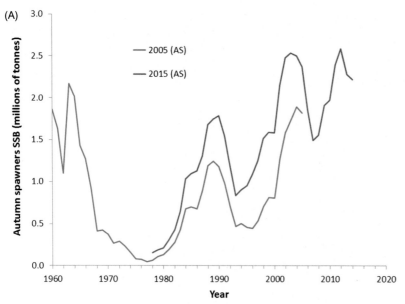

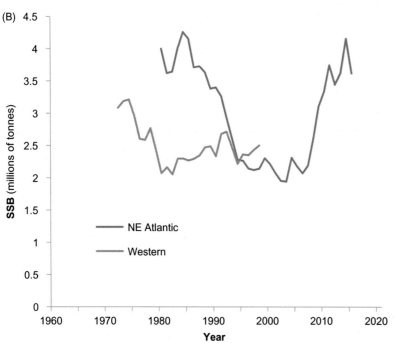

FIGURE 5.8 Herring (A) and mackerel (B) stock trends in the North Sea. *From ICES data.*

or jaws. These injuries are particularly associated with animals that have been caught in trawls, trapped within the net rather than entangled, and in trying to find an escape route they appear to try to push their way through the meshes with such force that fractures occur. They can also suffer extreme internal injuries including bruising, torn muscles, internal haemorrhaging, congestion of organs such as the liver, kidneys and spleen, and punctured and collapsed lungs. The processes that lead to these internal injuries are still a matter of conjecture. However, the muscular tears and haemorrhaging that are frequently recorded are consistent with extreme struggling or thrashing as the animal tries to escape the net, or convulses before death.

Almost any gear can cause entanglement but certain types are known to be more problematic than others. In Europe, five types of fishing gear are particularly identified as having a cetacean by-catch associated with them. These are midwater or pelagic trawls that are towed along either by one or a pair of vessels, static fishing gear such as

FIGURE 5.9 Lacerations over the beak of a common dolphin. *Photo: Marijke de Boer/WDC.*

bottom-set gillnets, driftnets, seine nets and pot lines (Kaschner 2003, Read, Dinker, and Northridge 2006; Reeves, McClellan, and Werner 2013; Dolman, Baulch, Evans, Read, and Ritter 2016).

Trawls (Fig. 5.10a) seem to catch dolphins in particular, such as common and striped dolphin. It is probable that dolphins come into contact with trawls in their pursuit of prey and find themselves engulfed by the gear. During the 1990s the winter pair trawl fishery, targeting bass in the Celtic Sea off southwest England and south of Ireland, was catching the two dolphin species in the low hundreds to low thousands every year (Tregenza, Berrow, Hammond, & Leaper, 1997a), as well as smaller numbers of bottlenose dolphin and long-finned pilot whale. The summer trawl fisheries targeting tuna west of Ireland and in the southern Bay of Biscay have caught mainly common dolphin but also striped dolphin, Atlantic white-sided dolphin and pilot whale. Some bottom trawls are operated with very high opening nets and these can also catch numbers of common dolphins, as has occurred particularly in and around the Bay of Biscay.

Bottom-set gillnets are fixed to the seabed by means of anchors, and are generally used to ensare fish that swim close to the bottom such as cod, turbot, lumpfish, plaice, sole and ray. When very loosely set, these nets are termed tangle nets. They tend to be set in deep water and wrap themselves around the fish. Gillnets may be set over shipwrecks. A further variation of the gillnet is the trammel net which comprises three layers and operates by trapping fish in a pocket of the inner mesh as they swim through from one side of the outer mesh to the other. It is thought that the animals do not notice the nylon mesh in their pursuit of prey, and become entangled. Juveniles seem to be particularly vulnerable, and the main species caught in NW Europe are harbour porpoise and common dolphin. Both species suffer by catch in the hake and pollack fisheries in the Celtic Sea and western English Channel, with the harbour porpoise also in the central and southern North Sea where gillnets targeted cod, hake, turbot, plaice and sole (Fig. 5.10b). Annual by-catches of porpoises in the early 1990s were estimated to average 2200 in the Celtic Sea (Tregenza, Berrow, Hammond, & Leaper, 1997b) (Fig. 5.11) and 8000 in the North Sea (Northridge & Hammond, 1999; Vinther, 1999; Vinther & Larsen, 2004), in both cases levels that were considered unsustainable. In the southernmost North Sea, the number of by-caught porpoises recorded has increased since 2000, possibly reflecting the southward shift in porpoises, with catches highest during late winter and spring. Gillnets are also responsible for by-catches of mainly common dolphin off the coast of Portugal.

Norway has a very large number of commercial small vessels (<15 m length), which operate a variety of gear types in coastal fisheries. These include a gillnet fishery for lumpsucker, a large-mesh net fishery for anglerfish, and in the north off the Lofoten Islands, a gillnet fishery for spawning cod. In the case of the latter two, these are believed to cause an annual by-catch numbering some thousands of porpoises (approximately 21,000 over the 3 years, 2006−08) (Bjørge, Skern-Mauritzen, & Rossman, 2013).

By-catches of porpoises have been reported also from gillnet, trammel net, and pound net fisheries operating in the Skagerrak and Kattegat as well as in the Belt Seas. In the inner Baltic Sea, porpoises have long experienced by-catch

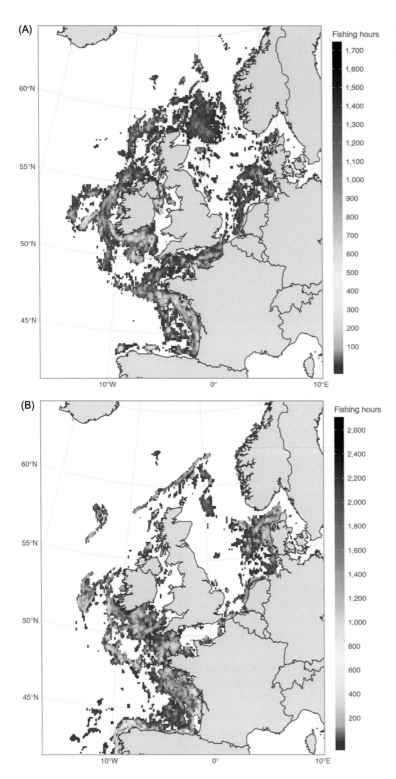

FIGURE 5.10 (A) Distribution of pelagic trawl fishing effort. (B) Distribution of gillnetting fishing effort. *From ICES data.*

from set gillnets, mainly in the western part, in Swedish, German and Polish waters. Some of these are set as semidrift-nets to catch cod or salmonids (as in Polish waters).

During the 1980s the use of large-scale driftnets was established in most oceans of the world, with nets up to 50 km in length regularly deployed in the Pacific. They resulted in very sizeable by-catches in many regions

FIGURE 5.11 Harbour porpoise by-catch was estimated to be causing mortality of 6% of the population in the Celtic Seas during the 1990s. *Photo: Sea Watch Foundation Photo Library.*

(IWC, 1994), including not only cetaceans but also seabirds, turtles, sharks and other nontarget fish species. In western Europe, there were major driftnet fisheries in the eastern North Atlantic for tuna (French and Spanish fisheries) and along the Atlantic coasts of Norway and Ireland for salmon (IWC, 1994). Harbour porpoises were the species mainly caught near-shore, and common, striped and Atlantic white-sided dolphins offshore. The French tuna driftnet fisheries in the Celtic Shelf and Bay of Biscay during 1992−93 had an estimated by-catch of mainly striped dolphins of between 1000 and 2000 per year (Goujon, 1996; Goujon, Antoine, Collet, & Fifas, 1993). The Irish tuna driftnet fishery between 1990 and 2000 was estimated to have killed nearly 12,000 common dolphins (Rogan & Mackey, 2007).

In response to widespread concerns, the United Nations imposed a moratorium on the use of all large-scale driftnets in the open ocean in 1992, and the European Commission responded with a series of resolutions leading to a total ban in European Atlantic waters in 2002.

In the 1980s and 1990s, purse seine nets were set around pods of dolphins associating with tuna in the eastern tropical Pacific and this became a serious conservation problem (Hall, 1998; Hall & Donovan, 2001; Northridge & Hofman, 1999). Fortunately, this particular mode of fishing has now ceased, and the most common form of interaction with purse seining in the North Atlantic is with killer whales taking herring or mackerel from the nets as they are being hauled in (Couperus, 1994).

The setting of pots or traps for fish and crustaceans can inadvertently catch cetaceans, particularly baleen whales such as humpback and minke whale (Lien, 1994; Lien et al., 1995; Northridge et al., 2010; Ryan et al., 2016).

They often become caught in the leader ropes rather than the traps themselves. The problem seems to be greatest in north and west Scotland, involving minke whales and humpbacks (Northridge et al., 2010; Ryan et al., 2016).

Fishing nets and lines that are cut loose and discarded may also entangle cetaceans. These are referred to as ghost netting, and have been known to entangle species ranging from minke whale to harbour porpoise.

ASCOBANS for the best knowledge of the interaction between populations of small cetaceans and fisheries

For conservation of cetacean populations, it is important to improve our knowledge on by-catch of cetaceans, since cetaceans can be caught in various fishing gears. ASCOBANS has therefore put the emphasis on the necessity to increase knowledge on this issue. Those regulations that the EU has taken, to implement the Habitats Directive in fisheries and to improve the assessment of by-catch, are in great part due to ASCOBANS advice. In the years 1995−2002, ASCOBANS pointed out that observers aboard fishing vessels were required to estimate the amount of by-catch of small cetaceans and particularly harbour porpoise, as there were many indications from strandings that fishing was involved as the main cause of death (Fig. 5.12).

FIGURE 5.12   Fishing activities particularly from trawls and set nets are the largest known human-derived source of mortality to small cetaceans in the 21st century. *Photo: Peter GH Evans.*

The knowledge on by-catch in set net fisheries and trawl fisheries in the European Union has increased a great deal during the last 20 years. Every year ASCOBANS has encouraged the Parties to publish their assessments on by-catch of small cetaceans inside their annual report. The impact of pelagic pair trawling (Fig. 5.13) has been investigated by various Parties to ASCOBANS, for example United Kingdom and France, and the numbers of common dolphins dying every year by this type of fishing activity is now estimated with a little more precision. The impact of set net fisheries on harbour porpoise populations is now better known in some areas. ASCOBANS has stimulated and encouraged the scientific observations at sea as it was the only way to assess the amount of by-catch. Whilst some progress has been made on by-catch estimates, the 812/2004 European Regulation contains a lot of gaps, which require urgent improvement. The impact of net fishing on porpoise populations has not been well investigated, as the present regulation does not require observations on vessels setting nets in the areas where the regulation requires that pingers be deployed. This is a dangerous gap as we do not really know if the pingers are being used efficiently by fishermen. The Plan for Conservation of Porpoises in the North Sea, which has been put forward by ASCOBANS, is a way to increase our knowledge of by-catch in all fishing activities including recreational activities. What is the present level of fishing pressure in terms of numbers of by-caught animals? Are they less or greater than 1.7% of the population, a level which is scientifically considered critical for biological conservation? The answer to this question may take some years to become clear. If no fishing has up to now been identified as exceeding this level, the total impact has to be considered. Without waiting more for the answer, mitigation has to be put forward to reduce by-catch as much as possible. Solutions for that have to be strongly encouraged. When working to find solutions it becomes easier to approach the fishing industry for discussions and to persuade them to accept observers on board.

(Continued)

(Continued)

FIGURE 5.13 Atlantic white-sided dolphin accidentally captured in a trawl in the northern North Sea. *Photo: Sea Watch Foundation Photo Library.*

A review of by-catch mitigation measures at a global level and within the EU Common Fisheries Policy was achieved at the joint ASCOBANS/ECS workshop held in March 2010 in Stralsund, Germany. This workshop was very positive and has permitted an understanding of why not all of these mitigation measures had been successful. It was felt that there was a need to improve communication with fishing communities. A guidance framework was suggested for cooperative projects that would bring together fishers, gear technologists and cetacean scientists in order to find solutions for mitigation. If pingers (acoustic alarms, acoustic deterrents) are proven to be scientifically efficient to reduce by-catch, the industry has to improve the technical reliability of the devices and their practical handling for them to be accepted by fishers. However, habitat exclusion and environmental noise are also serious questionable items regarding the possible negative effects of acoustic deterrent devices. These aspects are often highlighted during the annual ASCOBANS Advisory Committee meetings.

During the 25 years of its life, ASCOBANS has successfully made efforts to stimulate our knowledge of by-catch of small cetaceans in northeast Atlantic fisheries in order to measure the level of human pressure on the populations. The North Sea Conservation Plan should also improve our scientific knowledge on by-catch by filling gaps in the present European regulations. Discussions that have started between ASCOBANS and the fishing industry should result in the future in mitigation measures that should be more efficiently applied in the main areas of issue, areas that have to be defined for a better conservation of porpoises and other small cetaceans in accordance with the MSFD and the new CFP and its ecosystem management of fisheries.

Dr Yvon Morizur
Fisheries Advisor and Member of French Delegation, 2009—15

Chemical pollutants and other hazardous substances

We live in a world of synthesised products, which we use in industrial processes, food manufacturing and the production of consumer goods (Fig. 5.14). About 100,000 substances are on the European market and around 30,000 of these are estimated to have an annual production of more than one tonne per year (OSPAR Commission, 2010). Some of these are persistent, and may accumulate in living organisms, contaminating the marine environment, and causing harmful effects on marine life. Thus although only small amounts may occur in seawater, they can build up through the food chain from plankton through fish and squid to top predators like seabirds, seals and cetaceans.

The first major environmental effects were observed in the 1960s following the widespread use of organochlorine pesticides such as DDT and dieldrin, leading to breeding failure and phenomena such as eggshell thinning in top predators such as raptors and fish-eating birds. At the same time, PCBs (or polychlorinated biphenyls to give its full name), a by-product of the plastics industry, entered the environment, resulting in a marked increase in levels of this contaminant during the 1960s and 1970s.

Pollutants enter the body through the diet, and organic compounds made from carbon and hydrogen with at least one covalently bonded atom of an element such as chlorine or bromine, are readily soluble in lipid (fat). It is therefore not surprising that these may accumulate in the lipid-rich blubber of cetaceans. During pregnancy, some of this may be transferred from the mother to the foetus, but the majority is believed to pass to the firstborn calf during the first few weeks of lactation. In this way, the firstborn can have relatively high pollutant levels; in adult females those levels may actually decline with age whereas for males, they progressively increase.

A large number of organochlorine compounds such as DDT and PCBs disrupt hormones and the endocrine system, causing problems in reproduction (see, e.g., Murphy et al., 2015). They may also affect the immune system, and in Britain, strong links have been found between high levels of PCBs in the blubber of porpoises and mortality from infectious disease (Hall et al., 2006; Jepson, Bennett, Allchin et al., 1999; Jepson, Bennett, Deaville et al., 2005). Experimental studies on seals, otters and mink have indicated a threshold of 17 mg/kg of PCB lipid weight resulting in adverse health effects (Kannan, Blankenship, Jones, & Giesy, 2000).

The production of pesticides such as DDT and dieldrin has been completely banned throughout northwest Europe since the 1970s, and for PCBs since the mid-1980s. This has resulted in declines in levels of those contaminants in both

FIGURE 5.14 Milford Haven Oil Refinery, southwest Wales. The coasts of Europe have become increasingly industrialised in the 20th and 21st centuries. *Photo: Peter GH Evans.*

the terrestrial and marine environment. However, the decline in PCBs has been rather slow. It was estimated that in 2000 there remained a total accumulated world production of about two million tonnes, much of it contained in sealed systems. Releases occur from leakages, for example from sealants in buildings, accidental losses and spills, and emissions from PCB-containing materials and soils. EU regulations aimed at the complete phase out of PCBs by 2010, but not all PCBs in smaller applications, in particular in electrical equipment, are likely to have been removed within that period. PCBs emitted and deposited during the years of intensive production and use will remain a diffuse source to the global environment. Evaporation of PCBs from polluted soils and waters has been shown to be a significant source to the atmosphere. Once in the atmosphere, PCBs enter the global circulation and can be transported to remote places, entering the sea also from rivers.

OSPAR for many years has had a Coordinated Environmental Monitoring Programme, and these data show that levels of PCBs both in sediments and fish and shellfish remain unacceptably high for many parts of the British Isles and countries bordering the southern North Sea, Channel and even Bay of Biscay (OSPAR Commission, 2010, 2017; Fig. 5.15). PCB levels in harbour porpoises sampled in the United Kingdom showed a decline in the 1990s but since around 1998 they have remained relatively stable, and mean levels, particularly in males, are still above the 17 mg/kg ΣPCB lipid concentration thought to have adverse effects on health. Furthermore, a comparison of ΣPCB lipid concentrations in three other European species, bottlenose dolphin, striped dolphin and killer whale (Fig. 5.16), showed that they all had levels well above that threshold, and amongst the highest PCB contaminant burdens exhibited by those species anywhere in the world (Fig. 5.17a,b). Many female common dolphins from European seas have also had PCB levels above the threshold thought to cause adverse effects, with evidence suggesting that, although not actually inhibiting ovulation, conception or implantation, PCBs were negatively affecting the survival rates of foetuses or newborns, at least temporarily (Murphy et al., 2018).

Attention has been focused upon PCBs because of their widespread presence at raised levels in the northwest European marine environment and demonstrated negative effects upon cetaceans. However, other persistent organic pollutants (frequently abbreviated to POPs) remain present at unacceptable levels. Concentrations in the North Sea are still widely above background values for mercury, cadmium, lead and polyaromatic hydrocarbons and are rated

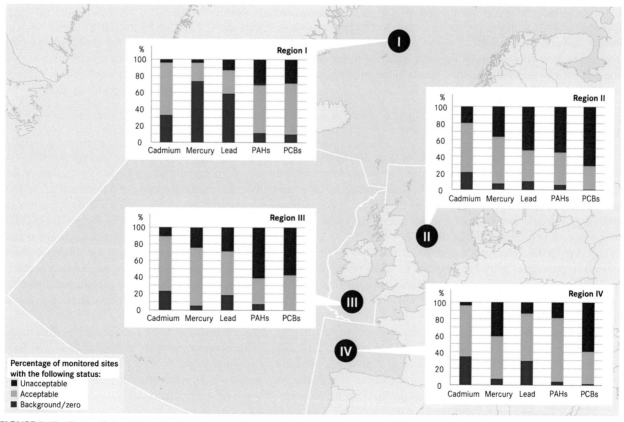

FIGURE 5.15 Status of chemical contamination in the OSPAR Region of western Europe. *OSPAR Quality Status Report 2010.*

FIGURE 5.16 As a top predator feeding upon other marine mammals, the killer whale is a species with disturbingly high concentrations of persistent organic pollutants such as PCBs. *Photo: John Irvine/Sea Watch Foundation.*

unacceptable in many, mostly coastal, areas. Overall, the situation is better for heavy metals, although more than 40% of sites monitored have shown unacceptable levels of lead in the North Sea, and mercury in the Bay of Biscay.

Various trace metals can have toxic effects on cetaceans, causing organ damage. These include methyl mercury, organotins, lead, cadmium, copper and zinc (see recent reviews in Evans 2014, Fossi and Panti 2018). Some such as mercury and cadmium can build up through the food chain. However, it appears that cetaceans are protected from the effects of many heavy metals by detoxification due to the presence of metallothioneins (a family of proteins rich in cysteine and of low molecular weight, found in the membrane of the Golgi apparatus). High levels of cadmium in some species tend to reflect the prey that they feed upon. Squid in particular can assimilate large amounts of cadmium.

Marine litter

In January and February 2016, 29 sperm whales stranded in the southern North Sea (Fig. 5.18a). Although probably the cause was simply that these had wandered into dangerously shallow waters, postmortems highlighted a feature of our modern world, revealing quantities of plastic including a 70 cm cover to a car engine and parts of a bucket (Fig. 5.18b). Our widespread use of plastics has led to the suffering of many marine animals. Sea turtles and seabirds appear to be particularly affected but several cetacean species may also ingest items of marine litter (Baulch & Perry, 2014; CBD, 2012; Fossi, Baini, Panti, & Baulch, 2018; IWC, 2013, 2014; Lusher et al., 2015; OSPAR Commission, 2014; Puig-Lozano et al. 2018). The main culprit has been plastics. Their effects can be by entanglement in plastic sheeting, which can lead to drowning, or by the ingestion of small plastic objects, which can lead to blockages in the stomach or intestines. The species of cetacean most affected appear to be those that feed upon squid and other cephalopods, such as beaked whales. They employ a suction mode of feeding and presumably take up these items when feeding over the seabed.

Most plastics are extremely durable materials and persist in the marine environment for a considerable period, possibly as much as hundreds of years (OSPAR Commission, 2014; Fig. 5.19). However, plastics also deteriorate and fragment in the environment as a consequence of exposure to sunlight in addition to physical and chemical deterioration. This breakdown of larger items results in numerous tiny plastic fragments, which, when smaller than 5 mm are called secondary microplastics. Other microplastics that can be found in the marine environment are categorised as primary microplastics due to the fact that they are produced either for direct use, such as for industrial abrasives or cosmetics, or for indirect use, such as preproduction pellets or nurdles (OSPAR Commission, 2014).

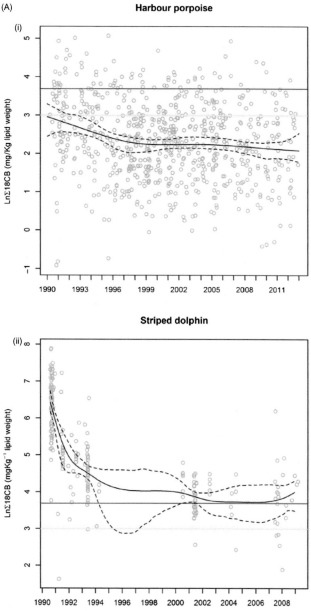

FIGURE 5.17   (A) Temporal trends in ΣPCBs in UK-stranded harbour porpoise and striped dolphins in the western Mediterranean Sea. (i) Ln ΣPCBs (sum 18−25CB) mg/kg lipid concentrations in United Kingdom harbour porpoise blubber against date for all data for 1990−2012 ($n = 706$). The continuous line represents the smoothed trend from a Generalised Additive Model fitted to the data. The trend is statistically significant ($P < .001$, $F = 11.76$, residual df = 701.97, trend df = 3.03) against the null hypothesis of no trend. The dashed lines represent the 95% bootstrapped Confidence Intervals. The yellow line represents ln ΣPCBs equivalent to 20.0 mg/kg lipid and the red line 40 mg/kg lipid. (ii) Ln ΣPCBs (sum 18−25CB) lipid concentrations in biopsied striped dolphin blubber from the Mediterranean Sea against date for all data for 1990−2009 ($n = 220$). Figure shows the natural logs (ln) of the whole data set plotted against the date found. The continuous line represents the smoothed trend from a Generalised Additive Model fitted to the data. The trend is statistically significant ($P < .001$, $F = 55.45$, residual df = 212.03, trend df = 6.97) against the null hypothesis of no trend. The dashed lines represent the 95% bootstrapped Confidence Intervals. The yellow line represents ln ΣPCBs equivalent to 20.0 mg/kg lipid and the red line 40 mg/kg lipid. *Adapted from Jepson, P. D., Deaville, R., Barber, J. L., Aguilar, A., Borrell, A., Murphy, S., … Law, R. J. (2016). PCB pollution continues to impact populations of orcas and other dolphins in European waters.* Nature Scientific Reports 6, 18573. Avaliable form: http://doi:org.10.1038/srep18573 (Jepson et al., 2016).

Microplastic ingestion in cetaceans has been found in fin whales in the Mediterranean (Fossi et al., 2014), in a humpback whale in the southern North Sea (Besseling et al., 2015) and two True's beaked whales in Ireland (Lusher et al., 2015).

Although a variety of European cetacean species on postmortem examination have had fragments of plastic in their stomachs, the percentage has generally been very low. Of 695 cetaceans, 22 sea turtles and two basking sharks

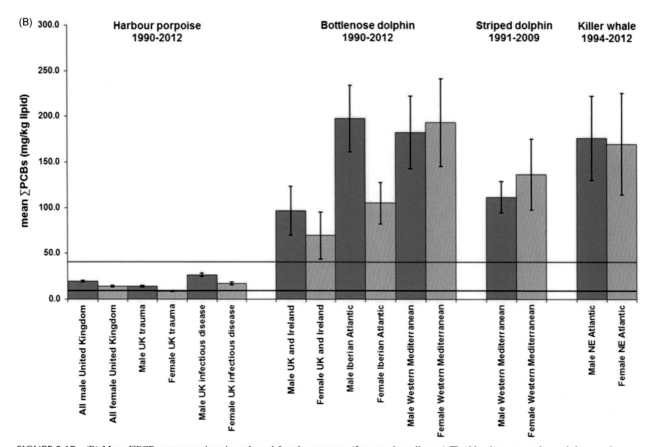

FIGURE 5.17 (B) Mean ΣPCBs concentrations in male and female cetaceans (four species; all ages) The blue bars are males and the grey bars are females. The lower line is the equivalent ΣPCBs concentrations threshold (9.0 mg/kg lipid) for onset of physiological effects in experimental marine mammal studies6. The upper line is the equivalent ΣPCBs concentrations threshold (41.0 mg/kg lipid) for the highest PCB toxicity threshold published for marine mammals based on marked reproductive impairment in ringed seals in the Baltic Sea. Mean ΣPCBs concentrations in male ($n = 388$) and female ($n = 318$) UK-stranded harbour porpoises (HPs) in 1990–2012. Mean blubber ΣPCBs (mg/kg lipid) concentrations in subsets of male ($n = 201$) and female ($n = 144$) UK-stranded HPs that died of acute physical trauma, and male ($n = 120$) and female ($n = 132$) HPs that died of infectious disease from the same 1990–2012 period. Mean blubber ΣPCBs (mg/kg lipid) concentrations (1990–2012) shown for stranded/biopsied male ($n = 29$) and female ($n = 17$) bottlenose dolphins (BNDs) from United Kingdom and Ireland; male ($n = 28$) and female ($n = 24$) BNDs from Atlantic coast of Spain and Portugal, and male ($n = 9$) and female ($n = 11$) BNDs from western Mediterranean Sea. Male ($n = 50$) and female ($n = 39$) striped dolphins from western Mediterranean Sea (1991–2009) and male ($n = 5$) and female ($n = 19$) KWs from NE Atlantic (1994–2012). Error bars = 1 Standard Error of the Mean (SEM). *Adapted from Jepson, P. D., Deaville, R., Barber, J. L., Aguilar, A., Borrell, A., Murphy, S., . . . Law, R. J. (2016). PCB pollution continues to impact populations of orcas and other dolphins in European waters. Nature Scientific Reports 6, 18573. Avaliable form: http://doi:org.10.1038/srep18573 (Jepson et al., 2016).*

examined in the United Kingdom between 2005 and 2010, only 16 cetaceans and four turtles were found with marine litter ingested (and one cetacean entanglement), and for none of the cetaceans was plastic ingestion thought to have been the cause of death. Nevertheless, this is a growing environmental concern for marine wildlife.

Noise disturbance

Living in an aquatic environment where vision, touch, smell and taste have severe limitations in effective range and speed of signal transmission, cetaceans rely heavily upon sound. Different species typically use different frequency bandwidths of sound, which may then overlap with the sounds produced by a variety of human activities. Loud sounds may cause injury, permanent threshold shifts (PTS), temporary threshold shifts (TTS), acoustic masking of communication or behavioural disturbance (Richardson, Greene, Malme, & Thomson, 1995, NRC 2005). Injury may take the form of damage to the hearing organ, haemorrhaging, or gas or fat emboli (Cox et al., 2006; Evans & Miller, 2004; Fernández, Arbelo, et al., 2004; Fernández, Edwards, et al., 2005). Both PTS and TTS represent actual changes in the ability of an animal to hear, usually at a particular frequency, whereby it becomes less sensitive at one or more

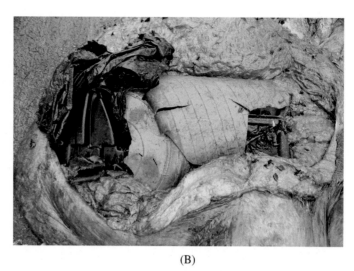

(B)

(A)

FIGURE 5.18 (A) Mass stranding of sperm whales along southern North Sea coasts in January–February 2016. (B) Sperm whale with quantities of plastic including a 70 cm cover of a car engine and parts of a bucket. *(A) Photo: Steve CV Geelhoed, (B). Photo: ITAW-TiHo, courtesy of U Siebert.*

frequencies as a result of exposure to sound (Cook et al., 2006; Finneran et al., 2000, 2005; Nachtigall, Pawloski, & Au, 2003; Nachtigall, Supin, Pawloski, & Au, 2004; Southall et al., 2007). Masking may occur when a sound overlaps with and then 'masks' a desired signal, making the latter more difficult to detect (Clark et al., 2009; Erbe, Reichmuth, Cunningham, Lucke, & Dooling, 2016). Finally, an animal may change its behaviour in response to a sound, such as altering its dive pattern, it might stop feeding or be disturbed from nursing its young, or simply move away from the source of the noise (DeRuiter et al., 2013; Nowacek et al., 2004; Tyack et al., 2011).

In most cases, however, we do not know whether or not the short-term reactions we might observe actually have long-term consequences. Just as we humans may move away from a loud noise, or be temporarily deafened by it, whether this increases the risk of death or reduces the number of offspring for individuals within a population is very difficult to establish. For this reason, we try to take a precautionary approach and aim to minimise potential risk of harm.

Before describing some of the effects it may be useful to explain some terms that one often sees in the literature. Sound has two components: a pressure component and a particle motion component. The former appears to be the most important for cetaceans whereas the latter is important for fish. There are two general sound types: (1) pulses which are brief, broadband, atonal, transient sounds and examples are explosions, seismic airgun pulses and pile driving strikes; these all have a relatively rapid rise from ambient pressure to a maximum followed by a decay period where the pressure may diminish or oscillate, and are the sounds more likely to cause physical injury; and (2) nonpulse sounds (which can be intermittent or continuous) which are tonal, broadband or both, and examples are vessels and machinery operations such as drilling or wind turbines, and many active sonar systems.

Sound pressure levels (SPLs) are given in decibels (dB) and a source level (SL) is the received level measured or estimated 1 m from the source, and related to units of Pascals, generally 1 microPascal (µPa). A peak sound pressure is the maximum absolute value of the instantaneous sound pressure during a specified time interval and is denoted P_{max} in units of Pascals. It is particularly useful for characterising sound pulses. Peak-to-peak sound pressure is the algebraic difference between the maximum positive and maximum negative instantaneous peak pressure. The root-mean-square (RMS) SPL is given as dB re: 1 µPa for underwater sound and dB re: 20 µPa for aerial sound.

The noise generated from human activities can be divided into particular types. The main ones are shipping, seismic, marine construction (including explosions) and active sonar. I shall briefly review each of these.

FIGURE 5.19 Beach at Hells Mouth, Llŷn Peninsula, North Wales. Prevailing currents can result in large quantities of marine litter coming ashore. *Photo: M. Czachur.*

Shipping

The introduction of the mechanical engine in the mid-19th century brought a new sound to the ocean. That sound has steadily increased as the number of vessels increased, along with their size and speed, resulting in ever larger engines. About 90% of world trade (in gross tonnage) depends on ship transport and, apart from declines during global economic downturns, the gross tonnage of goods transported by sea has steadily increased since the early 1970s. The global commercial shipping fleet expanded from about 30,000 vessels (of about 85 million gross metric tons) in 1950 to more than 85,000 vessels (about 525 million gross metric tons) in 1998 (NRC, 2003). Furthermore, a recent altimeter data analysis indicated a fourfold global increase in ship traffic between 1992 and 2012 (Tournadre, 2014).

Large vessels typically have sound SLs of 160−220 dB re 1 μPa @ 1 m over a bandwidth of 2−100 Hz, with peak energy around 25 Hz (NRC, 2003; Richardson et al., 1995), although it can vary between vessel type: bulk carriers may have higher SLs near 100 Hz, while container ship and tanker noise are predominantly below 40 Hz (McKenna, Ross, Wiggins, & Hildebrand, 2012). From 1950 to 2000, the shipping contribution to background sound at some locations increased by as much as 15 dB, corresponding to an average rate of increase of approximately 3 dB per decade (Andrew, Howe, & Mercer, 2011; Andrew, Howe, Mercer, & Dzieciuch, 2002; Chapman & Price, 2011; Frisk, 2012; Hildebrand, 2009). Shipping is probably the greatest single source of human-generated sound in the ocean (OSPAR 2009, Tyack et al., 2015; Fig. 5.20).

FIGURE 5.20 Large vessels such as tankers are probably the greatest source of human-generated noise in our oceans. *Photo: Peter GH Evans.*

The effects of shipping sound are largely unknown but it clearly has the potential to mask communication, particularly amongst baleen whales that vocalise at frequencies that overlap with shipping sound and which may rely upon their vocalisations being heard over great distances (Clark et al., 2009; Erbe et al., 2016). It is possible that species may have developed mechanisms to compensate for masking for example increasing the SL of their sounds when located in a noisy environment (Parks et al., 2011). However, even small cetaceans such as harbour porpoise have been shown to react negatively to shipping (Dyndo, Wisniewska, Rojano-Doñate, & Madsen, 2015; Wisniewska et al., 2018). There is also some evidence that underwater noise from shipping could have long-term effects exhibited through raised levels of stress hormones such as cortisol (Hunt, Rolland, Kraus, & Wasser, 2006; Rolland et al. 2012; Trumble et al., 2018).

Seismic

The second most pervasive source of noise from human activities is that of seismic, used for geophysical exploration for oil and gas (Fig. 5.21), which in Europe started in the North Sea in the late 1950s, and has continued here and in adjacent waters ever since (Fig. 5.22). Seismic surveys produce short duration broadband impulse sounds with peak SLs of approximately 220—255 dB re 1 μPa peak at 1 m (Evans & Nice, 1996; Nowacek, Thorne, Johnston, & Tyack, 2007; Richardson et al., 1995). The sound is directed downwards towards the sea floor with most energy lower than 300 Hz although high frequencies up to at least 15 kHz may also be produced. Several studies have shown negative reactions by baleen whales (e.g. bowhead whales, grey whales, blue whales and humpbacks), including deflections from migration routes (Malme et al. 1983, 1984; Malme & Miles 1985; Richardson & Malme 1993; McCauley, Jenner, Jenner, McCabe, & Murdoch, 1998), avoidance behaviour (Ljungblad, Wursig, Swartz, & Keene, 1988; McCauley et al., 2000; Richardson, Würsig, & Greene, 1986; Stone & Tasker, 2006; Stone, Hall, Mendes, & Tasker, 2017), cessation of feeding (Malme, Würsig, Bird, & Tyack, 1986, 1988; Miller et al., 2009), alterations in vocal behaviour such as calling rates (Blackwell et al., 2015; Castellote, Clark, & Lammers, 2012; Cerchio, Strindberg, Collins, Bennett, & Rosenbaum, 2014; Di Iorio & Clark, 2010), and changes in aerial or surfacing behaviour (Robertson et al., 2013), although there are also studies where behavioural reactions have been relatively small (Cato et al. 2013; Dunlop, Noad, McCauley, Kniest, Paton et al., 2015; Dunlop, Noad, McCauley, Kniest, Slade et al., 2017; Gailey, Würsig, & McDonald, 2007; Yazvenko, McDonald, Blokhin, Johnson, Meier et al., 2007; Yazvenko, McDonald, Blokhin, Johnson, Melton et al., 2007). Responses have varied both between and within species, some of which probably in part reflect the context in which the area is being used at the time (Abgrall, Moulton, & Richardson, 2008; Dunlop et al., 2013; Koski et al 2008, 2009; Miller et al. 2005; Richardson et al., 1995). Negative responses have been at received

FIGURE 5.21 Seismic vessel towing (A) a 2-D airgun array; and (B) a 3-D airgun array. *(Photos: PGS)*

sound levels varying between 130 and 160 dB re 1 μPa_{rms} @ 1 m, equivalent to a range of up to 30 km from the sound source.

Toothed whales and dolphins are often seen near operating airgun arrays. However, in general there seems to be a tendency for most smaller dolphins to show some limited avoidance of operating seismic vessels, in the order of 1 km or less (e.g. Holst, Smultea, Koski, & Haley, 2005; Moulton & Miller, 2005). Some species (e.g. beluga whale) even show long-distance avoidance of seismic vessels, in the order of 10−20 km (Harris, Elliot, & Davis, 2007; Miller et al., 2005). Sperm whales may continue vocalising but change some of their foraging behaviour in the presence of active airguns. They may avoid deep dives but not necessarily make larger changes to their behaviour such as swimming away (Madsen, Møhl, Nielsen, & Wahlberg, 2002; Madsen, Wahlberg, Tougaard, Lucke, & Tyack, 2006; Tyack et al., 2003). Ramping up the sound is often employed so as to give whales a chance to swim away before the sound may cause physical damage to their hearing. However, the fact that animals can remain in the vicinity throughout indicates that it may not be entirely effective (Dunlop et al., 2016; Jochens et al., 2006; Miller et al., 2009).

A recent study in the Moray Firth, northeast Scotland found group responses amongst harbour porpoises to airgun noise from a seismic survey over ranges of 5−10 km, at received peak-to-peak SPLs of 165−172 dB re 1 μPa

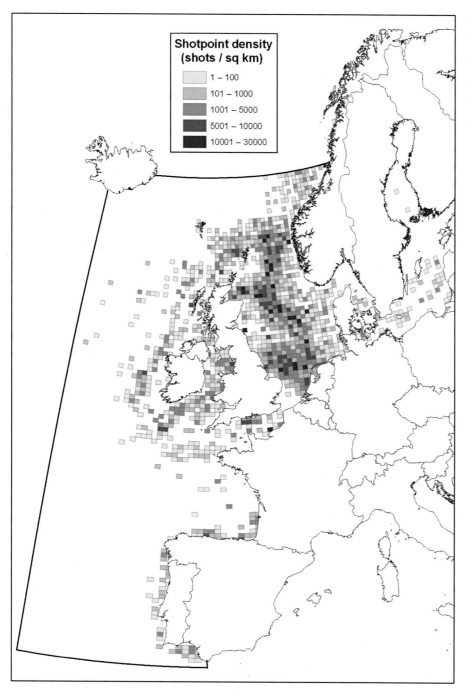

FIGURE 5.22 Map showing shot point densities from all 2-D and 3-D seismic surveys in NW Europe between 1960 and 2009. *Adapted from Evans, P. G. H., Baines, M. E., & Anderwald, P. (2016).* Cetacean stock assessment in Northwest Europe in relation to exploration and production industry sound. *Report to International Association of Oil & Gas Producers Joint Industry Program on Sound & Marine Life, 153 pp.*

(Thompson et al., 2013; Pirotta et al., 2014). However, animals were typically detected again at affected sites within a few hours, and the level of response declined through the 10-day survey. On the other hand, temporary hearing threshold shift has been found in a harbour porpoise after exposure to multiple airgun sounds (Kastelein et al., 2017).

A study of distribution changes of 10 cetacean species in northwest Europe over the last three decades in relation to distribution changes of seismic activity (expressed as shot point densities) found no relationship (Evans, Baines, & Anderwald, 2016). If there had been an effect it was either short-lived, occurring only in the first years of seismic exploration, or its effects were swamped by other factors influencing cetacean distribution patterns.

Marine construction: explosions, dredging and drilling

Marine construction activities, particularly explosions, have the potential to cause physical damage as well as behavioural disturbance (Anderwald et al., 2013; Culloch et al., 2016; Ketten, 1995; Nowacek et al., 2007; OSPAR Commission, 2009; Simmonds & Brown 2010).

Underwater explosions are used not only in construction but also to remove unwanted structures on the seabed, including military ordnance (Fig. 5.23). They are one of the strongest point sources of any man-made sound, starting with an initial shock pulse followed by a succession of oscillating bubble pulses (OSPAR Commission, 2009; Richardson et al., 1995). The SLs vary with the type and the amounts of explosives used, as well as the water depth at which the explosion occurs, and can range from 272 to 287 dB re 1 μPa zero to peak at 1 m distance (1−100 lb. TNT). Frequencies are rather low with most energy between 6−21 Hz and lasting <1−10 ms (NRC, 2003; Richardson et al., 1995). The disposal of unexploded military ordnance dumped during World War II in coastal areas of the southern North Sea has recently caused concern over potential impacts on marine mammals (Koschinski, 2011; Koschinski & Kock, 2015; von Benda-Beckmann et al., 2015).

There have been few studies of the impacts of explosions upon cetaceans, and although there are some poorly documented cases of injury and death of marine mammals which were thought to have been caused by explosions, most studies have detected little in the way of behavioural or acoustic reactions. However, these have generally been at received SPLs of 140−179 dB re 1 μPa$_{rms}$, @ 1 m, and it has been suggested that levels above 179 dB re 1 μPa$_{rms}$ @ 1 m are very likely to cause physical effects.

In July 2011, 70 long-finned pilot whales swam into the Kyle of Durness, a shallow tidal inlet east of Cape Wrath (north Scotland), Europe's largest live bombing range. Despite attempts to herd them back out to sea, 39 were left stranded by the tide, and 19 ultimately died. Three 1,000 lb bombs were detonated in the sea 24 hours before the mass stranding, and were concluded to be the most likely cause of the mass stranding (Brownlow et al., 2015).

Given the comparatively low SLs, injuries from either dredging or drilling operations are unlikely in cetaceans, unless they are very close to the source, but avoidance reactions have been recorded when received levels exceeded around 120 dB re. 1 μPa$_{rms}$ @ 1 m.

Since the first offshore wind farm was erected in Denmark in 1991, the global development of offshore wind energy (Fig. 5.24) has grown exponentially as countries work towards meeting the carbon reduction targets that have been developed to mitigate potential impacts of climate change. In Europe, these have focused upon the North Sea, Irish Sea and Baltic Sea where wind speeds tend to be greatest (Fig. 5.25). In the process of constructing a wind turbine, pile

FIGURE 5.23 A 300 kg bomb being exploded as part of a WWII military ordnance clearance exercise in the southern North Sea. *Photo: Klaus Lucke.*

FIGURE 5.24 Wind farm in the northeastern Irish Sea. *Photo: Peter GH Evans.*

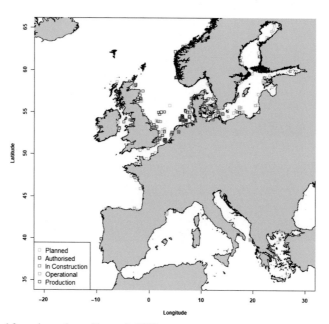

FIGURE 5.25 Map of offshore wind farms in northwest Europe in 2018.

driving is used, often for extended periods (Fig. 5.26). Development proposals for offshore wind farms have generally used large monopile designs with diameters of between 4 and 6 m. This size of driven pile has the potential to give rise to peak-to-peak SLs in excess of 250 dB re 1 μPa @ 1 m at peak frequencies of 100−500 Hz, although > 100 kHz can be produced (Dähne et al. 2013; Nedwell et al., 2008; OSPAR Commission, 2009; Tougaard et al. 2009). Shallow areas and pile driving are often preferred for economic reasons, so that the cetacean species most often likely to coincide at construction sites is the harbour porpoise.

Several studies have examined the effects of pile driving on marine mammals, particularly harbour porpoise (see Evans, 2008; Madsen et al., 2006; Mann & Teilmann, 2013; Teilmann, Tougaard, Carstensen, & Dietz, 2006;

FIGURE 5.26 The pile driving process used in the construction of a wind turbine (Nordsee Ost 1). (A) The jack up platform with the hydraulic hammer hanging from the crane, two monopiles on the platform and two behind the platform, two jacket foundations behind the platform and the orange ship, which is an anchor handler that is also used to lay out the bubble curtain. (B) The jack up platform with the hydraulic hammer hanging from the crane above a jacket foundation, which is installed on the sea floor via monopiles being driven into the ground through this jacket structure. Also visible are other foundations of the wind farm in the background and one still on the platform. *Photo: BioConsult SH, courtesy of Miriam Brandt.*

Thomsen, 2010 for reviews). In most cases they have been negative, with animals moving away from the sound source to distances of approximately 20 km and in the case of the large windfarm, Nysted, in the Danish Baltic, for a period exceeding 10 years. However, responses elsewhere have varied from only short-term effects (e.g. Horns Rev in the Danish North Sea) to no apparent effect (OWEZ in the Dutch North Sea) (Brandt, Diederichs, Betke, & Nehls, 2011; Carstensen et al. 2006; Carstensen et al., 2008; Scheidat et al. 2010; Teilmann & Carstensen, 2012; Tougaard, Carstensen, Bech, & Teilmann, 2006; Tougaard, Carstensen, Wisz, et al., 2006; Tougaard et al., 2009). These differences in responses could relate to a number of factors: differences in the noise exposure levels and duration, the experimental design and monitoring procedures, and perhaps most important of all, how the area is being utilised by the species. However, most studies show that porpoises return after the construction phase (Brandt et al. 2018).

Marine construction activities may have effects beyond noise disturbance. The development of marine renewable energy, particularly tidal turbines, may also pose collision risks to marine mammals and other marine megafauna (Wilson et al. 2007; Benjamins et al. 2015).

Wind farms − threat or sanctuary for porpoises?

As a potential alternative to fossil fuel and nuclear power, offshore renewable energy can be clean, safe and inexhaustible. In light of changing climate and growing energy demands, this is certainly desirable and necessary (Simmonds & Elliot, 2009). However, for these developments to be sustainable, any negative impacts on the marine ecosystem need to be considered as well.

Offshore wind farms (Fig. 5.27) already cover considerable areas of marine habitat in the North Sea and the plans for the future indicate that this is only the beginning. From what is currently known the main impact on small cetaceans occurs during construction when pile driving produces high noise levels (Weilgart, 2007). Displacements of harbour porpoises have been observed for tens of kilometres from the sound source (Tougaard, Carstensen, Teilmann Skov, & Rasmussen, 2009).

FIGURE 5.27 Wind farm in German waters of the North Sea. *Photo: Steve CV Geelhoed.*

FIGURE 5.28 The harbour porpoise is the species known to be most affected by piling activities used in the construction of wind farms. *Photo: Peter GH Evans.*

(Continued)

(Continued)

Particular concerns have been raised as this species could be negatively affected by noise pollution and habitat loss due to wind farms (Gilles, Scheidat, & Siebert, 2009; Fig. 5.28). How porpoises deal with operating wind farms is still an open question. In one case, at the Nysted wind farm in the Baltic Sea, 2 years after construction the animals had not returned to the site in the original numbers (Carstensen, Henriksen, & Teilmann, 2006; Teilmann & Carstensen, 2012). In the wind farm Horns Rev (North Sea, DK) porpoises returned in similar numbers (Teilmann, 2008) and a study done at the wind farm Egmond aan Zee (North Sea, NL) showed an increase in acoustic activity of porpoises in the park compared to outside, indicating an increased use of the park during operation (Scheidat et al., 2011). For all cases studied so far, no long-term follow-up studies have been conducted in operating wind farms. From these examples, it is obvious that each case is unique and that a large range of parameters, such as location and type of the wind farm, as well as density of animals and boat traffic, probably determine if and how their construction and operation will impact cetaceans. In some cases, porpoises might favour wind farms because of an increased prey availability (Lindeboom et al., 2011), or the limited vessel traffic in the park provides relative quiet conditions compared to outside the park (Scheidat et al., 2011). Because of this variability and our limited knowledge on harbour porpoise behaviour, it is impossible to use the results of one study as a generic blueprint for future planning and it is vital to apply a precautionary approach.

Wind farms at sea create a new type of unique habitat, and the increase of hard substrate from the wind pylons as well as the scouring around each mill could increase the local productivity. While at this moment most windfarms are exclusively used for generating power from wind, they also provide a potential location for multiuse by different human stakeholders. Introducing fishing activities, aquaculture and even windfarm tourism are being considered. On the one hand, some of these activities could pose a risk to small cetaceans, for example if gillnetting were to be allowed in the parks. On the other hand, this exploration of a new habitat also provides an opportunity to develop new innovative and economically viable activities that would not pose a threat to the wind farms or the animals occurring therein.

Over the next few decades the development and installation of renewable energy sources in the North Sea will expand. ASCOBANS with its long-standing international forum of experts has an important role to play in this development. ASCOBANS can provide advice to all Member States when it comes to standardising the monitoring methods used, and can work towards a common database of data collected in and around wind farms and prioritise the development of innovative "porpoise friendly" ways to use the North Sea. We still know very little about the cumulative effects of all current anthropogenic activities on cetaceans, such as by-catch, pollution, noise, overfishing and collision. ASCOBANS is the international forum in which all member states can work together to help research efforts to investigate the cumulative impact of humans on small cetaceans. Only by combining efforts across boundaries will we be able to assess impacts and develop and successfully implement conservation plans on highly migratory species such as the harbour porpoise.

Dr Meike Scheidat
Advisor to Dutch Delegation

Active sonar

Active sonar, operating with sound SLs of up to 235 dB re 1 μPa @ 1 m at frequencies mainly between 1 and 150 kHz, is frequently used for fish-finding, oceanography, charting and in military activities (e.g. locating submarines) (Fig. 5.29). Since the mid-1990s, concern has been expressed over the potential impact these sounds may have upon cetaceans, particularly deep-diving toothed whales such as the beaked whales (although recent studies suggest that other species including minke and humpback whale may also be affected). Postmortem studies of mass stranded animals in the Bahamas, Madeira and the Canaries have revealed widespread haemorrhaging and ear damage (see Bernaldo de Quirós et al., 2019; Cox et al., 2006; Evans & Miller, 2004, for reviews).

Several mass strandings of whales, particularly beaked whales (Fig. 5.30), have been linked to naval exercises, the most famous of which have occurred in Greece, the Bahamas, and the Canaries (Arbelo et al., 2013; Bernaldo de Quirós et al., 2019; Cox et al., 2006; D'Amico et al., 2009; Evans & England, 2001; Evans & Miller, 2004; Frantzis, 1998). Veterinary pathologists analysing whales from several of these strandings also identified decompression symptoms, suggesting that whales do not just die from stranding but may be injured or die at sea (Fernández, Arbelo, et al., 2004; Fernández, Edwards, et al., 2005; Jepson et al., 2005). In a series of careful studies where tagged whales were exposed to low levels of naval sonar, they were found to respond at levels of 98, 127 (DeRuiter et al., 2013) and 138 dB re 1 μPa (Tyack et al., 2011). Responses were unusually slow ascent rates, unusually long intervals between deep dives, and a premature stopping of echolocation used for foraging. They also included strong and prolonged directional movement away from the sound source. The theory is that the sonar disrupts the deep-diving behaviour of the beaked whales causing them to modify the recovery dives that are important to prevent gas bubbles (emboli) forming. Although beaked whales appear to be particularly susceptible, blue whales have also been recorded responding to

FIGURE 5.29 Type 23 frigate equipped with mid-frequency active sonar for submarine detection. *Photo: Peter GH Evans.*

FIGURE 5.30 Cuvier's beaked whale (part of a mass stranding on the coast of Almeria, Spain, Jan 2006) is one of the species that appears to be particularly vulnerable to the effects of mid-frequency active sonar use. *Photo: IUSA_ULPGC, courtesy of Antonio Fernández.*

mid-frequency naval sonar (Goldbogen et al., 2013), and behavioural response studies have yielded negative reactions by northern bottlenose whale (Wensveen et al. 2019) and minke whale (Kvadsheim et al. 2017), as well as by harbour porpoise in captivity (Kastelein et al. 2015).

Since active sonar is used widely in fisheries and oceanographic surveys, it is often questioned why these may not have similar effects to military sonar. However, their beam width is much narrower than those used for long-range detection.

A number of areas in northwest Europe have traditionally been used for military exercises, with mid-frequency sonar applied increasingly since the 1960s. However, so far there have been no cetacean deaths in the region that can unequivocally be linked to naval sonar. Between January and July 2008, 18 Cuvier's beaked whales, four Sowerby's beaked whales, five unidentified beaked whales and 29 long-finned pilot whales were reported stranded in Scotland, Ireland and Wales (Dolman et al., 2010), and people wondered if this might be linked to a military exercise. Most carcasses were too decomposed for postmortem examination, but the fact they occurred over a period of several months and over a very wide area argued against this, although the cause or causes remain undetermined.

In June 2008 there was a mass stranding of 26 common dolphins in the Fal Estuary, Cornwall, United Kingdom (Jepson & Deaville, 2009). All animals examined were in good condition although they had empty stomachs. There was no evidence of significant infectious disease or acute physical injury; levels of organochlorines, trace metals and butyltins were relatively low; the ears appeared normal (but mild decomposition prevented further investigation) and there were no signs of gas or fat emboli, or physical trauma. An international naval exercise using mid-frequency active sonar was conducted in the South Coast Exercise Area prior to the mass stranding. However, there were approximately 60 hours between the cessation of its use and the stranding and was therefore considered too far removed in time to have directly triggered the event, although it might have played a part in a behavioural response that ultimately led to the stranding. Nevertheless, a definitive cause of the mass stranding could not be identified.

More recently, between July and September of 2018, an unusually high number of beaked whale strandings occurred over a wide area of the northern North Atlantic. Around 80 beaked whales, mainly Cuvier's beaked, were reported stranded, in various states of decomposition, in the British Isles and Ireland, particularly along the west and north coasts of Ireland and the west of Scotland. Around the same time, there were 12 beaked whale strandings (Cuvier's beaked whale and northern bottlenose whale) in south and east Iceland, six in Svalbard and two on the west coast of Norway (northern bottlenose whale and Sowerby's beaked whale). To date, there is no definitive cause for these strandings although various countries were involved in military exercises in June around northwest Norway and in the Barents Sea, but it is not known if any took place further south that could have triggered the strandings. The problem with any military exercise, and even more so, any covert operations, is the difficulty obtaining information to establish whether or not there are causal links. Until both environmental and military authorities can work closer together nationally and internationally, it will be difficult to improve our understanding and take appropriate mitigation measures. Of potential concern is the recent development of near continuous active sonar that has been found to improve the detection of targets.

Risk assessment

We have seen that many human activities in our seas can have the potential to disturb and even physically harm cetaceans. There are three environmental assessment processes in Europe, each underpinned by European directives. The first is the Strategic Environmental Assessment (SEA) derived from the SEA Directive (European Directive 2001/42/EC) concerned with 'the assessment of effects of certain plans and programmes on the environment'; the second is the Habitats Regulations Appraisal or Assessment applied to European sites established to protect habitats listed in Annex I and species in Annex II (harbour porpoise and bottlenose dolphin) established within the Habitats Directive (Council Directive 92/43EEC); and the third is the Environmental Impact Assessment (EIA) derived from the EIA Directive (Council Directive 85/337/EEC, as amended and consolidated) 'on the assessment of the effects of certain public and private projects on the environment'. All of these play a role in measures taken by EU member states to assess whether particular human activities, including the introduction of underwater noise, could have a significant effect on the environment.

There remains much uncertainty on the effects of noise and under what circumstances mitigation actions are needed. In an attempt to address this a group of American scientists proposed noise exposure criteria giving received SPLs and sound energy levels (SEL) at which one might predict either hearing damage in the form of TTS or PTS, or behavioural disturbance (Southall et al., 2007). For all cetaceans, the received exposure level (SEL) for single and multiple pulses as well as nonpulses leading to TTS was estimated to be 230 dB re: 1 μPa_{peak} @ 1 m (flat) within a 24-hour period, and for behavioural disturbance it was estimated to be 224 dB re: 1 μPa_{peak} @ 1 m (flat) for single pulses. For multiple pulses and nonpulses the equivalent received exposure levels eliciting a behavioural response varied between the different hearing groups (low, medium and high frequencies) and tended to be anywhere from 160 to 90 dB re: 1 μPa_{peak} @ 1 m (flat).

The ability to hear sounds varies across a species' hearing range. Most mammal audiograms have a typical "U" shape, with frequencies at the bottom of the "U" being those to which the animal is more sensitive, in terms of hearing.

TABLE 5.1 Summary of TTS/PTS thresholds (SEL thresholds are in dB re 1 dB re: 1 μPa^2s and peak SPL thresholds are in dB re dB re: 1 μPa; SEL is weighted; peak SPL is unweighted).

Group	Nonimpulsive		Impulse			
	TTS threshold	PTS threshold	TTS threshold		PTS threshold	
	SEL	SEL	SEL	Peak SPL	SEL	Peak SPL
Low Freq.	179	199	168	213	183	219
Mid Freq.	178	198	170	224	185	230
High Freq.	153	173	140	196	155	202

Adapted from Finneran, J.J. (2016). *Auditory weighting functions and exposure functions for marine mammals exposed to underwater noise.* Technical Report 3026, SSC Pacific, San Diego, 59 pp Houser, D. S., Yost, W., Burkard, R., Finneran, J. J., Reichmuth, C., & Mulsow, J. (2017). A review of the history, development and application of auditory weighting functions in humans and marine mammals. Journal of the Acoustical Society of America, *141*, 1371−1413. Avaliable form: http://doi.org.10.1121/1.4976086.

Auditory weighting functions were therefore applied to assess the overall hazard of noise on hearing. Thus while these functions are based on regions of equal loudness and best hearing, they are meant to reflect the susceptibility of the ear to noise-induced threshold shifts.

After further research, several of these threshold values have been revised, and Table 5.1 summarises the current recommendations (Finneran, 2016; Houser et al., 2017), which have then been proposed by NMFS (2016). Low-frequency cetaceans are defined as those with a generalised hearing range of 7 Hz−35 kHz, which covers the baleen whales; mid-frequency cetaceans are those with a generalised hearing range of 150 Hz−60 kHz and include dolphins, toothed whales, beaked whales and bottlenose whales; and high-frequency cetaceans are those with a generalised hearing range of 275 H−160 kHz, and include the porpoises, pygmy and dwarf sperm whales (plus river dolphins and some Southern Hemisphere species − Commerson's, Hector's and Heaviside's dolphins, hourglass and Peale's dolphin).

Vessel strikes

Records of collisions between ships and cetaceans exist at least as far back as the early 17th century. However, the number of cases reported appears to have increased markedly from the 1950s onwards, corresponding to the period when vessels regularly attained speeds of 14−15 knots or more (Laist, Knowlton, Mead, Collet, & Podesta, 2001; IWC, 2008). Whales may be hit either by the bow (Fig. 5.31), the keel or any other part of a vessel's hull, or by its propeller. Hit whales at times may be stuck on the bow of large ships and are then brought into a harbour, sometimes after carrying the carcass over substantial distances (e.g. Laist et al., 2001; Pesante, Panigada, & Zanardelli, 2002).

The seas around western Europe contain some of the busiest waterways in the world. Almost half a million ship movements a year take place in the Strait of Dover, and other busy traffic routes include the dredged entrance route to Rotterdam/Europort and its connecting route through the Channel, the route between the North Sea and the Baltic via the Kiel Canal, across the southern Bay of Biscay, around the Iberian Peninsula, and the Strait of Gibraltar. Over the last 20 years, the numbers of shipping movements, sizes of vessels and their average speeds have all increased in the region. Approximately half the shipping activity in the North Sea consists of ferries and roll-on/roll-off vessels on fixed routes (Fig. 5.32).

A global review of ship strikes demonstrated that although all types and sizes of vessels can be involved, most lethal or severe injuries are usually caused by ships travelling 14 knots (26 km/h) or faster and of 80 m length or more (Laist et al., 2001), and a North American study found that the probability of a ship strike being lethal increases markedly as vessel speeds increase from 10 to 15 knots (Vanderlaan & Taggart, 2007).

Evidence of vessel collisions has been reported for at least 21 cetacean species (Evans, 2003), but the species that seem to be most vulnerable are large whales like the fin whale, humpback whale, North Atlantic right whale (Fig. 5.33), and sperm whale.

Since 1990, the United Kingdom has undertaken regular postmortem studies of cetaceans stranding around the British Isles as part of the Cetacean Strandings Investigation Programme (CSIP). Causes of death have been assessed, resulting in estimates of the proportions of postmortem examinations that can be attributed to physical trauma (believed to be caused by vessel strikes). Over the 20 years since 1990, between 15%−20% of baleen whales (fin whales and

FIGURE 5.31 Fin whale on the bow thruster of a large container ship. *Photo: Souffleur d'Ecume, courtesy of Morgane Ratel.*

FIGURE 5.32 High speed ferry crossing the Irish Sea. Vessels travelling above 10 knots pose a significantly higher risk of lethal strike. *Photo: New England Aquarium, courtesy of Scott Kraus.*

minke whales) and 4%−6% of small cetaceans (e.g. harbour porpoises and common dolphins) examined at postmortem appeared to have suffered mortality from physical trauma (Evans, Baines, & Anderwald, 2011; from data provided by Rob Deaville).

Shipping movements can be determined for large vessels by a variety of methods, the most common of which is AIS (Automatic Identification System). This is a VHF broadcast system that sends information at regular intervals including the identity of the vessel, its position, course and speed to other vessels and to shore receivers.

Since January 2005 the International Maritime Organization's International Convention for the Safety of Life at Sea has required AIS to be fitted aboard international voyaging ships with gross tonnage of 300 or more, and all passenger ships regardless of size. Within the EU fishing vessels with an overall length of more than 15 m were also required to

FIGURE 5.33 The endangered northern right whale is a species that is particularly vulnerable to ship strike. *Photo: Peter GH Evans.*

use AIS by the year 2014. It is estimated that more than 40,000 ships now carry AIS equipment. Anyone can follow the movements of vessels equipped with AIS if they go to the website www.marinetraffic.com//ais, which has approximately 200 AIS receivers within northwest Europe.

A study of the densities of shipping and the distribution of the more vulnerable whale species highlighted that although the areas having the highest shipping densities were the English Channel, southernmost North Sea, Kattegat and Danish Belt Seas, and western and central Baltic, the large whales are relatively uncommon in those areas. Instead, the areas at highest risk of ship strike were parts of the Celtic Sea, the Bay of Biscay and off northwest Spain (Evans et al., 2011).

Recreational disturbance

The coasts of Europe are popular for leisure and recreation, attracting both local people and tourists from inland and abroad. Activities include bathing, surfing, sailing, sea angling, water sports and wildlife watching (for whales, dolphins and seals, seabirds and basking sharks). Tourism in northwest Europe is distinctly seasonal, concentrated into the summer months, and over the last 30 years has been growing steadily (Fig. 5.34).

As recreational activities have increased in coastal zones around the world (Fig. 5.35), pressures upon a number of cetacean species have also increased, leading to concerns expressed in many regions (Bejder & Samuels, 2003; Constantine, Brunton, & Dennis, 2004; Evans, 1996; Higham, Bejder, & Williams, 2014; Higham, Bejder, Allen, Corkeron, & Lusseau, 2016; Lusseau, 2004; Lusseau & Higham, 2004; Lusseau, Slooten, & Currey, 2006; New et al., 2013; Williams et al., 2002, 2006; Würsig & Evans, 2001).

Whale watching in particular has increased dramatically in many parts of the world. The latest comprehensive estimate for Europe (O'Connor et al., 2009) found that numbers had doubled in less than 10 years since the previous estimate (Hoyt, 2001), averaging 7% growth per annum. Over that decade, whale watching in Europe had expanded to a total of 22 countries, generating annually nearly US$100 million in expenditure, from Cyprus to Greenland.

Whale watching in Norway and Iceland focuses upon the larger whales such as humpback, minke, fin and sperm whales, plus killer whales in certain localities such as the Lofoten Islands. Further south, there is whale watching for minke whales, common dolphins, Risso's dolphins, white-beaked dolphins and occasional killer whales in the Hebrides of west Scotland, bottlenose dolphins in the Moray Firth (east Scotland) and Cardigan Bay (west Wales), and harbour porpoise, common, bottlenose and Risso's dolphins in Pembrokeshire (southwest Wales) and Cornwall (southwest England). As humpback whale sightings increase at various localities in the British Isles, opportunistic whale watching trips follow, and sites such as the Farne Islands (Northumberland) and Whitby (Yorkshire) have recently developed whale-watching trips for a range of species from the diminutive porpoise to white-beaked and bottlenose dolphins, and

FIGURE 5.34 Bottlenose dolphins bow-riding a small motor vessel in North Wales. In this instance, there was no evidence that the vessel was disturbing the animals. *Photo: Peter GH Evans.*

FIGURE 5.35 Personal water craft such as jet skis and speedboats have become very popular in recent years but can pose a threat to cetaceans by their high speed and often unpredictable movements. *Photo: Peter GH Evans.*

baleen whales such as minke and fin whale. In the Republic of Ireland, there is whale watching off the southwest coast, aimed primarily at humpback whales, although other species frequently seen include minke whale, common dolphin and harbour porpoise. There is no regular whale watching in the southern North Sea, and rather little around the coasts of France, confined to some bottlenose dolphin watching on an opportunistic basis off the coasts of Normandy and Brittany.

Other recreational activities (sailing, kayaking, water sports) are much more widespread, occurring all along the coasts of northern Germany, the Netherlands and Belgium as well as the south coast of England, and at scattered localities in southwest England, Wales, Ireland and southwest Scotland. Marine recreation in eastern Britain takes place mainly in the southeast.

The presence of vessels can have both direct and indirect effects on cetaceans (Lusseau, 2006; Mattson, Thomas, & Aubin, 2005; Nowacek, Wells, & Solow, 2001). They can cause disturbance of feeding activities, separate calves from their mothers and interfere with acoustic communication, whilst physical contact may lead to injury or death.

Short-term effects include changing behavioural patterns such as increasing swim speeds, increasing dive intervals, avoidance, animals clumping together and reductions in resting behaviour. In the presence of vessels, cetacean acoustics may be masked or altered and this may affect group cohesion. Longer-term effects include changes in residency patterns, suppression of breeding capabilities, and disruptions to feeding leading to reduced energy intake. In highly disturbed areas, animals may actually abandon the use of favoured sites.

In western Europe, there is some evidence of negative effects upon bottlenose dolphins in Cardigan Bay, west Wales (Feingold & Evans, 2014; Pierpoint et al., 2009; Richardson, 2012; Thompson, 2012) and in the Moray Firth, northeast Scotland (Hastie, Wilson, Tufft, & Thompson, 2003; Janik & Thompson, 1996; New et al., 2013; Pirotta, Merchant, Thompson, Barton, & Lusseau, 2015).

Climate change

The Earth's climate has exhibited broad extremes over geological time but there is general consensus that temperatures have steadily increased through the last century and, since around 1980, the rate of global warming has been unprecedented in both terrestrial and marine environments (IPCC, 2007). Marine ecosystems are highly dependent on changes to both ocean climate and acidification, whilst storms and waves, sea level rise and coastal erosion pose clear threats to human life as well as to other creatures (MCCIP, 2010).

Both air and sea temperatures have risen sharply over the northeast Atlantic in the last 25 years, the largest increase in sea surface temperatures occurring in the southern North Sea and eastern English Channel, at a rate of between 0.6°C and 0.8°C per decade (Dye et al., 2013; Rayner et al., 2003). Temperatures do fluctuate annually but the first decade of the 2000s was the warmest on instrumental record (Hughes et al., 2010; IPCC, 2007), with further recorded highs since then (Hughes et al. 2017). The rate of change in ocean pH is also thought to be faster than anything experienced in the last 55 million years, with a 30% decrease in pH, and a 16% decrease in carbonate ion concentrations since 1750 (Caldeira & Wickett, 2003; Doney, Fabry, Feely, & Kleypas, 2009; IPCC, 2007).

In the Arctic, the past few decades have seen greater reductions in sea ice coverage during summer and increased melt from Greenland than have been observed over the past thousand years (ACIA, 2004; IPCC, 2007; Morison et al., 2000; Walsh, 2008).

Relative to the underlying warming trend that has taken place during the 20th century, the surface waters averaged across the North Atlantic have been cool in the period between 1900 and 1930, warm from 1930 to 1960, cool between the late 1960s and 1990, and then warm from 1990 to present (Dye et al., 2013). This is thought to be a natural pattern of variation and has been described as the North Atlantic Oscillation (NAO) and Atlantic Multidecadal Oscillation (AMO) (Knight, Allan, Folland, Vellinga, & Mann, 2005). Superimposed upon this have been the changes in the climate that are believed to be human induced.

Unstable weather patterns leading to increased frequency of cyclones and other types of storm also appear to be influenced at least in part by the North Atlantic Oscillation (the index of which is a measure of the difference in mean atmospheric pressure between high pressure in the Azores (or Gibraltar) and low pressure in Iceland). The recent strong trend in the NAO (towards stormier conditions) is apparently unique in its history, but added to this have been more general changes in westerly winds in the North Atlantic that result in greater wave heights and storminess around western Europe.

One of the consequences of increased sea surface temperatures has been extensive changes in plankton communities. In the North Sea, the cold-water copepod *Calanus finmarchicus* once dominated the plankton community. Since the 1960s, however, it has declined in biomass by 70%, replaced by species with warmer-water affinities (such as *Calanus helgolandicus*). Unfortunately these are present at much lower abundance. Such changes can have major repercussions on the marine ecosystem. For example the life cycle of the sandeel is timed to make use of the seasonal production of copepods, which in turn depend upon planktonic plant production. Not only has copepod abundance declined but the spring occurrence of copepods and fish larvae has become out of synchrony, resulting in far fewer young sandeels being produced. When this occurs, it can affect top predators such as seabirds (Daunt & Mitchell, 2013) and marine mammals (Evans & Bjørge, 2013), and this is what we are seeing now in northern Britain, the Faroes and western Norway. In the North Sea climate change impacts have been predicted in fish species like sand eel and sprat, the former being negatively affected whereas the latter may benefit from warming sea temperatures (Pinnegar & Heath, 2010). This might

FIGURE 5.36 Recent sightings of grey whale in the Mediterranean and off West Africa may be an unexpected positive consequence of climate change, with animals thought to have migrated into the North Atlantic from the Pacific via the Arctic Ocean as warming has resulted in ice-free areas, the North Atlantic population having gone extinct in the 17th century. *Photo: Peter GH Evans.*

explain the southwards shift in harbour porpoises in the western North Sea since the 1990s, given that sand eel breeding success has been poor in recent years (Hammond et al., 2008; Pinnegar & Heath, 2010).

Several reviews have been published recently of the possible effects of climate change upon marine mammals (Evans & Bjørge, 2013; Evans, Boyd, & MacLeod, 2010; Evans, Pierce, & Panigada, 2010; Frederiksen & Haug, 2016; Huntington & Moore, 2008; IWC, 1997, 2009; Laidre et al., 2008; Learmonth et al., 2006; MacLeod, 2009; Würsig, Reeves, & Ortega-Ortiz, 2001). Being warm-blooded vertebrates that can thermorgulate, marine mammals might be expected to cope well with most environmental variation predicted from climate change. They employ complex behavioural adaptations that can buffer them against wide changes in the environment. On the other hand, changes in the availability of their habitat (including food resources) may lead to changes in population size or distribution in particular cases (Fig. 5.36). An obvious example is the reduction in ice cover affecting ice-breeding seals such as the walrus, bearded, hooded, ribbon, harp or ringed seal, and its consequent effect upon Arctic predators such as the polar bear (Derocher, Lunn, & Stirling, 2004; Ferguson, Stirling, & McLoughlin, 2005; Huntington & Moore, 2008; Stirling, Lunn, & Iacozza, 1999).

Arctic cetaceans such as narwhal, beluga and bowhead whale are the species thought most likely to be affected by climate change as these animals associate closely with ice (Huntington & Moore, 2008; IWC, 2009; Laidre et al., 2008). In northwest Europe, however, one might expect actually an increase in the number of species occurring, many extending their range from subtropical and tropical regions as sea temperatures rise. This is indeed what we seem to be experiencing, with species such as the striped dolphin, pygmy sperm whale and Cuvier's beaked whale being recorded more regularly further north, and dwarf sperm whale being added to the French and British faunas (Evans & Bjørge, 2013; Evans, Boyd, et al., 2010; Evans, Pierce, et al., 2010). On the other hand, cold-water shelf species such as the harbour porpoise and white-beaked dolphin, and shelf edge species such as the Atlantic white-sided dolphin may find conditions less favourable, and their ranges could start to contract at the southern margins (Evans & Bjørge, 2013; Evans, Boyd, et al., 2010; Evans, Pierce, et al., 2010; MacLeod, 2009).

References

Hunting

Holt, S. J. (1977). Does the bottlenose whale have a sustainable yield and if so is it worth taking? *Reports of the International Whaling Commission,* 27, 206–208.

Jonsgård, A. (1977). Tables showing the catch of small whales (including minke whales) caught by Norwegians in the period 1938-75, and large whales caught in different North Atlantic waters in the period 1868-1975. *Reports of the International Whaling Commission,* 27, 413–426.

Tønnesen, J. N., & Johnsen, A. O. (1982). *The history of modern whaling.* London: C. Hurst & Co., 798 pp.

Zachariassen, P. (1993). Pilot whale catches in the Faroes, 1709−1992. *Report of the International Whaling Commission* (Special Issue 14), 69−88.

Reduction of prey from fishing

Deaville, R. (compiler) (2016). *UK cetacean strandings investigation programme* (76 pp.). Final Report to Defra for the Period 1st January−31st December 2015. (Contract numbers MB0111). London: Institute of Zoology.

Deaville, R. & Jepson, P. D. (compilers) (2011). *UK Cetacean strandings investigation programme* (98 pp). Final Report to Defra for the period 1st January 2005−31st December 2010. (Contract numbers CR0346 and CR0364). London: Institute of Zoology.

Evans, P. G. H. (1990). European cetaceans and seabirds in an oceanographic context. *Lutra, 33,* 95−125.

HELCOM (2013). HELCOM Red List of Baltic Sea species in danger of becoming extinct. *Baltic Sea Environmental Proceedings* No. 140. 106pp.

HELCOM (2018). *State of the Baltic Sea. Second HELCOM holistic assessment 2011-2016. Baltic Sea Environment Proceedings* No. 116B. 155pp.

OSPAR Commission. (2010). *Quality Status Report 2010.* London: OSPAR Commission, 176pp.

OSPAR Commission. (2017). *Intermediate Assessment 2017.* London: OSPAR Commission.Available from https://oap.ospar.org/en/ospar-assessments/intermediate-assessment-2017.

Incidental capture in fishing gear

Bjørge, A., Skern-Mauritzen, M., & Rossman, M. C. (2013). Estimated by-catch of harbour porpoise (*Phocoena phocoena*) in two coastal gillnet fisheries in Norway, 2006−2008. Mitigation and implications for conservation. *Biological Conservation, 161,* 164−173.

Couperus, A. S. (1994). Killer whales (*Orcinus orca*) scavenging on discards of freezer trawlers north-east of the Shetland islands. *Aquatic Mammals, 20,* 47−51.

Dolman, S., Baulch, S., Evans, P. G. H., Read, F., & Ritter, F. (2016). Towards an EU action plan on cetacean by-catch. *Marine Policy, 72,* 67−75.

Goujon, M. (1996). Captures accidentelles du filet maillant derivant et dynamique des populations de dauphins au large du Golfe de Gascogne. (Ph.D. thesis). École Nationale Superieure Agronomique de Rennes.

Goujon, M., Antoine, L., Collet, A., & Fifas S. (1993). *Approche de l'mpact de la ecologique de la pecherie thoniere au filet maillant derivant en Atlantique nord-est.* Rapport internese al Direction des Resources Vivantes de I'IFREMER. Plouzane: Ifremer, Centre de Brest.

Hall, M. A. (1998). An ecological view of the tuna-dolphin problem: Impacts and trade-offs. *Reviews in Fish Biology and Fisheries, 8,* 1−34.

Hall, M. A., & Donovan, G. P. (2001). Environmentalists, fishermen, cetaceans and fish: Is there a balance and can science help to find it? In P. G. H. Evans, & J. A. Raga (Eds.), *Marine mammals: Biology and conservation* (pp. 491−520). London: Kluwer Academic/Plenum Press, 630 pp.

IWC. (1994). Report of the workshop on mortality of cetaceans in passive fishing nets and traps. In W. F. Perrin, G. P. Donovan, & J. Barlow (Eds.), *International whaling commission, Special Issue, 15: Gillnets and cetaceans* (pp. 1−71). Cambridge: International Whaling Commission.

Kaschner, K. (2003). *Review of small cetacean by-catch in the ASCOBANS area and adjacent waters − current status and suggested future actions* (122 pp.). In: ASCOBANS Report MOP4/Doc22(s) presented at the 4th Meeting of the Parties to ASCOBANS, Esbjerg.

Lien, J. (1994). Entrapments of large cetaceans in passive inshore fishing gear in Newfoundland and Labrador (1979-1990). In W. F. Perrin, G. P. Donovan, & J. Barlow (Eds.), *Gillnets and cetaceans* (pp. 149−158). Cambridge: International Whaling Commission, IWC Special Issue No. 15.

Lien, J., Hood, C., Pittman, D., Ruel, P., Borggaard, D., Chisholm, C., ... Mitchel, D. (1995). Field tests of acoustic devices on groundfish gillnets: Assessment of effectiveness in reducing harbour porpoise by-catch. In R. A. Kastelein, J. A. Thomas, & P. E. Nachtigall (Eds.), *Sensory systems of aquatic mammals* (pp. 1−22). Woerdon: De Spil Publishers.

Northridge, S., Cargill, A., Coran, A., Mandleberg, L., Calderan, S., & Reid, R. J. (2010). *Entanglement of minke whales in Scottish waters: An investigation into occurrence, causes and mitigation* (57 pp.). Sea Mammal Research Unit, Final Report to Scottish Government CR/2007/49.

Northridge, S. P. & Hammond, P. S. (1999). *Estimation of porpoise mortality in uk gill and tangle net fisheries in the north sea and west of scotland.* Document SCl51lSM42. Cambridge: The Scientific Committee of the International Whaling Commission.

Northridge, S. P., & Hofman, R. J. (1999). Marine mammal interactions with fisheries. In J. R. Twiss, & R. R. Reeves (Eds.), *Conservation and management of marine mammals* (pp. 99−119). Washington and London: Smithsonian Institution Press.

Read, A. J., Dinker, P., & Northridge, S. (2006). By-catch of marine mammals in U.S. and global fisheries. *Conservation Biology, 20,* 163−169.

Reeves, R., McClellan, K., & Werner, T. (2013). Marine mammal by-catch in gillnet and other entangling net fisheries, 1990 to 2011. *Endangered Species Research, 20,* 71−97.

Rogan, E., & Mackey, M. (2007). Megafauna by-catch in drift-nets for albacore tuna (*Thunnus alalunga*) in the NE Atlantic. *Fisheries Research, 86,* 6−14.

Ryan, C., Leaper, R., Evans, P. G. H., Robinson, K. P., Haskins, G. N., Calderan, S., ... & Jack, A. (2016). Entanglement: An emerging threat to humpback whales in Scottish waters. In: *Presented to the Scientific Committee Meeting of the International Whaling Commission, 2016,* SC/66b/HIM/01.

Tregenza, N. J. C., Berrow, S. D., Hammond, P. S., & Leaper, R. (1997a). Common dolphin, *Delphinus delphis* L., by-catch in bottom set gillnets in the Celtic Sea. *Reports of the International Whaling Commission, 47,* 835−839.

Tregenza, N. J. C., Berrow, S. D., Hammond, P. S., & Leaper, R. (1997b). Harbour porpoise (*Phocoena phocoena* L.) by-catch in set gillnets in the Celtic Sea. *ICES Journal of Marine Science, 54,* 896−904.

Vinther, M. (1999). By-catches of harbour porpoises *Phocoena phocoena* (L.) in Danish set-net fisheries. *Journal of Cetacean Research and Management, 1,* 123−135.

Vinther, M., & Larsen, F. (2004). Updated estimates of harbour porpoise by-catch in the Danish bottom set gillnet fishery. *Journal of Cetacean Research and Management, 6*(1), 19−24.

Chemical pollutants and other hazardous substances

Evans, P. G. H. (Ed.) (2014). Chemical pollution and marine mammals. In: *Proceedings of the ECS/ASCOBANS/ACCOBAMS joint workshop, held at the European Cetacean Society's 25th Annual Conference* (93 pp), Cadiz, Spain, 20th March 2011. ECS Special Publication Series No. 55.

Fossi, M. C., & Panti, C. (Eds.), (2018). *Marine Mammal Ecotoxicology: Impacts of multiple stressors on population health*. New York: Academic Press, 486pp.

Hall, A. J., Hugunin, K., Deaville, R., Law, R. J., Allchin, C. R., & Jepson, P. D. (2006). The risk of infection from polychlorinated biphenyl exposure in harbour porpoise (*Phocoena phocoena*) − a case-control approach. *Environmental Health Perspectives, 114*, 704−711.

Jepson, P. D., Bennett, P. M., Allchin, C. R., Law, R. J., Kuiken, T., Baker, J. R., ... Kirkwood, J. K. (1999). Investigating potential associations between chronic exposure to polychlorinated biphenyls and infectious disease mortality in harbour porpoises from England and Wales. *Science of the Total Environment, 243−244*, 339−348.

Jepson, P. D., Bennett, P. M., Deaville, R., Allchin, C. R., Baker, J. R., & Law, R. J. (2005). Relationships between polychlorinated biphenyls and health status in harbour porpoises (*Phocoena phocoena*) stranded in the United Kingdom. *Environmental Toxicology and Chemistry, 24*(1), 238−248.

Jepson, P. D., Deaville, R., Barber, J. L., Aguilar, A., Borrell, A., Murphy, S., ... Law, R. J. (2016). PCB pollution continues to impact populations of orcas and other dolphins in European waters. *Nature Scientific Reports, 6*, 18573. Available from https://doi.org/10.1038/srep18573.

Kannan, K., Blankenship, A. L., Jones, P. D., & Giesy, J. P. (2000). Toxicity reference values for the toxic effects of polychlorinated biphenyls to aquatic mammals. *Human and Ecological Risk Assessment, 6*, 181−201.

Murphy, S., Barber, J. L., Learmonth, J. A., Read, F. L., Deaville, R., Perkins, M. W., ... Jepson, P. D. (2015). Reproductive Failure in UK Harbour Porpoises *Phocoena phocoena*: Legacy of Pollutant Exposure? *PLoS One, 10*(7), e0131085. Available from https://doi.org/10.1371/journal.pone.0131085.

Murphy, S., Law, R. J., Deaville, R., Barnett, J., Perkins, M. W., Brownlow, A., ... Jepson, P. D. (2018). Organochlorine contaminants and reproductive implication in cetaceans: A case study of the common dolphin. In M. C. Fossi, & C. Panti (Eds.), *Marine mammal ecotoxicology: Impacts of multiple stressors on population health* (pp. 3−38). New York: Academic Press, 486 pp.

OSPAR Commission. (2010). *Quality Status Report 2010*. London: OSPAR Commission, 176pp.

OSPAR Commission. (2017). *Intermediate Assessment 2017*. London: OSPAR Commission.Available from https://oap.ospar.org/en/ospar-assessments/intermediate-assessment-2017.

Marine litter

Baulch, S., & Perry, C. (2014). Evaluating the impacts of marine debris on cetaceans. *Marine Pollution Bulletin, 80*, 210−221.

Besseling, E., Foekema, E., Van Franeker, J., Leopold, M., Kühn, S., Bravo Rebolledo, E., ... Koelmans, A. (2015). Microplastic in a macro filter feeder: Humpback whale *Megaptera novaeangliae*. *Marine Pollution Bulletin, 95*, 248−252. Available from https://doi.org/10.1016/j.marpolbul.2015.04.007, Epub 24 April 2015.

CBD (2012). *Impacts of marine debris on biodiversity: Current status and potential solutions* (61pp.). Montreal: Technical Series No. 67. Convention on Biological Diversity.

Fossi, M., Baini, M., Panti, C., & Baulch, S. (2018). Impacts of marine litter on cetaceans: A focus on plastic pollution. In M. C. Fossi, & C. Panti (Eds.), *Marine mammal ecotoxicology: Impacts of multiple stressors on population health* (pp. 147−183). New York: Academic Press, 486 pp.

Fossi, M., Coppola, D., Baini, M., Giannetti, M., Guerranti, C., Marsili, L., ... Clo, S. (2014). Large filter feeding marine organisms as indicators of microplastics in the pelagic environment: The case studies of the Mediterranean basking shark (*Cetorhinus maximus*) and fin whale (*Balaenoptera physalus*). *Marine Environmental Research, 100*, 17−24.

IWC (International Whaling Commission) (2013). *Report of the IWC scientific committee workshop on marine debris*. (SC/65a/Rep06). International Whaling Commission.

IWC (International Whaling Commission) (2014). *Report of the IWC Workshop on mitigation and management of the threats posed by marine debris to cetaceans*. (IWC/65/CCRep04). International Whaling Commission.

Lusher, A., Hernandez-Milian, G., O'Brien, J., Berrow, S., O'Connor, I., & Officer, R. (2015). Microplastic and macroplastic ingestion by a deep diving, oceanic cetacean: The True's beaked whale *Mesoplodon mirus*. *Environmental Pollution, 199*, 185−191.

OSPAR Commission. (2014). *Regional Action Plan for Prevention and Management of Marine Litter in the North-East Atlantic*. London: OSPAR Commission, 18 pp.

Puig-Lozano, R., de Quirós, B., Diaz-Delgado, J., García-Álvarez, N., Sierra, E., De la Fuente, J., ... Arbelo, M. (2018). Retrospective study of foreign body-associated pathology in stranded cetaceans, Canary Islands (2000-2015). *Environmental Pollution, 243*, 519−527.

Noise disturbance
(a) General

Cook, M. L. H., Varela, R. A., Goldstein, J. D., McCulloch, S. D., Bossart, G. D., Finneran, J. J., ... Mann, D. A. (2006). Beaked whale auditory evoked potential hearing measurements. *Journal of Comparative Physiology A - Neuroethology Sensory Neural and Behavioral Physiology, 192*, 489−495.

Finneran, J. J. (2016). *Auditory weighting functions and exposure functions for marine mammals exposed to underwater noise* (59 pp.) (Technical report 3026). San Diego: SSC Pacific.

Finneran, J. J., Carder, D. A., Schlundt, C. E., & Ridgway, S. H. (2005). Temporary threshold shift in bottlenose dolphins (*Tursiops truncatus*) exposed to mid-frequency tones. *Journal of the Acoustical Society of America, 118*, 2696−2705.

Finneran, J. J., Schlundt, C. E., Carder, D. A., Clark, J. A., Young, J. A., Gaspin, J. B., & Ridgway, S. H. (2000). Auditory and behavioral responses of bottlenose dolphins (*Tursiops truncatus*) and a beluga whale (*Delphinapterus leucas*) to impulsive sounds resembling distant signatures of underwater explosions. *Journal of the Acoustical Society of America, 108*, 417−431.

Houser, D. S., Yost, W., Burkard, R., Finneran, J. J., Reichmuth, C., & Mulsow, J. (2017). A review of the history, development and application of auditory weighting functions in humans and marine mammals. *Journal of the Acoustical Society of America, 141*, 1371−1413. Available from https://doi.org/10.1121/1.4976086.

Nachtigall, P., Supin, A., Pawloski, J., & Au, W. (2004). Temporary threshold shifts after noise exposure in the bottlenose dolphin (*Tursiops truncatus*) measured using evoked auditory potentials. *Marine Mammal Science, 20*, 673−687.

Nachtigall, P. E., Pawloski, J. L., & Au, W. W. L. (2003). Temporary threshold shifts and recovery following noise exposure in the Atlantic bottlenosed dolphin (*Tursiops truncatus*). *Journal of the Acoustical Society of America, 113*, 3425−3429.

NMFS. (2016). *Technical guidance for assessing the effects of anthropogenic sound on marine mammal hearing: Underwater acoustic thresholds for onset of permanent and temporary threshold shifts.* U.S. Dept. of Commerce, NOAA, NOAA Technical Memorandum NMFS-OPR-55, 178 pp.

NRC. (2005). *Marine mammal populations and ocean sound: Determining when sound causes biologically significant effects.* Washington, DC: National Academy Press.

Nowacek, D., Johnson, M. P., & Tyack, P. L. (2004). North Atlantic right whales (*Eubalaena glacialis*) ignore ships but respond to alerting stimuli. *Proceedings of the Royal Society of London. Series B, Biological Sciences, 271*, 227−231.

Nowacek, D., Thorne, L. H., Johnston, D. W., & Tyack, P. L. (2007). Responses of cetacean to anthropogenic noise. *Mammal Review, 37*(2), 81−115.

OSPAR Commission. (2009). *Overview of the impacts of anthropogenic underwater sound in the marine environment.* Paris: OSPAR Commission, 134 pp.

Richardson, W. J., Greene, C. R., Malme, C. I., & Thomson, D. H. (1995). *Marine Mammals and Noise.* San Diego, CA: Academic Press.

Southall, B. L., Bowles, A. E., Ellison, W. T., Finneran, J. J., Gentry, R. L., Greene, C. R., ... Tyack, P. L. (2007). Marine mammal sound exposure criteria: Initial scientific recommendations. *Aquatic Mammals, 33*, 411−522.

Weilgart, L. S. (2007). The impacts of anthropogenic ocean noise on cetaceans and implications for management. *Canadian Journal of Zoology, 85*, 1091−1116. Available from https://doi.org/10.1139/Z07-101.

(b) Shipping

Andrew, R. K., Howe, B. M., & Mercer, J. A. (2011). Longtime trends in ship traffic noise for four sites off the North American West Coast. *Journal of the Acoustical Society of America, 129*(2), 642−651.

Andrew, R. K., Howe, B. M., Mercer, J. A., & Dzieciuch, M. A. (2002). Ocean ambient sound: Comparing the 1960s with the 1990s for a receiver off the California coast. *Acoustic Research Letters Online, 3*, 65−70.

Chapman, N. R., & Price, A. (2011). Low frequency deep ocean ambient noise trend in the Northeast Pacific Ocean. *Journal of the Acoustical Society of America, 129*(5), EL161−EL165.

Clark, C. W., Ellison, W. T., Southall, B. L., Hatch, L., Van Parijs, S. M., Frankel, A., & Ponirakis, D. (2009). Acoustic masking in marine ecosystems: Intuitions, analysis, and implication. *Marine Ecology Progress Series, 395*, 201−222.

Dyndo, M., Wisniewska, D. M., Rojano-Doñate, L., & Madsen, P. T. (2015). Harbour porpoises react to low levels of high frequency vessel noise. *Scientific Reports, 5*, 1−9. Available from https://doi.org/10.1038/srep11083.

Erbe, C., Reichmuth, C., Cunningham, K., Lucke, K., & Dooling, R. (2016). Communication masking in marine mammals: A review and research strategy. *Marine Pollution Bulletin, 103*, 15−38. Available from https://doi.org/10.1016/j.marpolbul.2015.12.007, Epub 18 December 2015.

Frisk, G. V. (2012). Noiseonomics: The relationship between ambient noise levels in the sea and global economic trends. *Nature Scientific Reports, 2*. Available from https://doi.org/10.1038/srep00437, Article number 437.

Hildebrand, J. A. (2009). Anthropogenic and natural sources of ambient sound in the ocean. *Marine Ecology Progress Series, 395*, 5−20.

Hunt, K. E., Rolland, R. A., Kraus, S. D., & Wasser, S. K. (2006). Analysis of fecal glucocorticoids in the North Atlantic right whale (*Eubalaena glacialis*). *General and Comparative Endocrinology, 148*, 260−272.

McKenna, M. F., Ross, D., Wiggins, S. M., & Hildebrand, J. (2012). Underwater radiated noise from modern commercial ships. *Journal of the Acoustical Society of America, 131*, 92−103.

NRC. (2003). *Ocean noise and marine mammals.* Washington, DC: National Academy Press, 204 pp.

Parks, S. E., Johnson, M., Nowacek, D., & Tyack, P. L. (2011). Individual right whales call louder in increased environmental noise. *Biology Letters, 7*, 33−35. Available from https://doi.org/10.1098/rsbl.2010.0451, PMID: 20610418.

Rolland, R. M., Parks, S. E., Hunt, K. E., Castellote, M., Corkeron, P. J., Nowacek, D. P., ... Krause, S. D. (2012). Evidence that ship noise increases stress in right whales. *Proceedings of the Royal Society B, 279*, 2363−2368. Available from https://doi.org/10.1098/rspb.2011.2429.

Tournadre, J. (2014). Anthropogenic pressure on the open ocean: The growth of ship traffic revealed by altimeter data analysis. *Geophysical Research Letters, 41*. Available from https://doi.org/10.1002/2014GL061786.

Trumble, S. J., Norman, S. A., Crain, D. D., Mansouri, F., Winfield, Z. C., Sabin, R., ... Usenko, S. (2018). Baleen whale cortisol levels reveal a physiological response to 20th century whaling. *Nature Communications, 9*, 4587. Available from https://doi.org/10.1038/s41467-018-07044-w.

Tyack, P., Frisk, G., Boyd, I., Urban, E., & Seeyave, S. (Eds) (2015). *International Quiet Ocean Experiment Science Plan.* Scientific Committee on Oceanic Research (SCOR) & Partnership for Observation of the Global Oceans. 103pp.

Wisniewska, D. M., Johnson, M., Teilmann, J., Siebert, U., Galatius, A., Dietz, R., & Madsen, P. T. (2018). High rates of vessel noise disrupt foraging in wild harbour porpoises (*Phocoena phocoena*). *Proceedings of the Royal Society B, 285*. Available from http://dx.doi.org/10.1098/rspb.2017.2314.

(c) Seismic

Abgrall, P., Moulton, V. D., & Richardson, W. J. (2008). *Updated review of scientific information on impacts of seismic survey sound on marine mammals, 2004−present.* LGL Rep. SA973-1. Report from LGL Limited, St. John's, NL and King City, ON, for Department of Fisheries and Oceans, Habitat Science Branch, Ottawa, ON. 27 pp. + appendices.

Blackwell, S. B., Nations, C. S., McDonald, T. L., Thode, A. M., Mathias, D., Kim, K. H., ... Macrander, A. M. (2015). Effects of airgun sounds on bowhead whale calling rates: Evidence for two behavioral thresholds. *PLoS One*, *10*(6), e0125720. Available from https://doi.org/10.1371/journal.pone.0125720.

Castellote, M., Clark, C. W., & Lammers, M. O. (2012). Acoustic and behavioural changes by fin whales *(Balaenoptera physalus)* in response to shipping and airgun noise. *Biological Conservation*, *147*(1), 115—122.

Cato, D. H., Noad, M. J., Dunlop, R. A., McCauley, R. D., Gales, N. J., Kent, C. P. S., ... Duncan, A. J. (2013). A study of the behavioural response of whales to the noise of seismic airguns: Design, methods and progress. *Acoustics Australia*, *41*(1), 88—97.

Cerchio, S., Strindberg, S., Collins, T., Bennett, C., & Rosenbaum, H. (2014). Seismic surveys negatively affect humpback whale singing activity off northern Angola. *PLoS One*, *9*(3), e86464. Available from https://doi.org/10.1371/journal.pone.0086464.

Di Iorio, L., & Clark, C. W. (2010). Exposure to seismic survey alters blue whale acoustic communication. *Biology Letters*, *6*(1), 51—54.

Dunlop, R. A., Noad, M. J., Cato, D. H., Kniest, E., Miller, P. J. O., Smith, J. N., & Stokes, M. D. (2013). Multivariate analysis of behavioural response experiments in humpback whales *(Megaptera novaeangliae)*. *The Journal of Experimental Biology*, *216*, 759—770. Available from https://doi.org/10.1242/jeb.071498.

Dunlop, R. A., Noad, M. J., McCauley, R. D., Kniest, E., Paton, D., & Cato, D. H. (2015). The behavioural response of humpback whale *(Megaptera novaeangliae)* to a 20 cubic inch air gun. *Aquatic Mammals*, *41*(4), 412—433. Available from https://doi.org/10.1578/AM.41.4.2015.412.

Dunlop, R. A., Noad, M. J., McCauley, R. D., Kniest, E., Paton, D., & Cato, D. H. (2016). Response of humpback whales *(Megaptera novaeangliae)* to ramp-up of a small experimental air gun array. *Marine Pollution Bulletin*, *103*, 72—83.

Dunlop, R. A., Noad, M. J., McCauley, R. D., Kniest, E., Slade, R., Paton, D., & Cato, D. H. (2017). The behavioural response of migrating humpback whales to a full seismic airgun array. *Proceedings of the Royal Society B*, *284*. Available from https://doi.org/10.1098/rspb.2017.1901, 20171901.

Evans, P. G. H., Baines, M. E., & Anderwald, P. (2016). *Cetacean stock assessment in north-west Europe in relation to exploration and production industry sound* (153 pp.). Report to International Association of Oil & Gas Producers Joint Industry Program on Sound & Marine Life.

Evans, P. G. H., & Nice, H. (1996). Review of the effects of underwater sound generated by seismic surveys on cetaceans. *Report to UKOOA*. Oxford: Sea Watch Foundation, 50 pp.

Gailey, G., Würsig, B., & McDonald, T. L. (2007). Abundance, behavior, and movement patterns of western gray whales in relation to a 3-D seismic survey, Northeast Sakhalin Island, Russia. *Environmental Monitoring and Assessment*, *134*(1-3), 75—91. Available from https://doi.org/10.1007/s10661-007-9812-1.

Harris, R.E., Elliot, T., & Davis, R.A. (2007). *Results of mitigation and monitoring program, Beaufort Span 2-D marine seismic program, open-water season 2006*. LGL Rep. TA4319-1. Rep. from LGL Ltd., King City, Ont., for GX Technology Corp., Houston, TX. 48 pp.

Holst, M., Smultea, M. A., Koski, W. R., & Haley, B. (2005, June). *Marine mammal and sea turtle monitoring during Lamont-Doherty EarthObservatory's marine seismic program off the Northern Yucatán Peninsula in the Southern Gulf of Mexico, January—February 2005* (96 pp.). LGL Report TA2822-31, prepared by LGL Ltd. environmental research associates, King City, ONT, for Lamont-Doherty Earth Observatory, Columbia University, Palisades, NY, and NMFS, Silver Spring, MD.

Jochens, A., Biggs, D., Engelhaupt, D., Gordon, J., Jaquet, N., Johnson, M., ... Würsig, B. (2006). *Sperm whale seismic study in the Gulf of Mexico; summary report, 2002—2004* (345 pp.) (OCS Study MMS 2006-034). MMS, Gulf of Mexico OCS Region, New Orleans, LA.

Kastelein, R. A., Helder-Hoek, L., Van de Voorde, L., von Benda-Beckmann, A. M., Lam, F.-P. A., Jansen, E., de Jong, C. A. F., & Ainslie, M. A. (2017). Temporary hearing threshold shift in a harbor porpoise *(Phocoena phocoena)* after exposure to multiple airgun sounds. *Journal of the Acoustical Society of America*, *142*, 2430. Available from https://doi.org/10.1121/1.5007720.

Ljungblad, D. K., Wursig, B., Swartz, S. L., & Keene, J. M. (1988). Observations on the behavioral responses of bowhead whales *(Balaena mysticetus)* to active geophysical vessels in the Alaskan Beaufort Sea. *Arctic*, *41*, 183—194.

Madsen, P. T., Johnson, M. P., Miller, P. J. O., Aguilar Soto, N., Lynch, J., & Tyack, P. (2006). Quantitative measures of air-gun pulses recorded on sperm whales *(Physeter macrocephalus)* using acoustic tags during controlled exposure experiments. *Journal of the Acoustical Society of America*, *120*, 2366—2379.

Madsen, P. T., Møhl, B., Nielsen, K., & Wahlberg, M. (2002). Male sperm whale behaviour during exposures to distant seismic survey pulses. *Aquatic Mammals*, *28*, 231—240.

Malme, C.J., & Miles, P.R. (1985). Behavioral responses of marine mammals (gray whales) to seismic discharges. In: G.D. Greene, F.R. Engelhardt and R.J. Paterson (editors) Proceedings of the Workshop on Effects of Explosives use in the Marine Environment, pp. 253—280. Jan 1985, Halifax, N.S. Technical Report 5. Canadian Oil & Gas Lands Administration Environmental Protection Branch, Ottawa, Ontario. 398pp.

Malme, C.I., Miles P.R., Clarke C.W., Tyack P., & Bird, J.E. (1983). Investigations of the potential effects of underwater noise from petroleum industry activities on migrating gray whale behavior. BBN Report 5366. Report by Bolt, Beranek, and Newman Inc, Cambridge, MA, for the U.S. Department of the Interior, Minerals Management Service, Anchorage, AK. NTIS PB86-174174.

Malme, C.I., Miles, P.R., Clark, C.W., Tyack, P., & Bird, J.E. (1984). Investigations of the potential effects of underwater noise from petroleum industry activities on migrating gray whale behavior/Phase II: January 1984 migration. BBN Report 5586. Report by Bolt, Beranek, and Newman Inc., Cambridge, MA, for the U.S. Department of the Interior, Minerals Management Service, Anchorage, AK. NTIS PB86-218377.

Malme, C. I., Würsig, B., Bird, J. E., & Tyack, P. (1986). *Behavioral responses of gray whales to industrial noise: Feeding observations and predictive modeling. Outer continental shelf environmental assessment program*, Final report, BBN Rep. 6265, OCS Study MMS 88-0048, prepared by BBN Labs Inc., Cambridge, MA, for NMFS and MMS, Anchorage, AK.

Malme, C. I., Würsig, B., Bird, J. E., & Tyack, P. (1988). Observations of feeding gray whale responses to controlled industrial noise exposure. In W. M. Sackinger, M. O. Jeffries, J. L. Imm, & S. D. Treacy (Eds.), *Port and ocean engineering under arctic conditions. Vol. II. Symposium on Noise 28 and marine mammals* (pp. 55—73). Fairbanks, AK: University of Alaska Fairbanks.

McCauley, R. D., Fewtrell, J., Duncan, A. J., Jenner, C., Jenner, M. N., Penrose, J. D., . . . McCabe, K. (2000). Marine seismic surveys – A study of environmental implications. *APPEA Journal, 40*, 692–708.

McCauley, R. D., Jenner, M. N., Jenner, C., McCabe, K. A., & Murdoch, J. (1998). The response of humpback whales (*Megaptera novaeangliae*) to offshore seismic survey: Preliminary results of observations about a working seismic vessel and experimental exposures. *APPEA Journal, 1998*, 692–706.

Miller, G. W., Moulton, V. D., Davis, R. A., Holst, M., Millman, P., MacGillivray, A., & Hannay, D. (2005). Monitoring seismic effects on marine mammals - southeastern Beaufort Sea, 2001-2002. In S. L. Armsworthy, P. J. Cranford, & K. Lee (Eds.), *Offshore oil and gas environmental effects monitoring/approaches and technologies* (pp. 511–542). Columbus, OH: Battelle Press.

Miller, P. J. O., Johnson, M. P., Madsen, P. T., Biassoni, N., Quero, M. E., & Tyack, P. L. (2009). Using at-sea experiments to study the effects of air-guns on the foraging behavior of sperm whales in the Gulf of Mexico. *Deep-Sea Research, 56*, 1168–1181.

Moulton, V. D. & Miller, G. W. (2005). Marine mammal monitoring of a seismic survey on the Scotian Slope, 2003. In: K. Lee., H. Bain, & G. V. Hurley (Eds.), *Acoustic monitoring and marine mammal surveys in the Gully and Outer Scotian Shelf before and during active seismic programs. (Environmental Studies Research Funds Report No. 151)* (pp. 29–40), 154 pp. + xx.

Pirotta, E., Brookes, K. L., Graham, I. M., & Thompson, P. M. (2014). Variation in harbour porpoise activity in response to seismic noise. *Biology Letters*. Available from https://doi.org/10.1098/rsbl.2013.1090.

Richardson, W. J., & Malme, C. I. (1993). Man made noise and behavioral responses. *The Bowhead Whale*, Special Publication 2, Society for Marine Mammology, Lawrence, KS. 787pp.

Richardson, W. J., Würsig, B., & Greene, C. R. J. (1986). Reactions of bowhead whales, *Balaena mysticetus*, to seismic exploration in the Canadian Beaufort Sea. *Journal of the Acoustical Society of America, 79*, 1117–1128.

Robertson, F. L., Koski, W. R., Thomas, T. A., Richardson, W. J., Würsig, B., & Trites, A. W. (2013). Seismic operations have variable effects on dive-cycle behaviour of bowhead whales in the Beaufort Sea. *Endangered Species Research, 549*, 243–262.

Stone, C. J., & Tasker, M. L. (2006). The effects of seismic airguns on cetaceans in UK waters. *Journal of Cetacean Research and Management, 8*(3), 255–263.

Stone, C. J., Hall, K., Mendes, S., & Tasker, M. L. (2017). The effects of seismic operations in UK waters: Analysis of Marine Mammal Observer data. *Journal of Cetacean Research and Management, 16*, 71–85.

Thompson, P. M., Brookes, K. L., Graham, I. M., Barton, T. R., Needham, K., Bradbury, G., & Merchant, N. D. (2013). Short-term disturbance by a commercial two-dimensional seismic survey does not lead to long-term displacement of harbour porpoises. *Proceedings of the Royal Society B, 280*. Available from https://doi.org/10.1098/rspb.2013.2001, 20132001.

Tyack, P., Johnson, M., & Miller, P. (2003). Tracking responses of sperm whales to experimental exposures of airguns. Pp. 115-120 in A.E. Jochens and D.C. Biggs, eds. Sperm whale seismic study in the Gulf of Mexico/Annual Report: Year 1. OCS Study MMS 2003-069. Prepared by Texas A&M University, College Station, TX, for MMS, Gulf of Mexico OCS Region, New Orleans, LA.

Yazvenko, S. B., McDonald, T. L., Blokhin, S. A., Johnson, S. R., Meier, S. K., Melton, H. R., . . . Wainwright, P. W. (2007). Distribution and abundance of western gray whales during a seismic survey near Sakhalin Island, Russia. *Environmental Monitoring and Assessment, 134*(1-3), 45–73.

Yazvenko, S. B., McDonald, T. L., Blokhin, S. A., Johnson, S. R., Melton, H. R., Newcomer, M. W., . . . Wainwright, P. W. (2007). Feeding of western gray whales during a seismic survey near Sakhalin Island, Russia. *Environmental Monitoring and Assessment, 134*(1-3), 93–106. Available from https://doi.org/10.1007/s10661-007-9810-3.

(d) Marine construction: explosions, dredging and drilling

Anderwald, P., Brandecker, A., Coleman, M., Collins, C., Denniston, H., Haberlin, M. D., . . . Walshe, L. (2013). Displacement responses of a mysticete, an odontocete, and a phocid seal to construction-related vessel traffic. *Endangered Species Research, 21*, 231–240.

Benda-Beckmann, A., Aarts, G., Sertlek, H. O., Lucke, K., Verboom, W. C., Kastelein, R. A., . . . Ainslie, M. A. (2015). Assessing the impact of underwater clearance of unexploded ordnance on harbour porpoises (*Phocoena phocoena*) in the Southern North Sea. *Aquatic Mammals, 41*(4), 503–523. Available from https://doi.org/10.1578/AM.41.4.2015.503.

Benjamins, S., Dale, A., Hastie, G., Lea, M., Scott, B. E., Waggitt, J. J., & Wilson, B. (2015). Confusion reigns? A review of marine megafauna interactions with energetic tidal features. *Oceanography and Marine Biology: An Annual Review, 53*, 1–54.

Brandt, M. J., Diederichs, A., Betke, K., & Nehls, G. (2011). Responses of harbour porpoise to pile driving at the Horns Rev II offshore wind farm in the Danish North Sea. *Marine Ecology Progress Series, 421*, 205–216.

Brandt, M. J., Dragon, A. C., Diederichs, A., Bellmann, M. A., Wahl, V., Piper, W., . . . Nehls, G. (2018). Disturbance of harbour porpoises during construction of the first seven offshore wind farms in Germany. *Marine Ecology Progress Series, 596*, 213–232.

Brownlow, A., Baily, J., Dagleish, M., Deaville, R., Foster, G., Jensen, S.-K., . . . & Jepson, P. (2015). *Investigation into the long-finned pilot whale mass stranding event, Kyle of Durness, 22nd July 2011* (60 pp.). Scottish Marine Animal Stranding Scheme, SRUC Wildlife Unit, Inverness.

Carstensen, J., Henriksen, O. D., & Teilmann, J. (2006). Impacts of offshore wind farm construction on harbour porpoises: Acoustic monitoring of echolocation activity using porpoise detectors (T-PODs). *Marine Ecology Progress Series, 321*, 295–308.

Culloch, R., Anderwald, P., Brandecker, A., Haberlin, D., McGovern, B., Pinfield, R., . . . Cronin, M. (2016). Effect of construction-related activities and vessel traffic on marine mammals. *Marine Ecology Progress Series, 549*, 221–242. Available from https://doi.org/10.3354/meps11686.

Dähne, M., Gilles, A., Lucke, K., Peschko, V., Adletr, S., Krügel, K., . . . Siebert, U. (2013). Effects of pile-driving on harbour porpoises (*Phocoena phocoena*) at the first offshore wind farm in Germany. *Environmental Research Letters, 8*(025002) 16pp. Available from https://doi.org/10.1088/1748-9326/8/2/025002.

Evans, P. G. H. (2008). Offshore wind farms and marine mammals: Impacts and methodologies for assessing impacts. In: *Proceedings of the ASCOBANS/ECS Workshop held San Sebastián, Spain, 22 April 2007, European Cetacean Society Special Publication Series* (pp. 1–68), 49.

Gilles, A., Scheidat, M., & Siebert, U. (2009). Seasonal distribution of harbour porpoises and possible interference of offshore wind farms in the German North Sea. *Marine Ecology Progress Series, 383*, 295–307.

Ketten, D. R. (1995). Estimates of blast injury and acoustic zones for marine mammals from underwater explosions. In R. A. Kastelein, J. A. Thomas, & P. E. Nachtigall (Eds.), *Sensory Systems of aquatic mammals* (pp. 391–407). Woerden, NL: De Spil Publishers.

Koschinski, S. (2011). Underwater noise pollution from munitions clearance and disposal, possible effects on marine vertebrates, and its mitigation. *Marine Technology Society Journal, 45*(6), 80–88. (9).

Koschinski, S. & Kock, C.-H. (2015). Underwater unexploded ordnance – Methods for a Cetacean-friendly removal of explosives as alternatives to blasting. ASCOBANS, AC22/Inf.4.6.e.

Lindeboom, H. J., Kouwenhoven, H. J., Bergman, M. J. N., Bouma, S., Brasseur, S., Daan, R., & Scheidat, M. (2011). Short-term ecological effects of an offshore wind farm in the Dutch coastal zone: a compilation. *Environmental Research Letters, 6*(035101) 13pp. Available from https://doi.org/10.1088/1748-9326/6/3/035101.

Madsen, P. T., Wahlberg, M., Tougaard, J., Lucke, K., & Tyack, P. (2006). Wind turbine underwater noise and marine mammals: Implications of current knowledge and data needs. *Marine Ecology Progress Series, 309*, 279–295.

Mann, J., & Teilmann, J. (2013). Environmental impact of wind energy. *Environmental Research Letters, 8*, 035001. Available from https://doi.org/10.1088/1748-9326/8/3/035001.

Nedwell, J. R., Parvin, S. J., Edwards, B., Workman, R., Brooker, A. G. & Kynoch, J. E. (2008). *Measurement and interpretation of underwater noise during construction and operation of offshore windfarms in UK waters.* (Subacoustech Report No. 544R0736), COWRIE Ltd, ISBN: 978-0-9554279-5-4.

Scheidat, M., Tougaard, J., Brasseur, S., Carstensen, J., Petel, T. V. P., Teilmann, J., & Reijnders, P. (2011). Harbour porpoises (*Phocoena phocoena*) and wind farms: A case study in the Dutch North Sea. *Environmental Research Letters, 6*. Available from https://doi.org/10.1088/1748-9326/6/2/025102.

Teilmann, J., & Carstensen, J. (2012). Negative long term effects on harbour porpoises from a large scale offshore wind farm in the Baltic - evidence of slow recovery. *Environmental Research Letters, 7*(4), 045101 10pp. Available from https://doi.org/10.1088/1748-9326/7/4/045101.

Teilmann, J., Tougaard, J., Carstensen, J., & Dietz, R. (2006). Marine mammals. In J. Kjær, J. K. Larsen, C. Boesen, H. H. Corlin, S. Andersen, S. Nielsen, A. G. Ragborg, & K. M. Christensen (Eds.), *Danish offshore wind - Key environmental issues* (pp. 80–93). Copenhagen: DONG Energy, Vattenfall, The Danish Energy Authority and The Danish Forest and Nature Agency, 142 pp.

Thomsen, F. (2010). Marine mammals. In J. Huddleston (Ed.), *Understanding the environmental impacts of offshore windfarms* (pp. 52–57). Cowrie, 138 pp.

Tougaard, J., Carstensen, J., Bech, N. I., & Teilmann, J. (2006). *Final report on the effect of nysted offshore wind farm on harbour porpoises.* Technical report to Energi E2 A/S. <http://www.ens.dk/graphics/Energiforsyning/Vedvarende_energi/Vind/havvindmoeller/vvm%20Horns%20Rev%202/Nysted/Nysted%20marsvin%20final.pdf>.

Tougaard, J., Carstensen, J., Teilmann, J., Skov, H., & Rasmussen, P. (2009). Pile driving zone of responsiveness extends beyond 20 km for harbor porpoises (*Phocoena phocoena* (L.)). *Journal of the Acoustical Society of America, 126*(1), 11–14. Available from https://doi.org/10.1121/1.3132523.

Tougaard, J., Carstensen, J., Wisz, M. S., Jespersen, M., Teilmann, J., Bech, N. I., & Skov, H. (2006). *Harbour porpoises on horns reef effects of the horns reef wind farm.* Final Report to Vattenfall A/S. <http://www.ens.dk/graphics/Energiforsyning/Vedvarende_energi/Vind/havvindmoeller/vvm%20Horns%20Rev%202/begge%20parker/Porpoises%20Horns%20Reef%202006%20final.pdf>.

Wilson, B., Batty, R. S., Daunt, F., & Carter, C. (2007). *Collision risks between marine renewable energy devices and mammals, fish and diving birds.* Report to the Scottish Executive, Scottish Association for Marine Science, Oban, Scotland, PA37 1QA.

(e) Sonar

Arbelo, M., Espinosa de los Monteros, A., Herraez, P., Andrada, M., Sierra, E., Rodriguez, F., Jepson, P. D., & Fernández, A. (2013). Pathology and causes of death of stranded cetaceans in the Canary Islands (1999–2005). *Diseases in Aquatic Organisms, 103*, 87–99. Available from https://doi.org/10.3354/dao02558.

Bernaldo de Quirós, Y., Fernandez, A., Baird, R. W., Brownell, R. L., Jr, Aguilar de Soto, N., Allen, D., ... Fahlman, A. (2019). Advances in research on the impacts of anti-submarine sonar on beaked whales. *Proceedings of the Royal Society B, 286*, 20182533. Available from https://doi.org/10.1098/rspb.2018.2533.

Cox, T. M., Ragen, T. J., Read, A. J., Vos, E., Baird, R. W., Balcomb, K., ... Benner, L. (2006). Understanding the impacts of anthropogenic sound on beaked whales. *Journal of Cetacean Research and Management, 7*(3), 177–187.

D'Amico, A. D., Gisiner, R., Ketten, D. R., Hammock, J. A., Johnson, C., Tyack, P., & Mead, J. (2009). Beaked whale strandings and naval exercises. *Aquatic Mammals, 35*, 452–472.

DeRuiter, S. L., Southall, B. L., Calambokidis, J., Zimmer, W. M. X., Sadykova, D., Falcone, E. A., ... Tyack, P. L. (2013). First direct measurements of behavioural responses by Cuvier's beaked whales to mid-frequency active sonar. *Biology Letters, 9*. Available from https://doi.org/10.1098/rsbl.2013.0223, 20130223.

Dolman, S. J., Pinn, E., Reid, R. J., Barley, J. P., Deaville, R., Jepson, P. D., ... Simmonds, M. P. (2010). A note on the unprecedented stranding of 56 deep-diving odontocetes along the UK and Irish coast. *Marine Biodiversity Records, 3*, 1–8.

Evans D.L., & England G.R. (2001) *Joint Interim Report Bahamas Marine Mammal Stranding Event of 15-16 March 2000*. Unpublished Report. 61pp. Available at www.nmfs.noaa.gov/prot_res/overview/Interim_Bahamas_Report.pdf.

Evans, P. G. H. & Miller, L. A. (Eds.) (2004). Active sonar and cetaceans. In: *Proceedings of workshop held at the ECS 17th annual conference, Las Palmas, Gran Canaria, 8th March 2003* (84 pp.). Kiel: European Cetacean Society.

Fernández, A., Arbelo, M., Deaville, R., Patterson, I., Castro, P., Baker, J., . . . Jepson, P. (2004). Pathology: Whales, sonar and decompression sickness. *Nature, 428*, 1−2.

Fernández, A., Edwards, J. F., Rodríguez, F., Espinosa de los Monteros, A., Herráez, P., Castro, P., . . . Arbelo, M. (2005). Gas and fat embolic syndrome involving a mass stranding of beaked whales (family Ziphiidae) exposed to anthropogenic sonar signals. *Veterinary Pathology, 42*, 446−457.

Frantzis, A. (1998). Does acoustic testing strand whales? *Nature (London), 392*, 29.

Goldbogen, J. A., Southall, B. L., DeRuiter, S. L., Calambokidis, J., Friedlaender, A. S., Hazen, E. L., . . . Tyack, P. L. (2013). Blue whales respond to simulated mid-frequency military sonar. *Proceedings of the Royal Society B, 280*. Available from https://doi.org/10.1098/rspb.2013.0657, 20130657.

Jepson, P. D. & Deaville, R. (compilers) (2009). *Investigation of the common dolphin mass stranding event in Cornwall, 9th June 2008*. Report to the Department for Environment, Food and Rural Affairs (under a variation to Contract CR0364). UK Cetacean Strandings Investigation Programme, Institute of Zoology, London, 30 pp. <http://randd.defra.gov.uk/Document.aspx?Document = WC0601_8031_TRP.pdf>

Jepson, P. D., Deaville, R., Patterson, I. A., Pocknell, R., Ross, H. M., Baker, J. R., . . . Cunningham, A. A. (2005). Acute and chronic gas bubble lesions in cetaceans stranded in the United Kingdom. *Veterinary Pathology, 42*, 291−305.

Kastelein, R. A., van den Belt, I., Helder-Hoek, L., Gransier, R., & Johansson, T. (2015). Behavioral Responses of a Harbor Porpoise (*Phocoena phocoena*) to 25-kHz FM Sonar Signals. *Aquatic Mammals, 41*(3), 311−326. Available from https://doi.org/10.1578/AM.41.3.2015.311.

Kvadsheim, P. H., DeRuiter, S., Sivle, L. D., Goldbogen, J., Roland-Hansen, R., Miller, P. J. O., . . . Southall, B. (2017). Avoidance responses of minke whales to 1-4kHz naval sonar. *Marine Pollution Bulletin, 121*(1-2), 60−68. Available from https://doi.org/10.1016/j.marpolbul.2017.05.037, Epub 2017 May 25.

Wensveen, P. J., Isojunno, S., Hansen, R. R., von Benda-Beckmann, A. M., Kleivane, L., van IJsselmuide, S., . . . Miller, P. J. O. (2019). Northern bottlenose whales in a pristine environment respond strongly to close and distant navy sonar signals. *Proceedings of the Royal Society B, 286*, 20182592. Available from https://doi.org/10.1098/rspb.2018.2592.

Simmonds, M. P., & Brown, V. C. (2010). Is there a conflict between cetacean conservation and marine renewable-energy developments? *Wildlife Research, 37*, 688−694.

Tyack, P. L., Zimmer, W. X. M., Moretti, D., Southall, B. L., Claridge, D. E., Durban, J., . . . Boyd, I. L. (2011). Beaked whales respond to simulated and actual navy sonar. *PLoS One, 6*(3), e17009. Available from https://doi.org/10.1371/journal.pone.0017009.

Vessel strikes

Evans, P. G. H. (2003). Shipping as a possible source of disturbance to cetaceans in the ASCOBANS region. In: *ASCOBANS 4th meeting of the parties, Esbjerg, Denmark, 19-22 August 2003* (88 pp.). Document MOP4/Doc. 17(S).

Evans, P. G. H., Baines, M. E., & Anderwald, P. (2011). *Risk assessment of potential conflicts between shipping and Cetaceans in the ASCOBANS region* (32 pp.). ASCOBANS AC18/Doc. 6-04 (S).

IWC (2008). *Third progress report to the conservation committee of the Ship Strike Working Group*. International Whaling Commission Scientific Committee IWC/60/CC3, International Whaling Commission.

Laist, D. W., Knowlton, A. R., Mead, J. G., Collet, A. S., & Podesta, M. (2001). Collisions between Ships and Whales. *Marine Mammal Science, 17*(1), 35−75.

Pesante, G., Panigada, S., & Zanardelli, M. (eds.) (2002). Collisions between cetaceans and vessels: Can we find solutions? *ECS Newsletter No. 40* (40 pp) (Special Issue).

Vanderlaan, A. S. M., & Taggart, C. T. (2007). Vessel collisions with whales: The probability of lethal injury based on vessel speed. *Marine Mammal Science, 23*(1), 144−156.

Recreational disturbance

Bejder, L., & Samuels, A. (2003). Evaluating the effects of nature based tourism on cetaceans. In N. Gales, M. Hindell, & R. Kirkwood (Eds.), *Marine mammals and humans: Towards a sustainable balance.* (pp. 229−256). Collingwood: CSIRO Publishing.

Constantine, R., Brunton, D. H., & Dennis, T. (2004). Dolphin-watching tour boats change bottlenose dolphin (*Tursiops truncatus*) behaviour. *Biological Conservation, 117*, 299−307.

Evans, P. G. H. (1996). Human disturbance of cetaceans. In N. Dunstone, & V. Taylor (Eds.), *The exploitation of mammals - Principles and problems underlying their sustainable use* (pp. 279−298). Cambridge: Cambridge University Press.

Feingold, D. & Evans, P. G. H. (2014). *Bottlenose dolphin and harbour porpoise monitoring in Cardigan Bay and Pen Llyn a'r Sarnau special areas of conservation 2011−2013* (124 pp) Natural Resources Wales Evidence Report Series No. 4.

Hastie, G. D., Wilson, B., Tufft, L. H., & Thompson, P. M. (2003). Bottlenose dolphins increase breathing synchrony in response to boat traffic. *Marine Mammal Science, 19*(1), 74−84.

Higham, J. E. S., Bejder, L., Allen, S. J., Corkeron, P. J., & Lusseau, D. (2016). Managing whale-watching as a non-lethal consumptive activity. *Journal of Sustainable Tourism, 24*(1), 73−90.

Higham, J. E. S., Bejder, L., & Williams, R. (Eds.), (2014). *Whale-watching: Sustainable tourism and ecological management.* Cambridge: Cambridge University Press, 418 pp.

Hoyt, E. (2001). *Whale Watching 2001: Worldwide Tourism Numbers, Expenditures, and Expanding Socioeconomic Benefits*. USA: IFAW, Yarmouth.

Janik, V. M., & Thompson, P. M. (1996). Changes in surfacing patterns of bottlenose dolphins in response to boat traffic. *Marine Mammal Science, 12*(4), 597−602.

Lusseau, D. (2004). The hidden cost of tourism: Detecting long-term effects of tourism using behavioral information. *Ecology and Society, 9*(2), 15 pp.

Lusseau, D. (2006). The short-term behavioural reactions of Bottlenose dolphins to interactions with boats in Doubtful Sounds, New Zealand. *Marine Mammal Science, 22*(4), 802−818.

Lusseau, D., & Higham, J. E. S. (2004). Managing the impacts of dolphin-based tourism through the definition of critical habitats: The case of bottlenose dolphins (*Tursiops* spp.) in Doubtful Sound, New Zealand. *Tourism Management, 25*, 657−667.

Lusseau, D., Slooten, E., & Currey, R. J. (2006). Unsustainable dolphin watching activities in Fiordland, New Zealand. *Tourism in Marine Environments, 3*, 173−178.

Mattson, M. C., Thomas, J. A., & Aubin, D. S. (2005). Effects of boat activity on the behaviour of bottlenose dolphins (*Tursiops truncatus*) in waters surrounding Hilton Head Island, South Carolina. *Aquatic Mammals, 31*(1), 133−140.

New, L. F., Harwood, J., Thomas, L., Donovan, C., Clark, J. S., Hastie, G., ... Lusseau, D. (2013). Modelling the biological significance of behavioural change in coastal bottlenose dolphins in response to disturbance. *Functional Ecology, 27*, 314−322.

Nowacek, S. M., Wells, R. S., & Solow, A. R. (2001). Short-term effects of boat traffic on bottlenose dolphins, *Tursiops truncatus* in Sarasota Bay, Florida. *Marine Mammal Science, 14*(4), 673−688.

O'Connor, S., Campbell, R., Knowles, T., & Cortez, H. (2009). *Whale Watching Worldwide: Tourism numbers, expenditures and expanding economic benefits*. USA: IFAW, Yarmouth.

Pierpoint, C., Allan, L., Arnold, H., Evans, P., Perry, S., Wilberforce, L., & Baxter, J. (2009). Monitoring important coastal sites for bottlenose dolphin in Cardigan Bay, UK. *Journal of the Marine Biological Association of the United Kingdom, 89*, 1033−1043.

Pirotta, E., Merchant, N. D., Thompson, P. M., Barton, T. R., & Lusseau, D. (2015). Quantifying the effect of boat disturbance on bottlenose dolphin foraging activity. *Biological Conservation, 181*, 82−89.

Richardson, H. (2012). *The effect of boat disturbance on the bottlenose dolphin (Tursiops truncatus) of Cardigan Bay in Wales*. (M.Sc. thesis) (71 pp.). University College London.

Thompson, K. (2012). *Variations in whistle characteristics of bottlenose dolphins (Tursiops truncatus) in Cardigan Bay, Wales*. (M.Sc. thesis) (62 pp.). University of Bangor.

Williams, R., Trites, A. W., & Bain, D. E. (2002). Behavioural responses of killer whales (*Orcinus orca*) to whale-watching boats: opportunistic observations and experimental approaches. *Journal of Zoology, 256*, 255−270.

Williams, R., Lusseau, D., & Hammond, P. S. (2006). Estimating relative energetic costs of human disturbance to killer whales (*Orcinus orca*). *Biological Conservation, 133*, 301−311.

Würsig, B., & Evans, P. G. H. (2001). Cetaceans and humans: Influences of noise. In P. G. H. Evans, & J. A. Raga (Eds.), *Marine mammals: Biology and conservation* (pp. 555−576). London: Kluwer Academic/Plenum Press, 630 pp.

Climate change

ACIA. (2004). *Impacts of a warming Arctic: Arctic climate impact assessment*. Cambridge: Cambridge University Press.

Caldeira, K., & Wickett, M. E. (2003). Anthropogenic carbon and ocean pH. *Nature, 425*, 365.

Daunt, F., & Mitchell, I. (2013). Impacts of climate change on seabirds. *MCCIP Science Review, 2013*, 125−133. Available from https://doi.org/10.14465/2013.arc14.125-133.

Derocher, E., Lunn, N., & Stirling, I. (2004). Polar bears in a warming climate. *Integrative and Comparative Biology, 44*, 163−176.

Doney, S. C., Fabry, V. J., Feely, R. A., & Kleypas, J. A. (2009). Ocean acidification: The other CO_2 problem. *Annual Review of Marine Science, 1*, 169−192.

Dye, S. R., Highes, S. L., Tinker, J., Berry, D. I., Holliday, N. P., Kent, E. C., ... Beszczynska-Möller. (2013). Impacts of climate change on temperature (air and sea). *Marine Climate Change Impacts Partnership (MCCIP) Science Review, 2013*. Available from https://doi.org/10.14465/2013.arc15.001-012, 1-12. Published online 28 November 2013.

Evans, P. G. H., & Bjørge, A. (2013). Impacts of climate change on marine mammals. *Marine Climate Change Impacts Partnership (MCCIP) Science Review, 2013*, 134−148. Available from https://doi.org/10.14465/2013.arc15.134-148, Published online 28 November 2013.

Evans, P. G. H., Boyd, I. L., & MacLeod, C. D. (2010). *Impacts of climate change on marine mammals*. Marine Climate Change Impacts Partnership (MCCIP) Annual Report Card 2009-2010 Scientific Review: 1−14.

Evans, P. G. H., Pierce, G. J., & Panigada, S. (2010). Climate change and marine mammals. *Journal of the Marine Biological Association of the United Kingdom, 90*, 1483−1488.

Ferguson, S., Stirling, I., & McLoughlin, P. (2005). Climate change and ringed seal (*Phoca hispida*) recruitment in western Hudson Bay. *Marine Mammal Science, 21*, 121−135.

Climate change and marine top predators. In M. Frederiksen, & T. Haug (Eds.), *Frontiers in ecology and evolution*. Lausanne: Frontiers Media. Available from https://doi.org/10.3389/978-2-88919-736-1.

Hammond, P. S., Macleod, K., Berggren, P., Borchers, D. L., Burt, M. L., Cañadas, A., ... Vázquez, J. A. (2013). Cetacean abundance and distribution in European Atlantic shelf waters to inform conservation and management. *Biological Conservation, 164*, 107−122.

Hughes, S.L., Holliday, N.P., Kennedy, J., Berry, D.I., Kent, E.C., Sherwin, T., ... Smyth, T. (2010). Temperature (Air and Sea) in MCCIP Annual Report Card 2010-11, MCCIP Science Review 2010: 1−16. http://www.mccip.org.uk/arc.

Hughes, S.L., Tinker, J., Dye, S., Andres, O., Berry, D.L., Hermanson, L., ... Smyth, T. (2017) Temperature. MCCIP Science Review 2017: 22–41. http://www.mccip.org.uk/media/1750/2017arc_sciencereview_003_tem.pdf.

Huntington, H. P., & Moore, S. E. (Eds.), (2008). Arctic marine mammals and climate change. *Ecological Applications, 18*(Supplement), S1–174.

IPCC. (2007). *Climate Change 2007. Intergovernmental panel on climate change (IPCC). fourth assessment report.* Cambridge and New York: Cambridge University Press. <https://www.ipcc.ch/>.

IWC. (1997). Report of the IWC workshop on climate change and cetaceans. *Reports of the International Whaling Commission, 47,* 293–313.

IWC. (2009). Report of the IWC workshop on cetaceans and climate change. *IWC/SC/61/Report, 4,* 1–31.

Knight, J. R., Allan, R. J., Folland, C. K., Vellinga, M., & Mann, M. E. (2005). A signature of persistent natural thermohaline circulation cycles in observed climate. *Geophysical Research Letters, 32,* L20708. Available from https://doi.org/10.1029/2005GL024233.

Laidre K. L., Stirling I., Lowry L., Wiig Ø., Heide-Jørgensen M. P., & Ferguson S. H. (2008) Quantifying the sensitivity of arctic marine mammals to climate-induced habitat change. In Huntington H.P. & Moore S.E. (Eds.) Arctic marine mammals and climate change. Ecological Applications, 18 (Supplement): S97–S125.

Learmonth, J. A., MacLeod, C. D., Santos, M. B., Pierce, G. J., Crick, H. Q. P., & Robinson, R. A. (2006). Potential effects of climate change on marine mammals. *Oceanography and Marine Biology: An Annual Review, 44,* 431–464.

MacLeod, C. D. (2009). Global climate change, range changes and potential implications for the conservation of marine cetaceans, a review and synthesis. *Endangered Species Research, 7,* 125–136.

MCCIP. (2010). Marine climate change impacts. In J. M. Baxter, J. P. Buckley, & C. J. Wallace (Eds.), *Annual Report Card 2010–2011. Summary Report.* Lowestoft: MCCIP, 12 pp.

Morison, J. K., Aagaard, K., & Steele, M. (2000). Recent environmental changes in the Arctic: a review. *Arctic, 53,* 359–371.

Pinnegar, J. K., & Heath, M. (2010). Impacts of climate change on fish in marine climate change impacts partnership (MCCIP) annual report card 2009-10. *MCCIP Science Review,* 1–23. Available from www.mccip.org.uk/arc.

Rayner, N. A., Parker, D. E., Horton, E. B., Folland, C. K., Alexander, L. V., Rowell, D. P., ... Kaplan, A. (2003). Global analyses of sea surface temperature, sea ice, and night marine air temperature since the late Nineteenth Century. *Journal of Geophysical Research, 108,* 4407. Available from https://doi.org/10.1029/2002JD002670.

Stirling, I., Lunn, N. J., & Iacozza, J. (1999). Long-term trends in the population ecology of polar bears in western Hudson Bay in relation to climate change. *Arctic, 52,* 294–306.

Walsh, J. E. (2008). Climate of the Arctic marine environment. *Ecological Applications, 18*(Supplement), S3–S22.

Würsig, B., Reeves, R. R., & Ortega-Ortiz, J. G. (2001). Global climate change and marine mammals. In P. G. H. Evans, & J. A. Raga (Eds.), *Marine mammals - biology and conservation* (pp. 589–608). New York: Kluwer Academic/Plenum Publishers, 630 pp.

Boxed contributions

ASCOBANS (2000). *Report of the 7th meeting of the advisory committee.* Bruges, 13–16 March 2000. <http://www.ascobans.org/pdf/ac7/ac7-finalreport.pdf>

ASCOBANS (2009). *Recovery* plan for baltic harbour porpoises *(Jastarnia Plan) - Revision.* ASCOBANS, Bonn. <http://www.service-board.de/ascobans_neu/files/MOP6-7-01_RevisionJastarniaPlan.pdf>

Carstensen, J., Henriksen, O. D., & Teilmann, J. (2006). Impacts of offshore wind farm construction on harbour porpoises: Acoustic monitoring of echolocation activity using porpoise detectors (T-PODs). *Marine Ecology Progress Series, 321,* 295–308.

Cooke, J. G., Deimer, P., & Schütte, H. J. (2006). *Opportunistic sightings of harbour porpoises (Phocoena phocoena) in the Baltic Sea: 3rd and 4th seasons (2004-2005) AC 13/Doc. 23(P).* ASCOBANS, Bonn.

Deimer, P., Schütte, H. J., & Wilhelms, S. (2003). *Opportunistic sightings of harbour porpoises (Phocoena phocoena) in the Baltic Sea. AC10/Doc.8 (P).* ASCOBANS, Bonn.

Gilles, A., Scheidat, M., & Siebert, U. (2009). Seasonal distribution of harbour porpoises and possible interference of offshore wind farms in the German North Sea. *Marine Ecology Progress Series, 383,* 295–307.

Hammond, P. S., Bearzi, G., Bjørge, A., Forney, K., Karczmarski, L., Kasuya,T. ... & Wilson, B. (2008). *Phocoena phocoena (Baltic Sea subpopulation). IUCN red list of threatened species.* <www.iucnredlist.org>

Herr, H., Siebert, U., & Benke, H. (2009). *Stranding numbers and by-catch implications of harbour porpoises along the German Baltic Sea Coast. AC16/Doc.62(P).* ASCOBANS, Bonn.

Koschinski, S., & Pfander, A. F. (2009). *By-catch of harbour porpoises (Phocoena phocoena) in the Baltic coastal waters of Angeln and Schwansen (Schleswig-Holstein, Germany): AC16/Doc.60(P),* ASCOBANS, Bonn.

Lindeboom, H. J., Kouwenhoven, H. J., Bergman, M. J. N., Bouma, S., Brasseur, S., Daan, R., & Scheidat, M. (2011). Short-term ecological effects of an offshore wind farm in the Dutch coastal zone: a compilation. *Environmental Research Letters, 6*(035101) 13pp. Available from https://doi.org/10.1088/1748-9326/6/3/035101.

Loos, P. (2009). *Opportunistic Sightings of harbour porpoises in the Baltic at large - Kattegat, Belt Sea, Sound, Western Baltic and Baltic Proper.* (Bachelor thesis). University of Hamburg. <www.gsm-ev.de>.

Scheidat, M., Tougaard, J., Brasseur, S., Carstensen, J., Van Polanen-Petel, T., Teilmann, J., & Reijnders, P. (2011). Harbour porpoises (*Phocoena phocoena*) and wind farms: A case study in the Dutch North Sea. *Environmental Research Letters, 6,* 025102.

Simmonds, M. P., & Elliot, W. J. (2009). Climate change and cetaceans: Concerns and recent developments. *Journal of the Marine Biological Association of the United Kingdom., 89*(1), 203–210.

Teilmann, J., & Carstensen, J. (2012). Negative long term effects on harbour porpoises from a large scale offshore wind farm in the Baltic - evidence of slow recovery. *Environmental Research Letters, 7*, 045101.

Teilmann, J., Tougaard, J., & Carstensen, J. (2008) Effects from offshore wind farms on harbour porpoises in Denmark. In: Proceedings of the ASCOBANS/ECS workshop: offshore wind farms and marine mammals: impacts and methodologies for assessing impacts. pp. 50−59. 21 April 2007, San Sebastian, Spain. (Editor P.G.H. Evans). ECS (European Cetacean Society) Special Publication Series No. 49. 68pp.

Teilmann, J., Sveegaard, S., & Dietz, R., (2011). *Status of a harbour porpoise population - evidence of population separation and declining abundance*. In Sveegard 2010. *Spatial and temporal distribution of harbour porpoises in relation to their prey*. (PhD thesis). University of Arhus.

Tougaard, J., Carstensen, J., Teilmann, J., Skov, H., & Rasmussen, P. (2009). Pile driving zone of responsiveness extends beyond 20km for harbour porpoises (*Phocoena phocoena*, (L.)). *Journal of the Acoustical Society of America, 126*, 11−14.

Tougaard, J., Carstensen, J., Wisz, M. S., Teilmann, J., Bech, N. I., & Skov, H. (2006). *Harbour porpoises on horns reef in relation to construction and operation of horns rev offshore wind farm*. Technical Report to Elsam Engineering A/S, Roskilde: National Environmental Research Institute.

Weilgart, L. (2007). The impacts of anthropogenic ocean noise on cetaceans and implications for management. *Canadian Journal of Zoology, 85*, 1091−1116.

Chapter 6

Conservation research

A conservation agreement cannot function without specific information about the animals it is designed to conserve. That may seem rather obvious but this fundamental requirement is actually rarely obtained because it is generally rather difficult to achieve. This applies particularly to marine mammals such as cetaceans that spend 95% of their lives out of sight of us.

Population structure

The first requirement is one of the most difficult of all — to define the population that is under investigation. Whereas it is comparatively easy to map the overall range of a species, it is much more difficult to draw boundaries around different populations unless there are obvious physical barriers. Island populations of land mammals such as mice frequently form distinct races because the sea around them represents an unfavourable habitat for them and so there is no mixing with other populations. Cetaceans on the other hand are not limited in the same way. It is usually only populations of those species that inhabit rivers or enclosed seas, which become isolated from one another. Thus porpoises inhabiting the Black Sea are genetically distinct from those in the Atlantic, since the species is largely absent from the Mediterranean Sea in between. The same applies to a lesser extent to porpoises living in the Baltic Sea compared with those from Danish waters or from the North Sea. Sometimes a species may have particular habitat preferences that ties it to localised areas. There is some indication, for example that beaked whales may form relatively isolated populations around deepwater canyons or basins, and coastal bottlenose dolphins may inhabit estuaries or bays for extended periods, isolating themselves from offshore populations. More often, however, it is ocean currents and the plankton, fish and cephalopod prey associated with these that influence the distribution of a particular cetacean species. Such currents may shift over time, consequently affecting all the organisms associated with them, but nevertheless there may be long-standing relationships with particular areas that can lead to populations becoming differentiated from one another.

A variety of methods are used to analyse for genetic differences. The most common markers that have been used include the mitochondrial DNA control region, which unlike other markers is exclusively maternally inherited; and microsatellites, which are repeating sequences of base pairs randomly distributed in the nuclear DNA. The mtDNA control region and nuclear microsatellite loci, generally found in noncoding regions, show higher levels of polymorphism than other markers due to a relatively high mutation rate, and so they tend to be more sensitive in detecting fine-scale population structure. Single nucleotide polymorphisms, which are single base pair substitutions among DNA sequences, form a new set of markers that offer much potential. They encompass the entire genome and provide higher quality and more efficient typing of the genome whilst the resulting data can be analysed relatively simply. In the past, genetic analyses were conducted only upon dead animals, using tissues or organs derived from strandings or bycaught specimens. Nowadays, samples for many species are routinely taken from skin and blubber biopsies using a long pole, or by firing a dart using a rifle or crossbow. And in some cases, DNA has been extracted using skin swabs, by collecting faeces, or even by sampling the blows of whales from an aerial drone.

One might think that for conservation management, population structure must best be studied through molecular genetics. However, this is not necessarily the case. First, any genetic differences observed may reflect evolutionary aspects of population separation involving tens, hundreds or thousands of generations rather than contemporary population structure. Second, there can often be sampling problems. Frequently, sample sizes used for comparisons are small (i.e. less than 30); they have been collected over varying periods of time; and their true origins may not be known (e.g. stranded or bycaught animals that have been washed ashore; and even live strandings may not reflect

European Whales, Dolphins, and Porpoises. DOI: https://doi.org/10.1016/B978-0-12-819053-1.00006-5

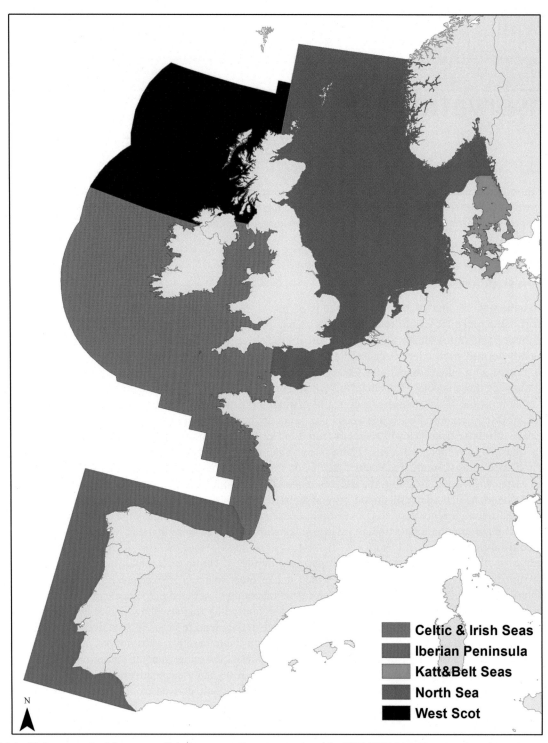

Celtic & Irish Seas
Iberian Peninsula
Katt&Belt Seas
North Sea
West Scot

FIGURE 6.1 Harbour porpoise Management Units in western Europe, as proposed by ICES WGMME. Note that these have since been further revised - see harbour porpoise species account in Chapter 4. *Courtesy of UK Joint Nature Conservation Committee.*

the area/population from which they originate). For all these reasons, the use of a suite of approaches has increasingly been advocated, incorporating both genetic and ecological information. The concept of the Management Unit has developed (see Evans & Teilmann, 2009; Moritz, 1994; Palsbøll, Bérubé, & Allendorf, 2007; Taylor & Dizon, 1999) (Figs 6.1 and 6.2). This focuses more upon defining populations that are demographically independent of one another,

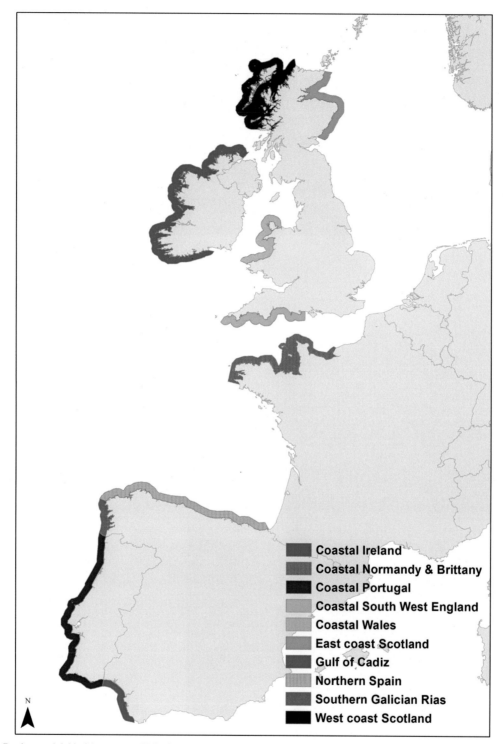

FIGURE 6.2 Bottlenose dolphin Management Units in western Europe, as proposed by ICES WGMME. *Courtesy of UK Joint Nature Conservation Committee.*

where population dynamics depend largely upon local birth and death rates rather than immigration. In this way the emphasis is upon the contemporary dispersal rate of individuals rather than the historical amount of gene flow.

A variety of ecological markers have been used to characterise separate populations. Radio telemetry (Fig. 6.3) and photoidentification of recognisable individuals (Fig. 6.4) allow one to examine ranging behaviour and establish whether

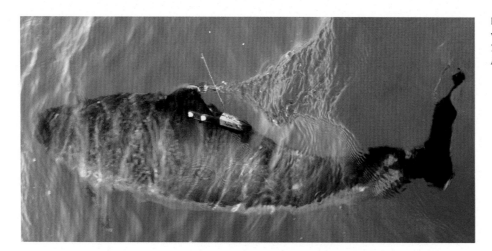

FIGURE 6.3 Harbour porpoise with satellite radio tag. *Photo: Jonas Teilmann, University of Aarhus, Denmark.*

FIGURE 6.4 Coastal populations of bottlenose dolphins may form discrete communities, as revealed by photoidentification studies of individual animals. *Photo: Peter GH Evans.*

groups of individuals mix with one another (Fig. 6.5). These can provide fine-scale information, although practical or resource limitations may mean that this information comes from only a small sample of animals. Many researchers have used other ecologically linked methods to distinguish between populations. These include taking measurements of skeletal characters, examining the diets through stomach contents analysis or from the blubber by deriving fatty acid profiles that can reflect the species of fish taken, or stable isotope signatures that can indicate the trophic level at which animals have been feeding. More indirect methods that have been used include examination of the parasite faunas of

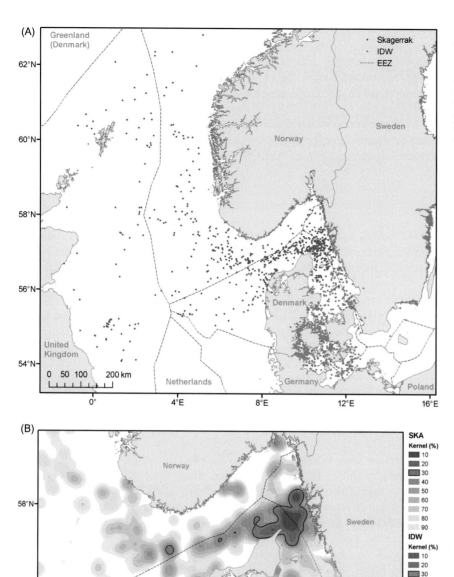

FIGURE 6.5 (A) Locations of tagged harbour porpoise. Green circles indicate those tagged in inner Danish waters; blue circles indicate those tagged in the Skagerrak. (B) Harbour porpoise hotspots as revealed from kernel analysis of the locations of tagged animals (*green areas* are animals tagged in inner Danish waters; *blue areas* are animals tagged in the Skagerrak). *Signe Sveegaard.*

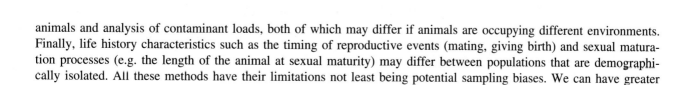

animals and analysis of contaminant loads, both of which may differ if animals are occupying different environments. Finally, life history characteristics such as the timing of reproductive events (mating, giving birth) and sexual maturation processes (e.g. the length of the animal at sexual maturity) may differ between populations that are demographically isolated. All these methods have their limitations not least being potential sampling biases. We can have greater confidence that differences between groups of animals are indeed reflecting separate management units when a variety of markers all indicate the same result.

Ideally, one should establish the relationship between the demographic characteristics and the population genetics dynamics of any target species since animals may mix together in the same areas, eat the same food, acquire the same parasite burdens, but not actually mate with one another so there is no gene exchange. That is often the greatest challenge and one that we rarely achieve.

In practice, most researchers operate over a limited area, which is considerably less than the range of the study population. And even when there is collaboration between researchers, these are often organised nationally, driven by political and funding infrastructures. A conservation agreement such as ASCOBANS therefore has a great opportunity to promote international collaborative projects.

Counting animals

A great deal of research conducted in the ASCOBANS Agreement Area involves surveys by boat or by plane, counting animals, and using the results to better understand distribution patterns and trends in abundance. The task is not an easy one, and although one aspires to obtain an accurate measure of absolute abundance of a population in a prescribed area, rarely can that be fully achieved.

The survey method most widely used to count cetaceans is the line-transect, and the platform for such surveys can be a vessel or an aircraft (Evans & Hammond, 2004; Hammond, 2010). Since it is rarely feasible to survey an entire area fully, lines are generally laid out using a random allocation scheme that provides representative coverage of the area for which a density or abundance estimate is desired (Buckland et al., 2001, 2004; Strindberg, Buckland, & Thomas, 2004). This enables one to extrapolate from a sample area by simply multiplying the sample density by the survey area. A survey design that gives representative coverage is one where each point in the area has an equal probability of being sampled. We do this because animals are not distributed randomly in space, and a computer program is commonly used to generate the lines, such as that which forms part of the Distance v.6 software that is routinely used to calculate abundance (Thomas et al., 2010).

The two survey designs most frequently used are sets of equally spaced parallel lines or a regular zigzag pattern. The former is generally used in aerial surveys and the latter in vessel surveys. If the objective is to produce estimates of average abundance/density, then new lines should be chosen each time. However, where the focus is upon measuring trends, then the same lines can be repeated each time (Thomas, Burnham, & Buckland, 2004). Transects should as far as possible run perpendicular to any density gradient (Buckland et al., 2001, 2004). Coastal surveys, for example typically have transects that run more or less perpendicular to the shore line or zigzag away from it.

Line-transects assume that all animals are detected along the track-line but of course that is rarely the case. Animals may be out of sight underwater, which creates what we call availability bias, or they may be at the surface but we miss them, perhaps because they were behind a wave or we simply did not notice them. We call that perception bias. The probability of detection along the track-line is referred to as g(0). In order to try to overcome these biases, if possible two observers or sets of observers are employed, usually one at a greater height than the other. We call the main observers primary observers and the others independent observers. By comparing the proportion of sightings seen by one set but not the other and those seen by both, we can calculate what proportion were missed and thus obtain a better estimate of the actual density. In a plane it is rarely possible to have an extra set of independent observers so one can either have two planes flying in tandem and compare the results of each (but this is expensive) or the plane can turn and go back over its track (termed the circle-back method) in order to get a second comparative sample. One compromise is to have a third observer on a rear seat recording sightings on the side which has the best sighting conditions, thus providing one double-counted strip on each transect (Diederichs et al., 2008; Grünkorn, Diederichs, & Nehls, 2005).

In order to calculate densities, we need to estimate the distance of the animal(s) from the track-line. This is referred to as the perpendicular distance and can be calculated by simple trigonometry by measuring or estimating the range and measuring the angle of the sighting to the vessel. By plotting these, one can determine the width over which one is effectively searching since the further away the animals are from the observer, the more difficult it is to see them. This also varies with the sea conditions. When the sea is very calm, one may spot animals one or more kilometres away whereas when it is a bit rougher, that can be reduced to just a few hundred metres. As soon as white caps appear on the waves (referred to as three on the Beaufort scale), it becomes much more difficult to distinguish animals surfacing and so one attempts to initiate line-transects only when the sea state is two or less on this scale.

Aerial surveys do not have quite the same constraints since one is looking directly down on the animals, and the width over which one is searching can be well-defined (often with markers on the wings) so long as one flies at a constant height. It is necessary to accurately measure the height of the plane throughout the survey using a radar altimeter. The angle of declination is collected using a handheld inclinometer to determine the perpendicular distance to the sighting from the track-line. An advantage of a plane is that animals usually do not have time to respond to it whereas with a vessel they have much more time to do so. If a cetacean hears the sound of a ship's engine a few kilometres away it may respond to it before the observer has seen it. Some species such as harbour porpoise tend to move away from vessels whereas others such as common and white-beaked dolphins tend to move towards vessels. In the case of the former occurring, this can negatively bias one's density estimates, whilst in the case of the latter it can cause a positive bias.

Large vessels (Fig. 6.6A) tend to be used for systematic line-transect surveys because they offer a stable and high platform, and sufficient space to house a large team, which is a particular advantage if the vessel is required to remain at sea for extended periods. They also offer a greater opportunity to operate double platforms for determining the probability of detection along the track-line, or at least, that which is affected by perception bias. Nevertheless, constraints of

FIGURE 6.6 (A) Larger vessels are most often used for line-transect surveys to estimate absolute abundance as they can generally provide higher observation platforms and have accommodation for teams to stay aboard overnight. (B) A small vessel used for regionally-based surveys. Note the double platform built upon the wheelhouse. *Photos: (A) Pia Anderwald; (B) Peter GH Evans.*

SCANS III / Nino Pierantonio

FIGURE 6.7 (A) A high-winged plane typically used for aerial surveys. (B) Observers aboard a plane in preparation for an aerial survey. *Photos: (A) Steve CV Geelhoed. (B) Nino Pierantonio/ SCANS III.*

habitat (e.g. shallow depth, complex coastlines) and resources (e.g. charter costs) often require the use of smaller vessels for surveys (Dawson, Wade, Slooten, & Barlow, 2008; Evans & Thomas, 2011) (Fig. 6.6B). Where possible, the observation height for the independent observer(s) should be at least 5 m as this allows them to see over a greater distance. This is particularly important for determining whether the animals are responding to the vessel.

Increasingly surveys are being conducted by plane (Fig. 6.7A,B) because although they can be expensive, they can cover a large area in a short space of time, making the best use of good weather conditions. Typically, boats survey at 8−10 knots (as it is important to travel faster than the cetaceans would be doing on average to avoid double counting) whereas planes travel at 90−110 knots (165−205 km/h) at a height of 500−600 ft. (150−180 m). The rapid travel across an area by a plane can introduce greater availability bias, particularly for those species that make long dives, and thus spend extended periods of time underwater. In areas of low density, aerial surveys may therefore not be very effective because they yield too small sample sizes. On the other hand, the availability process from a plane is quite simple: either the animal is at (or very near) the surface when the plane goes over, or it is not. This 'instantaneous' availability process is much easier to account for (assuming some knowledge of animal surfacing rates) than the 'intermittent' availability process that is more realistic in slower-moving surface vessels, where animals are intermittently either more-or-less continuously available at the surface for some periods of time and then below the surface for others. Hence, although availability bias is a greater problem for aerial surveys, it is easier to account for.

The choice of plane is very important. Partenavia PN68 aircraft are the preferred option, although larger twin otters, a two-motor Cessna, or a Britten Norman BN 2 Islander have been used in some surveys. The plane should have twin engines for safety and be high-winged, preferably with a bubble window beside each rear seat for ease of viewing (from abeam to the track-line). A belly port has the additional advantage of providing downward visibility (and, if necessary, deployment of a video camera), although most planes do not have this facility.

During surveys, it is important that observers do not become too tired as that will lead to loss in concentration. For this reason, one tends to have a team of observers and rotate them, and to have a separate person recording effort and the environmental conditions. This is obviously easier in a large vessel than a small one, and is further constrained in a plane, although in the latter case an aerial survey rarely lasts more than 5 hours, and observers can take a brief rest between transect lines if they are run parallel to one another.

Other forms of surveying cetaceans are increasingly being tested. These include the use of high-definition photography on planes, the deployment of aerial drones equipped with cameras, and even high-resolution satellite imagery from space.

I have mentioned a number of the potential biases that one tries to overcome or at least evaluate when conducting surveys of abundance. There are further issues to take into account. Assumptions are made that animals will be correctly identified, their numbers counted accurately, and the distances and angles measured accurately. These can be challenging, particularly when observing very large groups of cetaceans. Another issue that is difficult to address is that cetaceans move around in space and time. Ideally one should try to survey simultaneously the full range of the species (or its demographically distinct populations, or management units) but that is very difficult and expensive, and has never been fully undertaken. Snapshot wide-scale surveys such as North Atlantic sightings surveys (NASS), Small Cetacean Abundance in the North Sea (SCANS) and Cetacean Offshore Distribution and Abundance in European Atlantic waters (CODA), referred to earlier, are brave attempts but are never fully comprehensive. Their cost prohibits coverage at more frequent intervals than once every 10 years or so, and inevitably they provide no information on what is happening in the years and seasons in between surveys. For that, some more regular form of surveillance is required. Although northwest European countries have generally improved their effort, there are large gaps in space and time, making the results quite difficult to interpret. The precision of abundance estimates also depends heavily upon the number of encounters, which reflects how common the species is. Thus uncommon species such as killer whale, beaked whales and coastal bottlenose dolphin that may be at greatest conservation risk are not easily counted by this means.

Estimating the abundance of cetaceans in European Atlantic waters

Introduction

During the 1980s scientists and naturalists working in western Europe became increasingly aware of the extent to which harbour porpoises were being accidentally killed in fishing nets in the North Sea and adjacent waters. At that time there were no robust estimates of the magnitude of this mortality but the evident scale of bycatch was sufficient to cause considerable concern about its sustainability. Porpoises were being killed primarily in nylon bottom-set gill and tangle nets; fishing methods that had been widely adopted in the 1970s.

We now know that bycatch in Danish gillnet fisheries in the North Sea amounted to several thousand porpoises a year in the late 1980s, increasing to a peak estimated at more than 7000 animals in the early 1990s (Vinther & Larsen, 2004). More than 2000 porpoises per year were estimated to have been killed in English and Irish hake fisheries in the Celtic Sea at the same time (Tregenza, Berrow, Hammond, & Leaper, 1997). The problem was not restricted to European waters (e.g. Bravington & Bisack, 1996, and bycatch remains a major threat to small cetaceans worldwide (Read, Drinker, & Northridge, 2006).

Concerns about harbour porpoise bycatch and the need for coordinated action to obtain better information and to focus appropriate mitigation action, were drivers for the formation of the European Cetacean Society in 1987 (http://www.europeancetaceansociety.eu/) and the establishment of ASCOBANS in 1991 (http://www.ascobans.org/). They also featured in the development of the EU Habitats Directive (Council Directive 92/43/EEC).

Although it was widely believed that numbers of harbour porpoises were declining in European waters as a result of unregulated bycatch, this was based on limited information. There were no estimates of the size of the populations being impacted, which were known to range widely over shelf waters. In February 1992 a group of scientists from several European countries met in the Netherlands and put together a project proposal with the main objective of estimating the abundance of harbour porpoises in as large a part of their range as possible, thus allowing estimates of bycatch to be put in a population context. The proposal was supported by ASCOBANS, and the project was eventually funded by the European Commission's LIFE programme and the governments of Denmark, France, Germany, Ireland, the Netherlands, Norway, Sweden and the United Kingdom, and WWF Sweden. It became known as SCANS — Small Cetacean Abundance in the North Sea and adjacent waters.

(Continued)

(Continued)

FIGURE 6.8 Vessel used for SCANS survey. *Photo: Valeiras, Courtesy of Philip S Hammond.*

SCANS 1994

The late 1980s was a time of rapid development of methods to estimate the size of cetacean populations. Frameworks for data collection and analysis were being established for using line-transect sampling techniques on cetacean surveys (Hiby & Hammond, 1989) and for using mark-recapture methods applied to photoidentification data (Hammond, 1986). However, these methods were initially focused on large whales and there was a real need to develop survey methods suitable for small cetaceans. Harbour porpoises, in particular, are small and cryptic, which presents considerable challenges to obtaining robust estimates of abundance.

The SCANS project was a catalyst for developing methods that would allow unbiased estimates of abundance to be obtained from shipboard and aerial surveys. The main challenge was to find a way to correct for animals missed on the transect line, a key assumption of line-transect sampling. This involved estimating the proportion of animals actually seen on the transect line, known as g(0). This was done by building on earlier work by Buckland and Turnock (1992) to develop so-called 'double team' methods for shipboard survey (Borchers, Buckland, Goedhart, Clarke, & Hedley, 1998) and by developing an innovative method for two aircraft flying in tandem for aerial survey (Hiby & Lovell, 1998). The method developed for shipboard surveys (Fig. 6.8) also has the potential to correct for any movement of animals in response to the survey ship, another important assumption of the method. These new methods were trialled and modified on a pilot survey in April 1994. The methods developed during this period are still largely those in use today (Hammond, 2010).

The SCANS survey itself took place in July 1994, covering an area of 1 million km^2 with 27,000 km of transect line searched (Fig. 6.9A) and generating estimates of abundance, in round numbers, of 340,000 harbour porpoises, 8000 white-beaked dolphins and 8500 minke whales (Hammond et al., 2002).[1] These estimates have inherent uncertainty and it is equally important to present this information in the form of either coefficients of variation (0.14; 0.30 and 0.24, respectively) or as 95% confidence intervals (260,000–450,000; 4000–13,000 and 5000–13,500, respectively).

The large number of porpoises estimated to be in the area surveyed, including approximately 250,000 in the North Sea and 36,000 in the Celtic Sea, was a surprising result to many. However, the annual bycatch in these two areas as a percentage of these abundance estimates was approximately 3% in the North Sea and 6% in the Celtic Sea; mortality rates that were unlikely to be sustainable, particularly the latter.

The results of the SCANS survey made an immediate impact and have been widely used as a baseline for cetacean abundance in European Atlantic waters by EU Member States and a number of international organisations including the International Whaling Commission (IWC), International Council for Exploration of the Sea (ICES) and ASCOBANS.

(Continued)

---

1. These estimates of abundance have since been revised — see below.

(Continued)

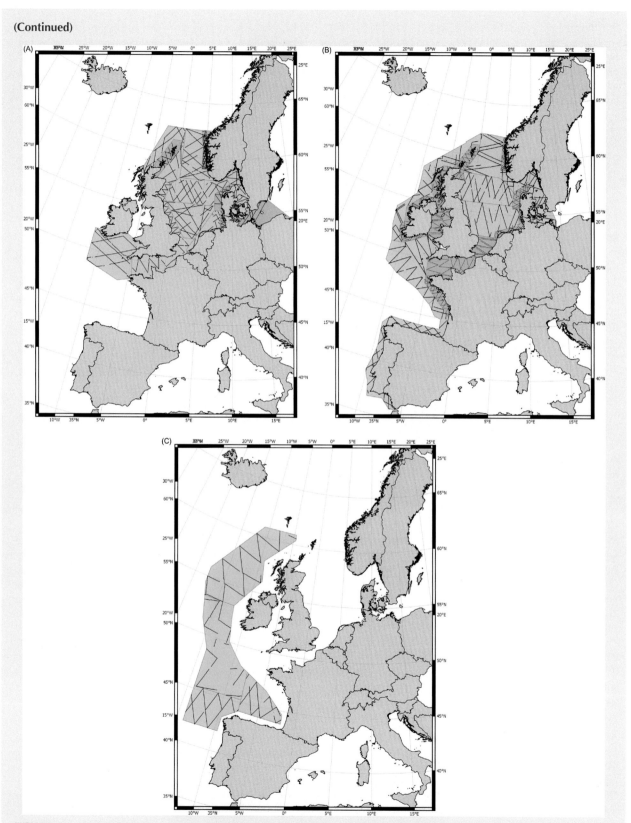

FIGURE 6.9 (A) SCANS Survey transects, July 1994. Blue, vessel; red, aerial. (B) SCANS-II Survey transects, July 2005. Blue, vessel; red, aerial. (C) CODA Survey transects, July 2007. Blue, vessel. *(A−C) Reproduced with permission from Philip S Hammond/SMRU.*

(Continued)

(Continued)

Gill and tangle net fishing in the North Sea declined during the late 1990s and early 2000s, and resulting estimates of harbour porpoise bycatch also declined. However, this continued to be a cause for concern (along with bycatch of common dolphins) and the European Council issued regulations 812/2004 and 814/2004, which made mandatory the monitoring of bycatch by observers in selected fleets and the use of acoustic devices (pingers) to reduce bycatch by vessels over 12 m in length involved in some gill or tangle net fisheries.

Around the same time, plans for a second major survey were taking shape, driven by the need to obtain up-to-date estimates of abundance for species subject to bycatch and to inform Habitats Directive reporting by EU Member States on the conservation status of all cetaceans. Planning was initially for a survey covering all European Atlantic waters but logistical realities meant that the Small Cetaceans in the European Atlantic and North Sea (SCANS-II) project focused only on continental shelf waters from 62°N to the Strait of Gibraltar (Fig. 6.9B). Additional project aims were to develop improved methods for monitoring, previously reviewed by Evans and Hammond (2004), and to develop a robust management framework for bycatch (Winship, 2009). SCANS-II was supported by the European Commission's LIFE Nature programme, and by the governments of Belgium, Denmark, France, Germany, Ireland, the Netherlands, Norway, Poland, Portugal, Spain, Sweden, and the United Kingdom.

SCANS-II 2005

The methods developed in the SCANS project had largely stood the test of time so preparatory work for SCANS-II focused mainly on improving certain aspects of data collection methodology. In particular, for shipboard surveys the methods for accurate measurement of angles and distances to sightings were improved and an integrated data collection system was developed (Gillespie, Leaper, Gordon, & Macleod, 2011). The tandem aircraft method used in SCANS was replaced by the more efficient single aircraft circle-back method (Hiby, 1999). A pilot survey in April was again invaluable in fine-tuning these methods prior to the survey in July 2005.

SCANS-II covered an area of 1.37 million km^2 with 36,000 km of transect line searched. Estimates of abundance obtained were, in round numbers, 380,000 harbour porpoises, 16,500 white-beaked dolphins, 56,000 common dolphins, 16,500 bottlenose dolphins and 19,000 minke whales (Hammond et al., 2013). Coefficients of variation of these estimates were 0.20, 0.30, 0.23, 0.42 and 0.35, respectively; 95% confidence intervals were: 260,000−550,000, 9000−30,000, 36,000−88,000, 7500−36,000 and 10,000−37,000, respectively.

The encouraging results were that there was no indication for a decline in abundance overall in any of the three species that were surveyed in both 1994 and 2005; these results have proved valuable to EU Member States for reporting on conservation status under the Habitats Directive. The distribution of harbour porpoises was markedly different between the two surveys. The main concentration of animals in the North Sea had shifted from the northwest in 1994 to the southwest in 2005, the high densities around coastal Denmark in 1994 had dissipated in 2005, and densities in the Celtic Sea were higher in 2005 than 1994 (Fig. 6.11A and B). The most likely major cause of this difference in distribution is a change in prey availability but interannual variability or the possibility of an impact of bycatch in some areas could not be excluded.

CODA 2007

Mindful of the need for information on cetaceans in deeper waters, project CODA was developed and supported by the governments of France, Ireland, Spain and the United Kingdom. A shipboard survey covered a large proportion of European Atlantic offshore waters, 3 million km^2, with 47,000 km of transect line searched (see Fig. 6.9C). The inshore boundary of the CODA survey area was designed to be the same as the offshore boundary of the SCANS-II survey area two years previously and the data from both these surveys, and from the Faroes block of the Trans North Atlantic Sightings Survey conducted in 2007, were analysed together to obtain estimates of abundance for cetaceans over the widest possible extent in European Atlantic waters.

Estimates of abundance obtained from this combined data analysis were 174,000 (CV = 0.27) common dolphins, 61,000 (CV = 0.93) striped dolphins, 36,000 (CV = 0.21) bottlenose dolphins, 124,000 (CV = 0.35) long-finned pilot whales, 27,000 (CV = 0.35) minke whales, 19,000 (CV = 0.24) fin whales, 2600 (CV = 0.26) sperm whales and 13,000 (CV = 0.21) beaked whales (all species). More details are given in Hammond et al. (2011) for baleen whales and Rogan et al. (2017) for deep-diving species.

SCANS-III 2016

More by chance than planning the third SCANS survey in summer 2016, maintained the interval of 11 years between these large-scale multinational collaborations. This time, as well as the need to consider current abundance in the context of ongoing fisheries bycatch and assessments of conservation status under the Habitats Directive, an additional driver was to provide information that could be used as a basis for the first assessment of environmental status under the EU's Marine Strategy Framework Directive, which came into force in 2008. The MSFD requires assessment at a regional scale and that Member States cooperate and coordinate − both features at the heart of the SCANS ambition.

(Continued)

(Continued)

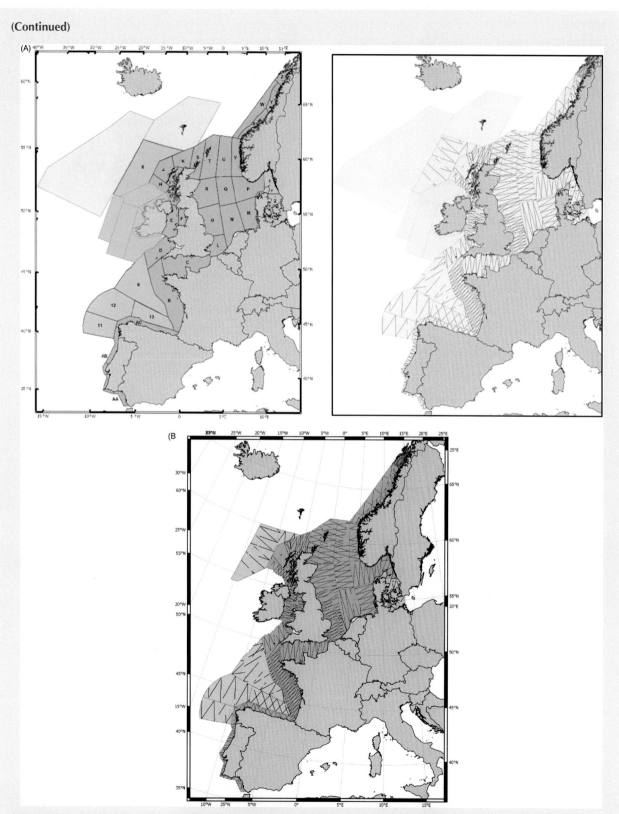

FIGURE 6.10 (A) Area covered by SCANS-III and adjacent surveys. (B) Track-lines used in SCANS-III: pink lettered blocks were surveyed by air; blue numbered blocks were surveyed by ship. Blocks coloured green to the south and west of Ireland were surveyed by the Irish ObSERVE project. Blocks coloured yellow were surveyed by the Faroe Islands as part of the North Atlantic Sightings Survey in 2015. *Reproduced with permission from Philip S Hammond/SMRU.*

(Continued)

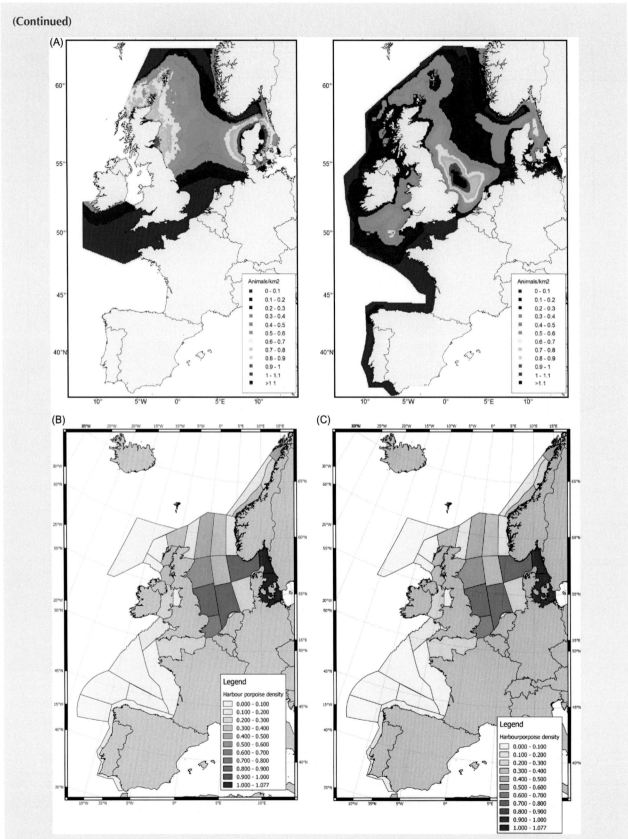

FIGURE 6.11 Harbour porpoise distribution in (A) 1994, (B) 2005 and (C) 2016. *Reproduced from Hammond et al., 2013, with permission from Philip S Hammond/SMRU.*

(Continued)

(Continued)

TABLE 6.1 Revised estimates of abundance from SCANS and SCANS-II (combined with CODA in 2007) compared to estimates for 2016.

Species	Region	1994		2005/07		2016	
		N	CV	N	CV	N	CV
Harbour porpoise	All	407,177	0.18	519,864	0.21	466,569	0.15
Bottlenose dolphin	All			35,936	0.21	27,697	0.23
White-beaked dolphin	All	23,716	0.30	37,689	0.36	36,287	0.29
Common dolphin	All			174,485	0.27	467,673	0.26
Striped dolphin	All			61,364	0.93	372,340	0.33
Unidentified common and/or striped dolphin	All			306,045	0.29	998,180	0.18
Pilot whale	All			123,732	0.35	20,656	0.40
Beaked whales (all species)	All			12,869	0.31	11,618	0.61
Sperm whale	All			2569	0.26	9599	0.41
Minke whale	All	9685	0.23	26,784	0.35	13,101	0.35
Fin whale	All			19,354	0.24	18,142	0.33
Harbour porpoise	North Sea	289,150	0.14	355,408	0.22	345,373	0.18
White-beaked dolphin	North Sea	22,619	0.23	29,010	0.35	20,453	0.36
Minke whale	North Sea	7495	0.15	9890	0.34	8854	0.24

SCANS-III covered a total of 1.8 million km^2 using the equivalent data collection methodology employed on previous surveys. For the first time, all continental shelf waters were surveyed by air, except for the Kattegat/Belt Seas area, enabling estimates of g(0) to be made from the circle-back method for dolphins and for minke whale, as well as harbour porpoise. The area surveyed was the same as for SCANS-II and CODA combined, with the addition of Norwegian coastal waters north to Vestfjorden but excluding Irish waters to the south and west of Ireland, which were surveyed as part of a separate project called ObSERVE (Fig. 6.10).

Estimates of abundance for the most frequently encountered species between 62°N and the Strait of Gibraltar are 440,000 (CV = 0.15) harbour porpoise, 36,000 (CV = 0.44) white-beaked dolphin, 470,000 (CV = 0.26) common dolphin, 370,000 (CV = 0.33) striped dolphin, 28,000 (CV = 0.23) bottlenose dolphin, 26,000 (CV = 0.35) long-finned pilot whale, 15,000 (CV = 0.33) minke whale, 18,000 (CV = 0.33) fin whale, 14,000 (CV = 0.41) sperm whale and 11,000 (CV = 0.50) beaked whales (all species). The distribution of porpoise densities remained broadly similar to that in 2005 (Fig. 6.11A,B,C).

To assess how abundance changes over time, it is important that it is estimated consistently. Improvements in methods of analysis over the years and the availability of new estimates of g(0) for aerial surveys for dolphins and minke whale, meant that analyses of data from SCANS and SCANS-II needed to be updated to ensure that comparisons with new estimates from SCANS-III were not confounded with differences in how data had been analysed. This revision has led to the updated estimates in Table 6.1.

The most striking difference in the results for 2016 compared to 2005/07 is the much larger estimates of abundance for common and striped dolphins totalling almost one million animals. Although much greater than in 2005/07, the 2016 estimates are nevertheless consistent with results from the French SAMM surveys in French waters of the Bay of Biscay and the Channel in summer 2012, which totalled almost 700,000 common and striped dolphins (Laran et al., 2017).

However, there are now three consistent estimates of abundance for 1994, 2005 and 2016 for the three main species occurring in the North Sea — harbour porpoise, white-beaked dolphin and minke whale, to which can be added an additional five estimates for minke whale from the Norwegian Independent Line-transect surveys. For these species in this region, the trend in abundance can be examined and power analysis can inform us of the annual rate of change that could be detected with high power from these data. There are also multiple consistent estimates of harbour porpoise abundance in the Skagerrak/Kattegat/Belt Seas area.

Fig. 6.12 shows trend lines fitted to these data and Table 6.2 gives details of the fitted lines and the results of calculations of the power to detect trends, given these data. The results of the trend analyses show that there is no statistical support for a change in abundance over the period covered by the surveys for any species/region. The results of the power calculations show that the data used have 80% power to detect annual rates of decline of 1.8% for harbour porpoise, 5% for white-beaked dolphin and 0.5% for minke whale in the North Sea and 3.7% for harbour porpoise in the Skagerrak/Kattegat/Belt Seas. The CVs used in these power calculations assume there is no "additional variance" resulting from any variation in the number of animals in the area in different years, independent of any trend. That is, it is assumed that there is a single discrete population that is not subject

(Continued)

(Continued)

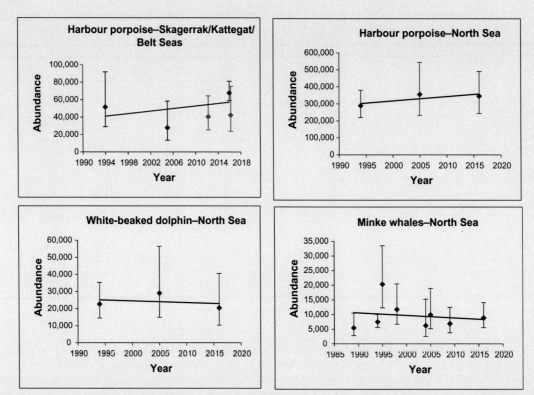

FIGURE 6.12 Estimates of abundance (error bars are log-normal 95% confidence intervals) for harbour porpoise in the Skagerrak/Kattegat/ Belt Seas area (*blue dots* and *line*) and Kattegat/Belt Seas ICES Assessment Unit (AU) (*red dots*) (top left); harbour porpoise in the North Sea AU (top right); white-beaked dolphin in the North Sea (bottom left); minke whale in the North Sea (bottom right). Trend lines are fitted to time series of more than two abundance estimates. All estimates are from SCANS surveys except Kattegat/Belt Seas in 2012 (Viquerat et al., 2014), and minke whale in the North Sea in 1989, 1994, 1998, 2004 and 2009 (Norwegian Independent Line-transect Surveys; Schweder, Skaug, Dimakos, Langas, & Øien, 1997; Skaug, Øien, Schweder, & Bøthun, 2004; Solvang, Skaug, & Øien, 2015).

TABLE 6.2 Estimated annual rates of change for species/regions where there are more than two estimates of abundance, and results of power calculations.

Species	Region	n	Estimated annual rate of change	P	CV	Annual rate of decline detectable at 80% power (%)
Harbour porpoise	Skagerrak/ Kattegat/Belt Seas	3	1.24% (95% CI −39; 67%)	0.81	0.30	3.7
Harbour porpoise	North Sea	3	0.8% (95% CI −6.8; 9.0%)	0.42	0.18	1.8
White-beaked dolphin	North Sea	3	−0.5% (95% CI −18; 22%)	0.82	0.36	5
Minke whale	North Sea	8	−0.25% (95% CI −4.8; 4.6%)	0.90	0.30	0.5

n is the number of abundance estimates. P is the probability of obtaining this rate of change by chance ($P > .05$ indicates no significant change). Power was calculated using the simplified inequality (Gerrodette, 1987): $r^2 n^3 > 12CV^2 (Z_{\alpha/2} + Z_\beta)^2$ where r is the rate of change estimated from the data. CV is the coefficient of variation of abundance, $Z_{\alpha/2}$ is the value of a standardised random normal variable for the probability of making a Type I error, α (set to 0.05), Z_β is the value of a standardised random normal variable for the probability of making a Type II error, β, and power is $(1 - \beta)$ expressed as a percentage.

(Continued)

(Continued)

to any movement of animals in or out of the area between years when estimates of abundance are available. If this does occur, the rates of change that would be detectable would be larger. For example, if the CV for harbour porpoise abundance in the North Sea were actually 0.27 (50% greater than the estimated 0.18), the annual rate of decline detectable would be around 3% instead of around 2%.

Concluding remarks — lessons learned from the SCANS experience

Overall the results from these large-scale international surveys have greatly expanded our knowledge of the distribution and abundance of cetacean species in the European Atlantic, enabling bycatch to be placed in a population context and giving a strong basis for assessments of conservation status. The information now available forms a good foundation for a large-scale time series for the coming decades. Although developed independently, the surveys have to some extent developed in parallel with ASCOBANS, with SCANS taking place in the year that ASCOBANS came into force and CODA preceding by a year the formal expansion of the Agreement area.

The data collection and analytical methodology used on SCANS surveys to account for animals missed on the transect line and for potential responsive movement prior to detection from ships is more complex and requires more resources than conventional surveys. Is this necessary? For assessing trends, it may be possible to assume that any biases are consistent across years and therefore to ignore any correction for this, although variation in survey conditions and survey vessels would likely result in more variable estimates of abundance. However, to assess the impact of bycatch or other anthropogenic pressures on cetacean populations, estimates of absolute (unbiased) abundance are required, at least periodically, and SCANS-type two-team survey methods are needed to achieve this.

SCANS-type surveys as stand-alone projects require considerable resources focused at one point in time. However, considering their decadal-scale frequency and the number of countries involved (up to 10), the annual cost per country is small. Even if the frequency were increased to match EU Directive reporting cycles of 6 years, they should be readily affordable. Although there have been three successful SCANS projects, they do not form a programme of surveys; each one has been developed from scratch by a team of dedicated scientists. If European Atlantic Range States value the information provided by SCANS it would be more appropriate for future surveys to be driven by government agencies responsible for implementing national and European policy. ASCOBANS is well placed to be a focus for coordinating a future SCANS programme perhaps in a way similar to the role played by its sister Agreement in driving and coordinating the ACCOBAMS Survey Initiative in the Mediterranean and Black Seas.

Philip Hammond
Sea Mammal Research Unit, University of St Andrews

Photoidentification

Since the 1980s the technique of photoidentification (known as Photo-ID) and mark—recapture has been applied widely to a variety of cetacean species (Hammond, 1986, 2002, 2010; Hammond, Mizroch, & Donovan, 1990; Würsig & Jefferson, 1990). Not only can Photo-ID provide estimates of population size, and thus over a period of time, population trends, it can also reveal useful information on movements, home range size and use, habitat use, life history parameters and social structure. The method when applied to cetaceans usually relies upon the reliable identification of individuals from natural markings such as nicks in the dorsal fin (Fig. 6.13) or pigmentation patterns on the body or tail of an animal. However, a number of species are not particularly amenable to the technique either because they are difficult to see clearly and photograph, or they are poorly marked, or identifiable markings are not long-lasting.

The mark—recapture (or more accurately, capture-mark-recapture, often abbreviated to CMR) technique as applied to cetaceans usually means 'capturing' an image using a digital single-lens reflex camera equipped with a telephoto lens. In that image should be recognisable features that are unique to that individual. This information is collected on encountering a group of cetaceans, and then further groups are sampled, and the proportion of already identified versus new animals is used to calculate the population size from multiple encounters.

As with line-transect distance sampling a number of assumptions are made and these can also be challenging to meet. Those most relevant to cetaceans are that: (1) a marked animal will always be recognised if it is seen again; (2) samples of individuals must be representative of the population being estimated (in other words there should be full mixing between sampling occasions) and (3) within any one sampling occasion, every animal in the population should have the same probability of being captured.

The most common violations of these assumptions in cetacean Photo-ID studies are that: (1) there is insufficient information in the mark to guarantee no duplicates, marks change or fade, or are erroneously recorded; (2) individuals have different capture probabilities (termed heterogeneity of capture probabilities) because they do not mix randomly due to differences in preference for particular areas, some may be more difficult to photograph, or are more distinctively marked (and therefore more easily recognised) than others. As a result, some individuals will be seen more often

FIGURE 6.13 Bottlenose dolphin individuals are typically identified by unique patterns of nicks in their dorsal fin. Depicted here are two examples (A, B) of different individuals from the community inhabiting Cardigan Bay, west Wales. *Photos: (A) Pia Anderwald; (B) Peter GH Evans.*

than expected and others less often and (3) for what we term 'closed' population models, the population is actually not closed, with births and deaths occurring, and either permanent or nonrandom temporary immigration or emigration.

The software program CAPTURE has models that can take account of heterogeneity of capture probabilities (Rexstadt & Burnham, 1991), whilst if animals are believed to emigrate temporarily from the study area, there are also methods for taking this into account in the analysis (Kendall, Nichols, & Hines, 1997; Whitehead, 1990).

If sampling occurs on two occasions close together in time, between which one can assume that births, deaths, immigration and emigration will be negligible, a closed population model can be applied. If the study has multiple sampling occasions, a time series of estimates can be obtained and there is more flexibility in analysis, and one can use open or closed population models (Hammond, 1986). Pollock's robust design model (Kendall, Pollock, & Brownie, 1995) for an open population assumes that no mortality, immigration or emigration takes place within sampling sessions. The probability of an animal being captured at least once in any year can be estimated solely from the data collected during that year, using closed population models, whereas the longer intervals between years allow estimation of survival, temporary emigration from the area sampled, and immigration of marked animals back to that area.

To minimise some of the problems of violation of model assumptions, where possible efforts should be made to sample photographically as many individuals as possible, obtain good photographs (in good light, left and right sides for fin notches, and at right angles), grade images according to quality and use only the best. This requires choosing sampling sessions carefully and using good photographers with the appropriate equipment.

In northern Europe, Photo-ID has been used mainly on bottlenose dolphin (Fig. 6.13), Risso's dolphin, common dolphin, white-beaked dolphin, long-finned pilot whale, northern bottlenose whale, killer whale, sperm whale, humpback whale (Fig. 6.14), minke whale and fin whale.

The identification of individual animals not only enables one to estimate numbers, it can be used for many other useful scientific investigations. Birth rates can be calculated from known mothers together with growth and survival rates of their young. The movements of individuals can be tracked by comparing catalogues of images to identify matches, and with lots of resightings one can start to build up a picture of the range of known individual animals. The social structure of groups can be investigated by determining who associates with whom. Thus just by carefully taking high quality photographs allowing individual identification, a rich amount of information can be gained.

Acoustics

Acoustic monitoring of cetaceans has the advantage that animals can potentially be detected in all weathers and during both day and night. Automated data collection reduces personnel costs whilst providing continuous monitoring in a standardised manner for extended periods. Because detection ranges are limited, at least for dolphins and porpoises that

FIGURE 6.14 Humpback whales may show distinct individual patterns on the underside of their tail flukes when diving. Depicted here are two examples (A, B) of different individuals photographed at the same time in the Hebrides, Scotland. *Photos: (A) Peter GH Evans; (B) Nick Davies.*

produce high frequency sounds (whistles in the case of the former, and clicks in both cases), monitoring is likely to be applicable to only a small area. Static acoustic monitoring (SAM) provides a cost-effective means for establishing trends in how particular areas (e.g. within a marine protected area, or in conjunction with certain human activities such as pile driving) are being used by cetaceans − between years, seasonally and over shorter time periods (daily and tidal cycles). Towed hydrophones are used in conjunction with a dedicated vessel survey or may be mounted on the hull of a vessel like a ferry.

By its nature, acoustic monitoring requires animals to be vocal and detectable. This works well with species like the harbour porpoise that echolocates more or less continuously, or the common dolphin that whistles frequently, but it may not necessarily provide a representative account of the presence of a species such as minke whale, which appears to vocalise more at particular seasons.

Estimates of absolute abundance from acoustic detections remain difficult to obtain, particularly when animals occur in groups, although much progress has been made in recent years with the development of hydrophone arrays, algorithms for better classification and location of sounds, and statistical approaches for density estimation (Marques et al., 2013; Marques, Thomas, Ward, DiMarzio, & Tyack, 2009; Mellinger, Stafford, Moore, Dziak, & Matsumoto, 2007; Moretti, Casey, & Mellinger, 2009; Thomas & Marques, 2012). The method has been applied successfully to derive an abundance estimate for the small, endangered population of harbour porpoise in the inner Baltic Sea (the SAMBAH project − see Carlén et al., 2018; SAMBAH, 2016 for full details).

SAM systems can be fixed cable hydrophones, radio-linked hydrophone systems or fixed autonomous devices (Mellinger et al., 2007; Moretti et al., 2009; Norris, Oswald, & Sousa-Lima, 2010). In Europe PODs, Pop-Ups and Sound Traps are the main SAM units that have been deployed to monitor cetaceans. In deeper waters (up to 4000 m), Pop-Ups have generally been used (see, e.g. Clark, Gillespie, Nowacek, & Parks, 2007; Swift et al., 2002), mainly to detect baleen whales and sperm whales. The hydrophone records sounds at frequencies of between 10 Hz and 2 kHz, and is surrounded by a perforated PVC tube that shields it from water motion (Fig. 6.15A). Pop-Ups can be deployed for periods up to 3 months, singly or in an array, and are not restricted to deepwater; they can also be deployed on or near the bottom in very shallow water. Sound Traps (Fig. 6.15B) have a spherical transducer element that can operate over bandwidths of 20 Hz to 150 kHz. They are compact self-contained units and can be either deployed statically for periods that range from 2 weeks to 2 months depending upon their sampling rate, or towed on a rope. In a recent study west of Ireland, autonomous multichannel acoustic recorders, with 32 and 2 kHz sampling rate, channels, were deployed close to the seafloor at depths approaching 2000 m to monitor the presence of beaked whales (Kowarski et al., 2018). Analyses of tonal sounds such as whistles are generally undertaken using a sound spectrograph (Fig. 6.16).

For monitoring small cetaceans, particularly harbour porpoise and bottlenose dolphin, PODs have been widely used in Europe (see, e.g. Bailey et al., 2010; Carstensen, Henriksen, & Teilmann, 2006; Kyhn et al., 2012; Leeney & Tregenza, 2006; Nuuttila, Courtene-Jones, Baulch, Simon, & Evans, 2017; Nuuttila et al., 2013; Simon et al., 2010; Verfuß et al., 2007, 2008), including for impact studies on harbour porpoises in conjunction with wind farm construction and operation (Brandt, Diederichs, Betke, & Nehls, 2011; Carstensen et al., 2006; Koschinski et al., 2003; Tougaard, Carstensen, Teilmann, Skøv, & Rasmussen, 2009; Tougaard, Henriksen, & Miller, 2009), effects of pingers (Leeney et al., 2007), boat disturbance (Pirotta, Merchant, Thompson, Barton, & Lusseau, 2015) and habitat use

FIGURE 6.15 (A) A towed fixed cable hydrophone set-up. (B) A sound trap passive acoustic monitoring set-up. *Photos: (A) Katrin Lohrengel/SWF; (B) Gemma Veneruso.*

(Nuuttila et al., 2013, 2017; Philpott, Englund, Ingram, & Rogan, 2007; Schaffeld et al., 2016; Senior, 2006; Simon et al., 2010; Verfuß et al., 2008).

The timed porpoise detector (T-POD) is an acoustic self-contained data logger comprising a hydrophone, a filter and digital memory, which logs the time of detections and durations of sequences of echolocation clicks (Fig. 6.17A). Memory capacity is 128 MB and battery life is approximately 6—10 weeks. To select echolocation clicks of cetaceans, the T-POD compares the sound energy picked up by a pair of bandpass filters with adjustable bandwidth, one of which (the target filter) is set to the click frequency of the species of interest. The energy picked up by the target filter has to be a certain amount higher than the energy picked up by the reference filter to cause a registration of the presented sound. A target filter of 50 kHz and reference filter of 70 kHz is used to detect bottlenose dolphin click trains, and a

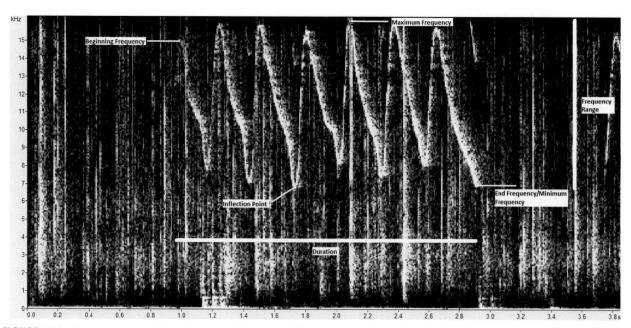

FIGURE 6.16   Sound spectrogram of a bottlenose dolphin whistle. The parameters often used to characterise a sound signal include beginning and end frequencies, minimum and maximum frequencies, the inflection point, frequency range and duration. The vertical axis is the frequency in kilo-hertz and the horizontal axis is time in seconds. *Bird, 2012, reproduced with permission from Anna Bird.*

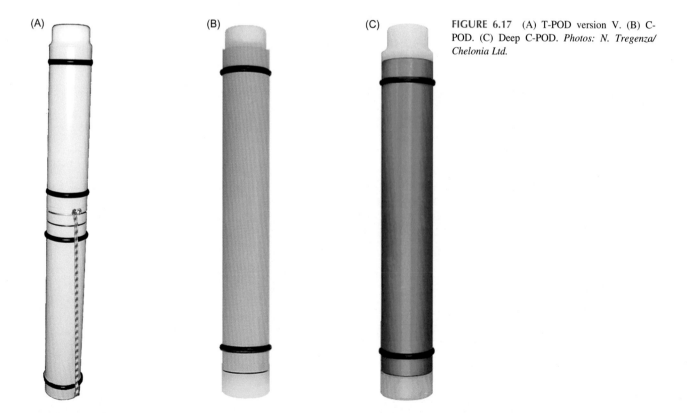

FIGURE 6.17   (A) T-POD version V. (B) C-POD. (C) Deep C-POD. *Photos: N. Tregenza/ Chelonia Ltd.*

target filter of 130 kHz and reference filter of 92 kHz, to detect porpoise click trains (Simon et al., 2010). Since T-POD units can vary individually in their sensitivities, it is necessary to conduct calibration tests in a tank and then validate the settings by placing units together in a cage in the sea for a period of time (Kyhn et al., 2008; Simon et al., 2010; Teilmann, Henriksen, Carstensen, & Skov, 2002; Verfuß et al., 2007).

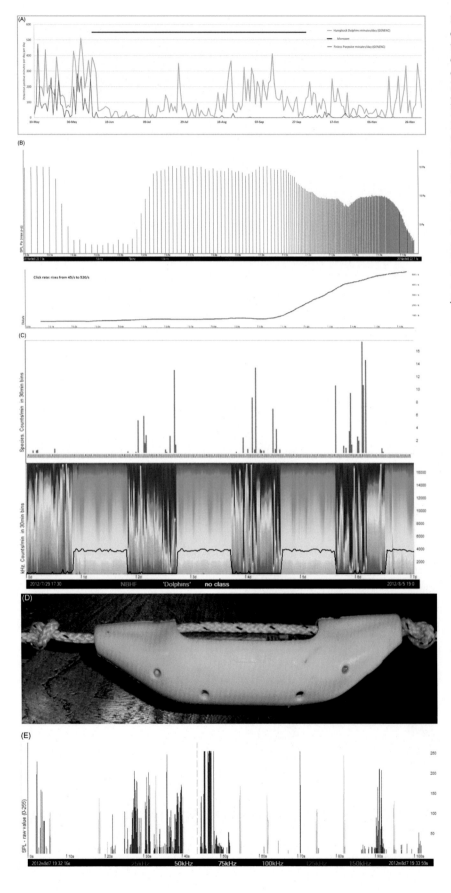

FIGURE 6.18 (A) Example of a C-POD data file (0.6 GB) collected over 206 days which generated 3 million clicks in trains C clicks in total, showing automated detections of humpback dolphins and finless porpoise, Konkan, India, 2016. (B) A C-POD harbour porpoise click train from the United Kingdom, with click rate below. Each vertical line is a click. The height shows the amplitude and the colour represents the frequency. (C) Example of harbour porpoise click detections during a trial to test the effectiveness of the banana pinger to mitigate fishing net entanglement. (D) Side view of a banana pinger used to alert porpoises to the presence of a net. (E) Harbour porpoise clicks at the onset of pinging by a banana pinger., *Photo: (A) Reproduced with permission from Nick Tregenza/Chelonia Ltd, with thanks to the Konkan research team for provision of data shown. (B−E) Reproduced with permission from Nick Tregenza/Chelonia Ltd.*

The T-POD was widely used in Europe during the 2000s, but around 2010 was superseded by the C-POD (Fig. 6.17B, C) — a digital version that uses waveform characterisation in order to select cetacean clicks, logging the time, centre frequency, sound pressure level, duration and bandwidth of each click. The false positive rate is low, and the wide frequency range, 20−160 kHz enables them to continuously log all small and medium-sized toothed whales, dolphins and porpoises (see Fig. 6.18A−E for examples of their uses). Because of their low power requirements, they can run for around 6 months whereas the batteries in T-PODs rarely would last more than 10 weeks. They also can hold a lot of memory, allowing large volumes of data (16−32 GB) to be collected, enabling data downloading to be undertaken at sea without the need for a computer. Individual units also show less variability. Although porpoises and bottlenose dolphins can be confidently distinguished from one another, discrimination between different species of dolphins remains a problem. However, a new version, the C-POD-F, is currently being trialled and shows promise for use also on other dolphin species (N. Tregenza/Chelonia, personal communication).

Often a problem with deployment of SAM devices is keeping them moored at a particular location. Heavy anchoring gear may need to be used to combat strong currents and displacement by heavy vessel traffic or fishing activities. In some cases these might require several anchors (weighing up to four tons), a cardinal buoy and one or more smaller buoys. Inevitably there is a loss rate but with care this can be minimised. In the case of the ambitious SAMBAH project, almost 70% of the total amount of data possible to acquire was obtained.

SAMBAH − Static Acoustic Monitoring of the Baltic Sea Harbour Porpoise

The LIFE+SAMBAH project was carried out from 2010 to 2015, involving all EU countries around the Baltic Sea (Table 6.3). The major aims were to estimate the size of the Baltic Proper harbour porpoise population, provide distribution maps, identify important areas and areas with higher risk of conflicts with human activities, and to increase the knowledge and awareness about the population among policymakers, managers, stakeholders, users of the marine environment and the general public. The hope was that the results should serve as a thorough basis for future designations of protected areas within the Natura 2000 network, as well as for other conservation measures for the Critically Endangered (CR) Baltic Proper harbour porpoise population.

The study area included the waters between 5 and 80 m depth, stretching from the Darss and Drogden underwater sills in the southwest to the northern border of the Åland and Archipelago Seas in the north. This includes a wide variety of habitats, from almost marine to brackish water, from open sea to sheltered archipelago areas, and with winter ice conditions ranging from metre-thick fast ice to no ice at all. Harbour porpoise click detectors called C-PODs (Fig. 6.19) were deployed at a total of 304 locations, following a regular grid with a random starting point and a random tilt. However, in a few instances when the primary positions could not be used due to military restrictions or major shipping lanes, the stations were either moved up to a couple of hundred metres, or one of four secondary positions located in between the primary positions was randomly selected and used instead. The field survey lasted for 2 full years, from May 2011 to April 2013.

All C-PODs were deployed approximately 2 m above the seafloor, but the anchoring systems varied among and within countries depending on local conditions. Some had surface buoys, others were retrieved by acoustic release mechanisms and yet others had double anchors with a line in between which was caught and hauled using a grapple. In Poland, a trawl-resistant anchoring system was developed and successfully used. The weight of the anchors used ranged from a few kilos to 1 ton. The C-PODs had to be lifted to the surface for data upload and battery replacement every 3−6 months.

Some data were lost due to delayed servicing caused by harsh weather conditions, including ice cover, wherefore the batteries ran out before the C-POD could be retrieved (Fig. 6.20). In other cases C-PODs were lost due to trawling, and in the beginning a programming error caused the C-PODs to switch off when half of the battery power was used. Much effort was put into communication with fishermen to prevent loss due to trawling, and the C-PODs were reprogrammed to use their full battery power. In total, 377 years' worth of data were retrieved from 298 stations corresponding to 62% of the total possible effort.

To estimate absolute abundance from station-based click rate data, auxiliary data on harbour porpoise click rates and the C-PODs' likelihood of detecting echolocating harbour porpoises at various distances were collected. Data on click rates were collected by deploying acoustic recorders on wild harbour porpoises incidentally and benignly caught in herring weirs in inner Danish waters. After instrumentation the animals were released to the wild again. Data on the C-PODs' likelihood of detecting echolocating harbour porpoises, that is the C-PODs' detection function, were collected in an experiment where wild harbour porpoises were acoustically tracked by a three-dimensional multihydrophone array above C-PODs anchored 2 m off the seafloor. This too was carried out in inner Danish waters where the harbour porpoise density is much higher than in the Baltic Proper.

Based on the estimated harbour porpoise density at the C-POD stations and environmental variables such as spatial coordinates, month, depth and topographic complexity, spatial modelling was used to produce monthly predictions of harbour porpoise distribution (Carlén et al., 2018). These predictions showed that during the half-year from May to October, the harbour porpoises were separated in two main clusters, divided by a line east of Bornholm in the southern Baltic Proper. The southwestern cluster increased in density to the west, indicating that these were Belt Sea animals utilising the southwestern part of the

(Continued)

(Continued)

TABLE 6.3 Number of C-PODs per country, project members and their role within the SAMBAH project.

Country	No. of C-PODs	Organisation	Role
Sweden	99	Kolmården Wildlife Park, Project Coordinator	Coordinator
		Swedish Agency for Marine and Water Management	Partner
		AquaBiota Water Research	Subcontractor
Finland	46	Turku University of Applied Sciences	Partner
		Ministry for the Environment	Partner
		Särkänniemi Adventure Park	Partner
Estonia	40	Pro Mare	Subcontractor
Latvia	34	Latvian Institute for Aquatic Ecology	Subcontractor
Lithuania	9	Coastal Research and Planning Institute, Klaipeda University	Subcontractor
Poland	39	University of Gdansk	Partner
		Institute of Meteorology and Water Management	Partner
		Chief Inspectorate for Environmental Protection	Partner
Germany	16	German Oceanographic Museum	Collaborator
Denmark	21	Aarhus University	Partner
		Danish Nature Agency	Partner
United Kingdom		Centre for Research into Ecological and Environmental Modelling, St Andrews University	Subcontractor
		Chelonia Ltd	C-POD manufacturer

FIGURE 6.19   Mats Amundin holding a C-POD used in the SAMBAH Project. *Photo: Cinthia Tiberi Ljungqvist, SAMBAH Project.*

(Continued)

(Continued)

FIGURE 6.20 Deploying a C-POD from the ice in the Baltic Sea. *Photo: Turku University of Applied Sciences, SAMBAH Project.*

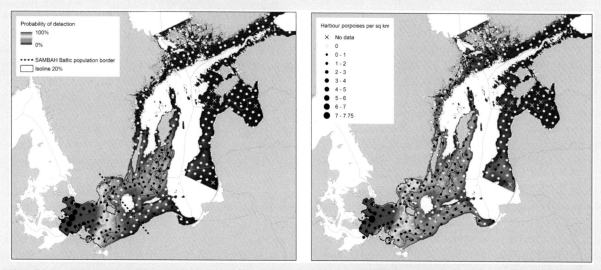

FIGURE 6.21 Predicted probability of detection of harbour porpoises per month in the SAMBAH project area during May−October (left) and November−April (right). The black line indicates 20% probability of detection, approximately equivalent to the area encompassing 30% of the population, often used to define high-density areas. The dots or crosses show the calculated density of porpoises at the SAMBAH survey stations. The border indicates the spatial separation between the Belt Sea and Baltic harbour porpoise populations during May−October. *Reproduced with permission from SAMBAH Project.*

(Continued)

(Continued)

Baltic Proper. Northeast of this line was another cluster around the Hoburg's Bank and the Northern and Southern Mid-sea Banks, south of Gotland and southeast of Öland. This area was previously virtually unknown to accommodate harbour porpoises. As calving and mating takes place during summer, the predictions indicate that these offshore banks and the surrounding areas are the major breeding site for the Baltic Proper population. From November to April, there was no spatial separation within the study area, suggesting that the two populations mix during this time of the year (Fig. 6.21). During winter animals were more spread out with detections throughout Polish waters, along the coasts of Sweden, Latvia and Lithuania, and in Finnish offshore waters, although the offshore banks were still of importance. During the spatial separation in summer, the size of the Baltic Proper population was estimated at 497 individuals (95% CI 80–1091) (SAMBAH, 2016).

The estimated size of the Baltic Proper population confirms that the population is critically endangered and in urgent need of efficient conservation measures. The monthly spatial predictions provide a powerful basis for designation of protected areas and other conservation measures. Based on the SAMBAH results, the Swedish Government has proposed a new Natura 2000 site (pSCI) for harbour porpoises, encompassing Hoburgs Bank, the Northern Mid-sea Bank and a large area around them. The site covers more than 1 million ha (more than 10,000 km^2) of the Baltic Proper, being the second largest for harbour porpoises in Europe, after Dogger Bank. A management plan is now to be developed for the site, with the potential of providing protection from bycatches and anthropogenic disturbance.

Dr Julia Carlström, MSc Ida Carlén and Dr Mats Amundin
Project managers and co-ordinators of the SAMBAH Project

Postmortem examinations and other approaches to investigating health status

Several European countries administer coordinated stranding schemes, whereby members of the public report the finding of cetaceans coming ashore either alive or dead. In the United Kingdom this started in 1913, based out of the London Natural History Museum, and has continued ever since, with coordination of postmortem examinations from the Institute of Zoology in London since 1990 when the Cetacean Strandings Investigation Programme (CSIP) was initiated (Fig. 6.22A, B). In the Netherlands annual reporting of strandings started in 1942, continued until 1965, and then after a gap resumed from 1970 to the present, coordinated first by Amsterdam Museum, then by Leiden Museum [which later became the National Museum of Natural History and then Naturalis Biodiversity Center]. In Belgium strandings data were routinely collected in conjunction with the University of Antwerp from the 1960s, then from the 1970s in collaboration with the Royal Belgian Institute of Natural Sciences and later with the University of Liège. In France a stranding scheme was started from the Musée Océanographique in La Rochelle in 1970 and has been run from La Rochelle ever since (the stranding network is known as Réseau National d'Echouages or RNE). When the Musée Océanographique was integrated within the University of La Rochelle, RNE was coordinated by the Centre de Recherche sur les Mammifères Marins. In Germany it was started on the coast of Schleswig-Holstein and Mecklenburg-Western Pomerania in 1990. Animals are collected and necropsied by the Research and Technology Centre West Coast (FTZ) in Büsum, for Schleswig-Holstein, and by the German Oceanographic Museum, Stralsund, for Mecklenburg-Western Pomerania. Recording of strandings from Lower Saxony is more incidental than in the other two regions.

Further south, stranding records have been compiled occasionally in northwest Spain (Galicia) since 1980, and have been systematically recorded since 1990, organised by the Coordinadora para o Estudio dos Mamiferos Mariños with help from the public and local authorities. In Portugal strandings have been systematically recorded since 1975, and are collated by the Instituto da Conservaçao da Natureza in Lisbon.

There is no formalised scheme reporting strandings in Norway, Finland, Poland or the eastern Baltic States. Reporting of strandings in Denmark has existed to a varying degree since the 1990s. Strandings when reported either go to the Fisheries and Maritime Museum in Esbjerg, the Zoological Museum of the University of Copenhagen, or the nearest forest districts of the Danish Nature and Forest Agency. In Sweden a project started in 1988 with the aim of collecting all dead harbour porpoises found in Sweden, either from strandings or as bycatches from fishing gear. Material was collected by the Museum of Natural History in Göteborg. Since 1990, postmortem investigations have been carried out on all fresh stranded and bycaught small cetaceans found in Swedish waters by the Swedish Museum of Natural History in Stockholm.

There is a great deal of useful information that can be derived from examining cetaceans after death. In addition to basic measurements and weights taken, the age of an animal can be ascertained by sectioning a tooth in the case of

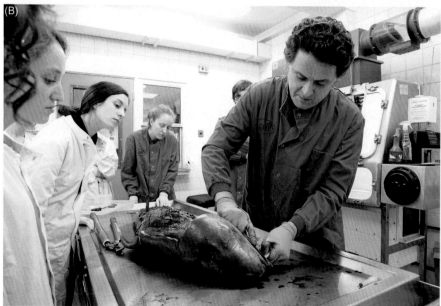

FIGURE 6.22 Dissections of (A) short-beaked common dolphin and (B) harbour porpoise, undertaken at the Institute of Zoology, London. *Photos: CSIP-ZSL.*

toothed whales, dolphins and porpoises, or taking an earplug from a baleen whale, and counting annual layers that are laid down rather as occurs with the growth rings of a tree (Bowen & Northridge, 2010). The reproductive system can be examined to establish the breeding status, and in the case of females, the number of births (and number of abortions) that have taken place. The age and length at sexual maturity can be determined. Differences between groups of animals can be found by examining skulls and other skeletal material. In recent years, geometric morphometrics have largely replaced traditional morphometrics based on length measurements. Geometric morphometrics employ the capture of two- or three-dimensional coordinates from previously defined morphological landmarks from biological specimens to get an approximation of shape.

Taking the stomach contents, one can extract fish otoliths (Fig. 6.23) and squid beaks, and by use of reference collections, identify the prey species and their sizes (see Hyslop, 1980; Pierce & Boyle, 1991; Tollit, Pierce, Hobson, Bowen, & Iverson, 2010, for reviews). Further information on diet can be derived by examining the fatty acids that are

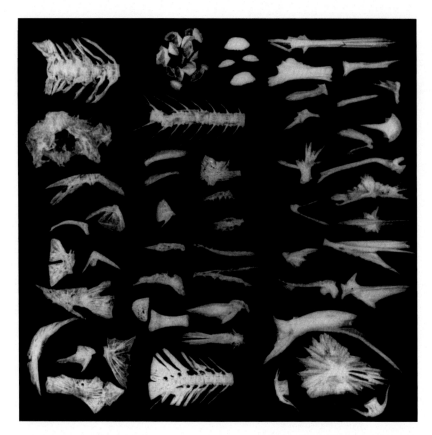

FIGURE 6.23 Otoliths of various fish species taken from a dolphin stomach. *Photo: Graham J Pierce.*

stored in the blubber (Iverson, 2009; Iverson, Field, Bowen, & Blanchard, 2004). Some of these are synthesised by the animal itself but others have to be derived from the diet, having been passed from phytoplankton, etc., upwards through the food chain. The relative composition (the % contribution by weight) of individual fatty acids found in a tissue is known as the fatty acid profile or signature (Fig. 6.24) . Different prey species have different fatty acid profiles so a comparison can be made. Although the blubber profile is influenced by the profiles of dietary species eaten, the relationship between them nevertheless is quite complex. Generally, it is better to take samples from the inner area of the blubber.

For a longer-term and more general profile of the diet of an animal, one can analyse stable isotopes within samples (e.g. skin which has a turnover rate measured in weeks, bone or teeth which present a permanent record) (West, Bowen, Cerling, & Ehleringer, 2006). Stable isotopes are elements with unique atomic masses (same number of protons, but different numbers of neutrons) that do not undergo radioactive decay; and most elements have more than one stable isotope. The ratio of those isotopes for an element is the relative amount of the common to the rare isotope; generally, the lighter isotope is more common in biological systems. Carbon isotope ratios ($\delta^{13}C$) help to characterise the habitat in which the animal was living through environmental effects; nitrogen isotope ratios ($\delta^{15}N$) provide dietary and trophic-level information; sulphur isotope ratios ($d^{34}S$) distinguish between benthic and pelagic producers, whilst hydrogen ($\delta^{2}H$) and oxygen isotope ratios ($\delta^{18}O$) record water-related dynamics in plants and animals. Together, these provide an 'isotopic signature', that is used to trace the movements of nutrients, compounds, particles and organisms across landscapes and between components of the biosphere, and to reconstruct aspects of dietary, ecological and environmental histories.

Information on the health status of a cetacean can be gained in many ways (Hall, Gulland, Hammond, & Schwacke, 2010). The thickness of the blubber layer indicates whether it is in good nutritive condition. Tests can be made to identify the presence of bacteria or viruses. Microscopic examination of tissues or organs can indicate whether the animal is diseased or has a heavy parasitic infection. Parasites, additionally, can be extracted from the bronchial, digestive or excretory systems, identified, and counted (Aznar, Balbuena, Férnandez & Raga, 2001). Samples of skin, blubber and tissue (e.g. muscle), or organ (e.g. liver or kidney) can be taken and sent away to a laboratory for DNA testing or to test for contaminants.

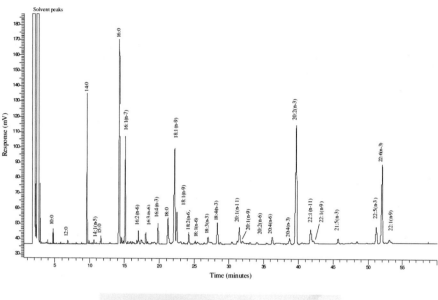

FIGURE 6.24 Fatty acid profile (above) from a blubber sample (below). *Photo: Jenny Learmonth.*

Both external and internal injuries can help to identify the cause of death and distinguish cases of bycatch in fishing gear, physical trauma (often called blunt trauma) from a vessel strike, or an attack from another species (e.g. a harbour porpoise killed by a bottlenose dolphin). Damage to the ear apparatus or haemorrhaging in various parts of the body may indicate loud noise as a likely cause of death (either direct or indirect).

Protocols have been developed to standardise how necropsies are undertaken (Geraci & Lounsbury, 2005; Kuiken & García Hartmann, 1993), and to help with diagnosis of causes of death (Dierauf & Gulland, 2001; García Hartmann, 2001; IJsseldyk & Brownlow, 2018; Kuiken, 1996).

All of these types of study are possible if the material is fresh, and some may even be feasible on long dead specimens. Some of the information can also be gained from live animals without causing long-term harm. Biopsies of skin and blubber can be taken remotely from cetaceans for DNA analysis or stable isotope analysis; hormone assays can be undertaken from sampling the blows of whales; and much can be learned about behaviour, dive patterns, and movements by temporarily attaching telemetry devices using suction cups or other methods to attach to the back of animals (Noren & Mocklin 2012). Using satellite tags, very high frequency transmitters or GPS mounted devices, animals can then be tracked, and time-depth recorders used to measure dive profiles (McConnell, Fedak, Hooker, & Patterson, 2010).

References

Population structure

Evans, P.G.H., & Teilmann, J. (Eds.) (2009). *Report of ASCOBANS/HELCOM Small Cetacean Population Structure Workshop* (140 pp). Bonn: ASCOBANS/UNEP Secretariat.

Moritz, C. (1994). Defining 'evolutionarily significant units' for conservation. *Trends in Ecology and Evolution, 9*, 373−375.

Palsbøll, P. J., Bérubé, M., & Allendorf, F. W. (2007). Identification of management units using population genetic data. *Trends in Ecology and Evolution, 22*, 11−16.

Taylor, B. L., & Dizon, A. E. (1999). First policy then science: Why a management unit based solely on genetic criteria cannot work. *Molecular Ecology, 8,* S11−S16.

Counting animals

Buckland, S. T., Anderson, D. R., Burnham, K. P., Laake, J. L., Borchers, D. L., & Thomas, L. (2001). *Introduction to Distance Sampling.* Oxford: Oxford University Press, 432 pp.

Buckland, S. T., Anderson, D. R., Burnham, K. P., Laake, J. L., Borchers, D. L., & Thomas, L. (Eds.), (2004). *Advanced distance sampling: Estimating abundance of biological populations.* Oxford: Oxford University Press.

Dawson, S., Wade, P., Slooten, E., & Barlow, J. (2008). Design and field methods for sighting surveys of cetaceans in coastal and riverine habitats. *Mammal Review, 38*(1), 19−49.

Diederichs, A., Nehls, G., Dähne, M., Adler, S., Koschinski, S., & Verfuß, U. (2008). *Methodologies for measuring and assessing potential changes in marine mammal behaviour, abundance or distribution arising from the construction, operation and decommissioning of offshore windfarms* (90 pp.). BioConsult SH and Deutsches Meeresmuseum Stralsund. Report Commissioned by COWRIE Ltd., ENG-01-2007.

Evans, P. G. H., & Hammond, P. S. (2004). Monitoring cetaceans in European waters. *Mammal Review, 34*(1), 131−156.

Evans, P. G. H., & Thomas, L. (2011). *Estimation of costs associated with implementing a dedicated cetacean surveillance scheme in UK.* Joint Nature Conservation Committee Commissioned Report, 45 pp.

Grünkorn, T., Diederichs, A., & Nehls, G. (2005). Aerial surveys in the German Bight − estimating *g(0)* for harbour porpoises (*Phocoena phocoena*) by employing independent double counts. In F. Thomsen, F. Ugarte, & P. G. H. Evans (Eds.), *Estimation of g(0) in line-transect surveys of cetaceans* (pp. 25−34). European Cetacean Society Special Issue No. 44.

Hammond, P. S. (2010). Estimating the abundance of marine mammals. In I. L. Boyd, W. D. Bowen, & S. J. Iverson (Eds.), *Marine mammal ecology and conservation. A handbook of techniques* (pp. 42−67). Oxford: Oxford University Press, 450 pp.

Strindberg, S., Buckland, S. T., & Thomas, L. (2004). Chapter 7: Design of distance sampling surveys and Geographic Information Systems. In S. T. Buckland, D. R. Anderson, K. P. Burnham, J. L. Laake, D. L. Borchers, & L. Thomas (Eds.), *Advanced distance sampling.* Oxford: Oxford University Press.

Thomas, L., Buckland, S. T., Rexstad, E. A., Laake, J. L., Strindberg, S., Hedley, S. L., … Burnham, K. P. (2010). Distance software: Design and analysis of distance sampling surveys for estimating population size. *Journal of Applied Ecology, 47,* 5−14.

Thomas, L., Burnham, K. P., & Buckland, S. T. (2004). Chapter 5: Temporal inferences from distance sampling surveys. In S. T. Buckland, D. R. Anderson, K. P. Burnham, J. L. Laake, D. L. Borchers, & L. Thomas (Eds.), *Advanced distance sampling.* Oxford: Oxford University Press.

Photoidentification

Hammond, P. S. (1986). Estimating the size of naturally marked whale populations using capture-recapture techniques. In *Behaviour of whales in relation to management* (pp. 253−282, Special issue 8). Report of the International Whaling Commission.

Hammond, P. S. (2002). The assessment of marine mammal population size and status. In P. G. H. Evans, & J. A. Raga (Eds.), *Marine mammals: Biology and conservation* (pp. 269−291). London and New York: Kluwer Academic.

Hammond, P. S. (2010). Estimating the abundance of marine mammals. In I. L. Boyd, W. D. Bowen, & S. J. Iverson (Eds.), *Marine mammal ecology and conservation. A handbook of techniques* (pp. 42−67). Oxford: Oxford University Press, 450 pp.

Hammond, P. S., Mizroch, S. A., & Donovan, G. P. (Eds.). (1990). *Individual recognition of cetaceans: Use of photo-identification and other techniques to estimate population parameters* [Special issue 12]. Report of the International Whaling Commission.

Kendall, W. L., Nichols, J. D., & Hines, J. E. (1997). Estimating temporary emigration using capture-recapture data with Pollock's robust design. *Ecology, 78,* 563−578.

Kendall, W. L., Pollock, K. H., & Brownie, C. (1995). A likelihood-based approach to capture-recapture estimation of demographic parameters under the robust design. *Biometrics, 51,* 293−308.

Rexstadt, E., & Burnham, K. (1991). *User's guide for interactive program CAPTURE.* Fort Collins, CO: Colorado Cooperative Fish and Wildlife Research Unit, Colorado State University.

Whitehead, H. (1990). Mark-recapture estimates with emigration and re-immigration. *Biometrics, 46,* 473−479.

Würsig, B., & Jefferson, T. A. (1990). Methods of photo-identification for small cetaceans. In P. S. Hammond, S. A. Mizroch, & G. P. Donovan (Eds.), *Individual recognition of cetaceans: Use of photo-identification and other techniques to estimate population parameters* (pp. 43−52, Special issue 12). Report of the International Whaling Commission.

Acoustics

Bailey, H., Clay, G., Coates, E. A., Lusseau, D., Senior, B., & Thompson, P. M. (2010). Using T-PODS to assess variations in the occurrence of bottlenose dolphins and harbour porpoises. *Aquatic Conservation, 20,* 150−158.

Bird, A. (2012) *Geographic Variation in the Whistle Characteristics of Bottlenose Dolphins* (Tursiops truncatus) *between Cardigan Bay, Wales, the Shannon Estuary, Ireland, the Molène Archipelago, France and the Sado Estuary, Portugal.* MSc Thesis, University of Bangor, Wales. 61pp.

Brandt, M. J., Diederichs, A., Betke, K., & Nehls, G. (2011). Responses of harbour porpoise to pile driving at the Horns Rev II offshore wind farm in the Danish North Sea. *Marine Ecology Progress Series, 421,* 205−216.

Carlén, I., Thomas, L., Carlström, J., Amundin, M., Teilmann, J., Tregenza, N., … Koblitz, J. C. (2018). Basin-scale distribution of harbour porpoises in the Baltic Sea provides basis for effective conservation actions. *Biological Conservation, 226,* 42−53.

Carstensen, J., Henriksen, O. D., & Teilmann, J. (2006). Impacts of offshore wind farm construction on harbour porpoises: Acoustic monitoring of echolocation activity using porpoise detectors (T-PODs). *Marine Ecology Progress Series, 321,* 295−308.

Clark, C. W., Gillespie, D., Nowacek, D. P., & Parks, S. E. (2007). Listening to their world: Acoustics for monitoring and protecting right whales in an urbanized ocean. In S. D. Kraus, & R. M. Rolland (Eds.), *The urban whale: North Atlantic right whales at the crossroads* (pp. 333−357). Cambridge, MA: Harvard University Press, 543 pp.

Koschinski, S., Culik, B. M., Henriksen, O. D., Tregenza, N., Ellis, G., Jansen, C., & Kathe, G. (2003). Behavioural reactions of free-ranging porpoises and seals to the noise of a simulated 2 MW windpower generator. *Marine Ecology Progress Series, 265,* 263−273.

Kowarski, K., Delarue, J., Martin, B., O'Brien, J., Made, R., Ó Cadhla, O., & Berrow, S. (2018). Signals from the deep: Spatial and temporal acoustic occurrence of beaked whales off western Ireland. *PLoS ONE, 13*(6), e0199431. Available from https://doi.org/10.1371/journal, pone.0199431.

Kyhn, L. A., Tougaard, J., Teilmann, J., Wahlberg, M., Jørgensen, P. B., & Bech, N. I. (2008). Harbour porpoise (*Phocoena phocoena*) static acoustic monitoring: Laboratory detection thresholds of T-PODs are reflected in field sensitivity. *Journal of the Marine Biological Association of the United Kingdom, 88,* 1085−1091.

Kyhn, L. A., Tougaard, J., Thomas, L., Duve, L. R., Stenback, J., Amundin, M., ... Teilmann, J. (2012). From echolocation clicks to animal density: Acoustic sampling of harbour porpoises with static dataloggers. *Journal of the Acoustical Society of America, 131,* 550−560.

Leeney, R. H., & Tregenza, N. J. C. (Eds.). (2006). Static acoustic monitoring of cetaceans. In *Proceedings of workshop held at 26th annual conference of European Cetacean Society, Gdynia, Poland, 2 April 2006.*Kiel: European Cetacean Society Special Issue No. 46.

Leeney, R. H., Berrow, S., McGrath, D., O'Brien, J., Cosgrove, R., & Godley, B. J. (2007). Effects of pingers on the behaviour of bottlenose dolphins. *Journal of the Marine Biological Association of the United Kingdom, 87*(1), 129−133. Available from https://doi.org/10.1017/S0025315407054677.

Marques, T. A., Thomas, L., Martin, S. W., Mellinger, D. K., Ward, J. A., Moretti, D. J., ... Tyack, P. L. (2013). Estimating animal population density using passive acoustics. *Biological Reviews, 88,* 287−309.

Marques, T. A., Thomas, L., Ward, J., DiMarzio, N., & Tyack, P. L. (2009). Estimating cetacean population density using fixed passive acoustic sensors: An example with Blainville's beaked whales. *Journal of the Acoustical Society of America, 125*(4), 1982−1994.

Mellinger, D. K., Stafford, K. M., Moore, S. E., Dziak, R. P., & Matsumoto, H. (2007). An overview of passive acoustic observation methods for cetaceans. *Oceanography, 20,* 36−45.

Moretti, D., Casey, T., &Mellinger, D. K. (2009). *Measuring the health of the field: Fixed passive acoustic marine mammal monitoring for estimating species abundance and mitigating the effect of operations on the marine environment* (56 pp). Final Report to the International Association of Oil & Gas Producers Joint Industry Programme on E&P Sound and Marine Life. Contract Reference #: JIP22 07-09.

Norris, T. F., Oswald, J. N., & Sousa-Lima, R. S. (2010). *A review and inventory of fixed installation passive acoustic monitoring methods and technology* (214 pp). Final Report to the International Association of Oil & Gas Producers Joint Industry Programme on E&P Sound and Marine Life. Contract Reference #: JIP22 08-03.

Nuuttila, H. K., Courtene-Jones, W., Baulch, S., Simon, M., & Evans, P. G. H. (2017). Don't forget the porpoise: Acoustic monitoring reveals fine scale temporal variation between bottlenose dolphin and harbour porpoise in Cardigan Bay SAC. *Marine Biology, 164,* 50. Available from https://doi.org/10.1007/s00227-017-3081-5.

Nuuttila, H. K., Meier, R., Evans, P. G. H., Turner, J. R., Bennell, J. D., & Hiddink, J. G. (2013). Identifying foraging behaviour of wild bottlenose dolphins (*Tursiops truncatus*) and harbour porpoises (*Phocoena phocoena*) with static acoustic dataloggers. *Aquatic Mammals, 39*(2). Available from https://doi.org/10.1578/AM.39.2.2013.

Philpott, E., Englund, A., Ingram, S., & Rogan, E. (2007). Using T-PODs to investigate the echolocation of coastal bottlenose dolphins. *Journal of the Marine Biological Association of the United Kingdom, 87,* 11−17.

Pirotta, E., Merchant, N. D., Thompson, P. M., Barton, T. R., & Lusseau, D. (2015). Quantifying the effect of boat disturbance on bottlenose dolphin foraging activity. *Biological Conservation, 181,* 82−89.

SAMBAH. (2016). Final report for LIFE + project SAMBAH LIFE08 NAT/S/000261 covering the project activities from 01/01/2010 to 30/09/2015 (80 pp.). Reporting date 29/02/2016.

Schaffeld, T., Bräger, S., Anja Gallus, A., Dähne, M., Krügel, K., Herrmann, A., ... Koblitz, J. C. (2016). Diel and seasonal patterns in acoustic presence and foraging behaviour of free-ranging harbour porpoises. *Marine Ecology Progress Series, 547,* 257-−2272.

Senior, B. (2006). Using T-PODs in areas with dolphins and porpoises. In R. H. Leeney & N. J. C. Tregenza (Eds.), *Static acoustic monitoring of cetaceans. Proceedings of workshop held at 25th annual conference of European Cetacean Society, Gdynia, Poland, 2 April 2006* (pp. 43−44). Kiel: European Cetacean Society Special Issue No. 46.

Simon, M., Nuuttila, H., Reyes-Zamudio, M. M., Ugarte, F., Verfuß, U., & Evans, P. G. H. (2010). Passive acoustic monitoring of bottlenose dolphin and harbour porpoise in Cardigan Bay, Wales, with implications for habitat use and partitioning. *Journal of the Marine Biological Association of the United Kingdom, 90,* 1539−1546.

Swift, R. J., Hastie, G. D., Barton, T. R., Clark, C. W., Tasker, M. L., & Thompson, P. M. (2002). *Studying the distribution and behaviour of cetaceans in the northeast Atlantic using passive acoustic techniques.* Report for the Atlantic Frontier Environmental Network, Aberdeen.

Teilmann, J., Henriksen, O. D., Carstensen, J., & Skov, H. (2002). *Monitoring effects of offshore windfarms on harbour porpoises using PODs (porpoise detectors).* Technical Report for the Ministry of the Environment, Denmark.

Thomas, L., & Marques, T. (2012). Passive acoustic monitoring for estimating animal density. *Journal of the Acoustical Society of America, 8*(3), 35−44.

Tougaard, J., Carstensen, J., Teilmann, J., Skøv, H., & Rasmussen, P. (2009). Pile driving zone of responsiveness extends beyond 20 km for harbor porpoises (*Phocoena phocoena* (L.)). *Journal of the Acoustical Society of America, 126,* 11−14. Available from https://doi.org/10.1121/1.3132523.

Tougaard, J., Henriksen, O. D., & Miller, L. A. (2009). Underwater noise from three types of offshore wind turbines: Estimation of impact zones for harbor porpoise and harbor seals. *Journal of the Acoustcal Society of America, 125*, 3766−3773.

Verfuß, U. K., Honnef, C. G., Medling, A., Dähne, M., Adler, S., Kilian, A., & Benke, H. (2008). The history of the German Baltic Sea harbour porpoise acoustic monitoring at the German Oceanographic Museum. In K. Wollny-Goerke, & K. Eskildsen (Eds.), *Marine warm-blooded animals in North and Baltic Seas, MINOS − Marine mammals and seabirds in front of offshore wind energy* (pp. 41−56). Wiesbaden: Teubner Verlag.

Verfuß, U. K., Honnef, C. G., Medling, M., Dähne, M., Mundry, R., & Benke, H. (2007). Geographical and seasonal variation of harbour porpoise (*Phocoena phocoena*) in the German Baltic Sea revealed by passive acoustic monitoring. *Journal of the Marine Biological Association of the United Kingdom, 87*, 165−176.

Postmortem examinations and other approaches to investigating health status

Aznar, F.J., Balbuena, J.A., Férnandez, M., & Raga, J.A. (2001) Living together: The parasites of marine mammals. Pp. 385−423. In: P.G.H. Evans & J.A. Raga (Eds.), *Marine Mammals: Biology and Conservation*. New York: Kluwer Academic/Plenum Publishers, 630 pp.

Bowen, W. D., & Northridge, S. (2010). Morphometrics, age estimation, and growth. In I. L. Boyd, W. D. Bowen, & S. J. Iverson (Eds.), *Marine mammal ecology and conservation. A handbook of techniques* (pp. 98−118). Oxford: Oxford University Press, 450 pp.

Dierauf, L. A., & Gulland, F. C. (2001). *CRC handbook of marine mammal medicine*. (2nd ed.). Boca Raton, FL: CRC Press.

García Hartmann, M. (2001). Lung pathology. In *Proceedings of the third European Cetacean Society workshop on Cetacean Pathology* (60 pp, Special issue). ECS newsletter 37.

Geraci, J. R., & Lounsbury, V. R. (2005). *Marine mammals ashore: A field guide for strandings* (2nd ed.). Baltimore, MD: National Aquarium in Baltimore, 305 pp.

Hall, A. J., Gulland, F. M. D., Hammond, J. A., & Schwacke, L. H. (2010). Epidemiology, disease, and health assessment. In I. L. Boyd, W. D. Bowen, & S. J. Iverson (Eds.), *Marine mammal ecology and conservation. A handbook of techniques* (pp. 144−164). Oxford: Oxford University Press, 450 pp.

Hyslop, E. J. (1980). Stomach contents analysis − A review of methods and their application. *Journal of Fish Biology, 17*, 411−429.

IJsseldijk, L., &Brownlow, A. (Eds.), *Cetacean pathology: Necropsy technique & tissue sampling* (27 pp.), 2018. ASCOBANS AC24/Inf.2.5.a. Available from <https://www.ascobans.org/sites/default/files/document/AC24_Inf._2.5.a_Cetacean%20Pathology%20Necropsy%20Sampling.pdf>.

Iverson, S. (2009). Tracing aquatic food webs using fatty acids from qualitative indicators to quantitative determination. In M. T. Arts, M. T. Brett, & M. J. Kainz) (Eds.), *Lipids in aquatic ecosystems* (pp. 281−307). New York: Springer.

Iverson, S., Field, C., Bowen, W. D., & Blanchard, W. (2004). Quantitative fatty acids signature analysis: A new method of estimating predator diets. *Ecological Monographs, 74*, 211−235.

Kuiken, T. (Ed.). (1996). Diagnosis of by-catch in cetaceans. In *Proceedings of the second European Cetacean Society workshop on cetacean pathology* (43 pp., Special issue). ECS Newsletter 26.

Kuiken, T., & García Hartmann, M. (Eds.). (1993). Dissection techniques and tissue sampling. In *Proceedings of the first European Cetacean Society workshop on cetacean pathology* (39 pp., Special issue). ECS Newsletter 17.

McConnell, B., Fedak, M., Hooker, S., & Patterson, T. (2010). Telemetry. In I. L. Boyd, W. D. Bowen, & S. J. Iverson (Eds.), *Marine mammal ecology and conservation. A handbook of techniques* (pp. 222−242). Oxford: Oxford University Press, 450 pp.

Noren, D. P., & Mocklin, J. A. (2012). Review of cetacean biopsy techniques: Factors contributing to successful sample collection and physiological and behavioral impacts. *Marine Mammal Science, 28*(1), 154−199.

Pierce, G. J., & Boyle, P. R. (1991). A review of methods for diet analysis in piscivorous marine mammals. *Oceanography and Marine Biology: An Annual Review, 29*, 409−486.

Tollit, D. J., Pierce, G. J., Hobson, K. A., Bowen, W. D., & Iverson, S. J. (2010). Diet. In I. L. Boyd, W. D. Bowen, & S. J. Iverson (Eds.), *Marine mammal ecology and conservation. A handbook of techniques* (pp. 191−221). Oxford: Oxford University Press, 450 pp.

West, J. B., Bowen, G. J., Cerling, T. E., & Ehleringer, J. R. (2006). Stable isotopes as one of nature's ecological recorders. *Trends in Ecology and Evolution, 21*(7), 408−414.

Boxed contributions

Borchers, D. L., Buckland, S. T., Goedhart, P. W., Clarke, E. D., & Hedley, S. L. (1998). Horvitz-Thompson estimators for double-platform line-transect surveys. *Biometrics, 54*, 1221−1237.

Bravington, M. V., & Bisack, K. D. (1996). Estimates of harbour porpoise bycatch in the Gulf of Maine sink gillnet fishery, 1990-1993. *Reports of the International Whaling Commission, 46*, 567−574.

Buckland, S. T., & Turnock, B. J. (1992). A robust line transect method. *Biometrics, 48*, 901−909.

Carlén, I., Thomas, L., Carlström, J., Amundin, M., Teilmann, J., Tregenza, N., ... Tougaard, J. (2018). Basin-scale distribution of harbour porpoises in the Baltic Sea provides basis for effective conservation actions. *Biological Conservation, 226*, 42−53.

Evans, P. G. H., & Hammond, P. S. (2004). Monitoring cetaceans in European waters. *Mammal Review, 34*(1), 131−156.

Gerrodette, T. (1987). A power analysis for detecting trends. *Ecology, 68*, 1364−1372.

Gillespie, D. M., Leaper, R., Gordon, J. C. D., & Macleod, K. (2011). A semi-automated, integrated, data collection system for line-transect surveys. *Journal of Cetacean Research and Management, 11*(3), 217−227.

Hammond, P. S. (1986). *Estimating the size of naturally marked whale populations using capture-recapture techniques* (pp. 253−282, Special issue 8). Reports of the International Whaling Commission.

Hammond, P. S. (2010). Estimating the abundance of marine mammals. In I. L. Boyd, W. D. Bowen, & S. J. Iverson (Eds.), *Marine Mammal Ecology and Conservation. A Handbook of Techniques* (pp. 42–67). Oxford: Oxford University Press, 450pp.

Hammond, P. S., Berggren, P., Benke, H., Borchers, D. L., Collet, A., Heide-Jørgensen, M. P., . . . Øien, N. (2002). Abundance of harbour porpoise and other cetaceans in the North Sea and adjacent waters. *Journal of Applied Ecology, 39*, 361–376.

Hammond, P. S., Macleod, K., Berggren, P., Borchers, D. L., Burt, M. L., Cañadas, A., . . . Vázquez, J. A. (2013). Cetacean abundance and distribution in European Atlantic shelf waters to inform conservation and management. *Biological Conservation, 164*, 107–122.

Hammond, P.S., MacLeod, K., Burt, L., Cañadas, A., Lens, S., Mikkelsen, B., . . . and Vazquez, J.A. (2011) Abundance of baleen whales in the European Atlantic. Paper SC/63/RMP24 presented to the Scientific Committee of the International Whaling Commission. 22pp.

Hiby, L. (1999). The objective identification of duplicate sightings in aerial survey for porpoise. In G. W. Garne, S. C. Armstrup, J. L. Laake, B. F. J. Manly, L. L. McDonald, & D. G. Robertson (Eds.), *Marine Mammal Survey and Assessment Methods* (pp. 179–189). Rotterdam: Balkema.

Hiby, A. R., & Hammond, P. S. (1989). *Survey techniques for estimating abundance of cetaceans* (pp. 47–80, Special issue 11). Reports of the International Whaling Commission.

Hiby, A. R., & Lovell, P. (1998). Using aircraft in tandem formation to estimate abundance of harbour porpoise. *Biometrics, 54*, 1280–1289.

Laran, S., Authier, M., Blanck, A., Dorémus, G., Falchetto, H., Monestiez, P., . . . Ridoux, V. (2017). Seasonal distribution and abundance of cetaceans within French waters: Part II: The Bay of Biscay and the English Channel. *Deep-Sea Research II, 14*, 31–40.

Read, A. J., Drinker, P., & Northridge, S. (2006). Bycatch of marine mammals in U.S. and global fisheries. *Conservation Biology, 20*, 163–169.

Rogan, E., Cañadas, A., Macleod, K., Santos, M. B., Mikkelsen, B., Uriarte, A., . . . Hammond, P. S. (2017). Distribution, abundance and habitat use of deep diving cetaceans in the North East Atlantic. *Deep Sea Research Part II: Topical Studies in Oceanography, 141*, 8–19. Available from https://doi.org/10.1016/j.dsr2.2017.03.015.

Sambah (2016) Final report for LIFE + project SAMBAH LIFE08 NAT/S/000261 covering the project activities from 01/01/2010 to 30/09/2015. Reporting date 29/02/2016, 80pp.

Schweder, T., Skaug, H. J., Dimakos, X. K., Langas, M., & Øien, N. (1997). *Abundance of northeastern Atlantic minke whales, estimates for 1989 and 1995* (pp. 453–483). Reports of the International Whaling Commission, 47.

Skaug, H. J., Øien, N., Schweder, T., & Bøthun, G. (2004). Abundance of minke whales (*Balaenoptera acutorostrata*) in the northeastern Atlantic. *Canadian Journal of Fisheries and Aquatic Sciences, 61*, 870–886.

Solvang, H. K., Skaug, H. J., & Øien, N. (2015). *Abundance estimates of common minke whales in the Northeast Atlantic based on survey data collected over the period 2008-2013.* Paper SC/66a/RMP8 presented to the IWC Scientific Committee.

Tregenza, N. J. C., Berrow, S. D., Hammond, P. S., & Leaper, R. (1997). Harbour porpoise, *Phocoena phocoena* L., bycatch in set gill nets in the Celtic Sea. *ICES Journal of Marine Science, 54*, 896–904.

Vinther, M., & Larsen, F. (2004). Updated estimates of harbour porpoise (*Phocoena phocoena*) bycatch in the Danish North Sea bottom-set gillnet fishery. *Journal of Cetacean Research and Management, 6*, 19–24.

Viquerat, S., Herr, H., Gilles, A., Peschko, V., Siebert, U., Sveegaard, S., & Teilmann, J. (2014). Abundance of harbour porpoises (*Phocoena phocoena*) in the western Baltic, Belt Seas and Kattegat. *Marine Biology, 161*, 745–754.

Winship, A. J. (2009). *Estimating the impact of bycatch and calculating bycatch limits to achieve conservation objectives as applied to harbour porpoise in the North Sea* (Unpublished Ph.D. thesis). St Andrews: University of St Andrews.

Chapter 7

Conservation actions

All the legislative conservation agreements and directives in the world count for nothing if the animals or environment they serve to protect are in no better state as a consequence. Will the intensive discussions held in a United Nations or government building over a number of years lead to actions that a porpoise, dolphin or whale might notice? In order to have some insight on this question, I will review the conservation actions that have been taken over the last quarter of a century or are being developed. I will take each of the main conservation threats in turn and summarise actions taken by the Parties to the ASCOBANS Agreement, starting with the oldest threat of all, that of direct human exploitation.

Hunting

The hunting of whales, dolphins and porpoises has persisted across Europe for hundreds, and in some cases, thousands of years. Besides important welfare and cultural considerations, undoubtedly it has had impacts on the populations of many cetacean species. It was not until the middle of the 20th century that human society woke up to the damage they were doing and attitudes changed from whale killing to whale watching.

'Save the Whale' campaigns started in the 1960s by environmental groups around the world. This public pressure led to an international moratorium on whaling recommended in 1972 by the United Nations Conference on the Human Environment in Stockholm, Sweden, and eventually adopted by the International Whaling Commission (IWC) 10 years later. The IWC adopted a New Management Procedure, which was designed to bring whale stocks to an optimum level, and protected stocks at 54% of their initial estimated population size. However, this assumed we had good knowledge of the dynamics of the different populations, which rarely was the case. The moratorium on whaling came into effect in 1986 establishing zero commercial catch limits and led to a 'comprehensive assessment' of whale stocks – the stock sizes, current population trends, carrying capacity, and productivity. In an attempt to incorporate uncertainty and take a more precautionary approach, the IWC Scientific Committee introduced a Revised Management Procedure.

Since the formation of the ASCOBANS Agreement, none of its Parties have actually engaged in hunting. However, as noted in Chapter 5, Conservation threats, minke whales continue to be hunted by Norway (with an average of 600–800 taken annually), Iceland has resumed commercial whaling with annual catches of approximately 150 fin whales (Fig. 7.1) and small numbers of minkes, whilst the Faroe Islands kill an average of approximately 700 long-finned pilot whales annually along with occasional captures of other species such as Atlantic white-sided dolphin and Risso's dolphin. In these three cases, the animals killed occupy waters beyond those countries' Exclusive Economic Zones (EEZs) so that Parties to ASCOBANS have a direct interest. Although it is unlikely that hunting at current levels will lead to population declines in those species since catch rates represent a relatively small fraction of the overall numbers estimated in the region, my own hope is that with the reduced economic need for direct exploitation, and the inevitable cultural changes in attitudes that are taking place in a modern Europe, those countries will turn increasingly to less destructive methods to sustain employment and their economy, such as wildlife tourism.

Reduction of prey from fishing

Whereas hunting is no longer the serious conservation concern that it was half a century ago, the depletion of prey resources from overfishing continues to be a major issue. Fisheries scientists repeatedly provide advice to the European Union through the International Council for Exploration of the Sea (ICES) on the setting of catch quotas for different stocks of fish, cephalopods, and crustaceans. Repeatedly, these recommendations are modified by governments as they jostle to satisfy the interests of domestic politics. Most of the animals commercially taken by fisheries produce large quantities of eggs or young upon which there may be a high mortality. The fluctuations in this recruitment to the

European Whales, Dolphins, and Porpoises. DOI: https://doi.org/10.1016/B978-0-12-819053-1.00007-7

FIGURE 7.1   Fin whale being brought in by a Faroese whaling ship in the 1970s. No fin whales have been taken in the Faroes since then. However, since 2006, the Icelandic Whaling Company, Hvalur H/F, has intermittently hunted fin whales in Icelandic waters, taking 155 in 2017 and 146 in 2018. *Photo: Peter GH Evans.*

population are influenced greatly by environmental conditions so that although fishing may have an important effect, the overall impact may be either dampened down or exacerbated by environmental change, and, in recent decades, not least by climate change. Thus the effects of overfishing are not always clear, and the implications for top predators such as cetaceans even less clear.

Nevertheless, the years of protection afforded to herring and mackerel after their stocks had crashed in the 1960s do seem to be yielding positive results, and as those stocks increase, they appear to be benefiting cetacean species that feed upon them, such as harbour porpoise, minke whale and humpback whale.

Incidental capture in fishing gear

One species above all has dominated the ASCOBANS Agreement and that is the harbour porpoise. This is perhaps not surprising since it occurs in the waters of all the Range States (albeit only rarely now in many of the eastern Baltic countries). When the Agreement came into force, the conservation issue that was the focus of attention was bycatch, and particularly of harbour porpoise. Two of the ASCOBANS Parties in particular, Denmark and the United Kingdom, gathered data on bycatch rates with the help of independent observers on fishing vessels. In May 1995 the IWC endorsed its Scientific Committee's advice that an estimated annual bycatch of 1% of estimated population size indicates that further research should be undertaken immediately to clarify the status of the stocks and that an estimated annual bycatch of 2% may cause the population to decline and requires immediate action to reduce bycatch.

The ASCOBANS aim, as set out at the 2nd Meeting of Parties in Bonn in November 1997 was 'to restore and/or maintain biological or management stocks of small cetaceans at the level they would reach when there is the lowest possible anthropogenic influence - a suitable short-term practical subobjective is to restore and/or maintain stocks/populations to 80% or more of the carrying capacity' (ASCOBANS, 1997).

A small IWC/ASCOBANS Working Group on harbour porpoises advised that 'the maximum annual by-catch that achieves the ASCOBANS interim objective over an infinite time horizon, assuming no uncertainty in any parameter, should be 1.7% of the population size in that year'. This advice was subsequently endorsed by the IWC in 1999 at its 51st meeting. If uncertainty is considered, such as measurement error in estimating population size, maximum annual bycatch must be less than 1.7% to ensure a high probability of meeting the ASCOBANS objective (ASCOBANS, 2000a).

At the 3rd Meeting of Parties in Bristol in July 2000, Parties agreed that an estimated annual bycatch of 2% was too high and that further research indicated that a total anthropogenic removal above 1.7% of a harbour porpoise population must be considered unacceptable. It also underlined that the intermediate precautionary objective should be to reduce bycatches to less than 1% of the best available population estimate (ASCOBANS, 2000b).

Earlier, Parties to the Agreement had pledged financial support for a wide-scale abundance survey (SCANS) in July 1994 with particular attention to the harbour porpoise. This generated an abundance estimate of 170,000 harbour porpoises in the central and southern North Sea which when compared with a minimum estimated annual bycatch of 4450 harbour porpoises in this area, indicated that mortality from this source alone was 2.6% and thus well exceeded the threshold set. Similarly an abundance estimate of 36,000 harbour porpoises compared with a minimum estimate annual bycatch of 2200 animals on the Celtic Shelf, yielded a mortality rate of 6.1%. Clearly action was needed.

During the 1990s a project initiated by biologists in Denmark with collaboration from others in the United Kingdom and Sweden investigated the potential to deploy pingers (Fig. 7.2) on bottom-set gillnets to deter porpoises from close

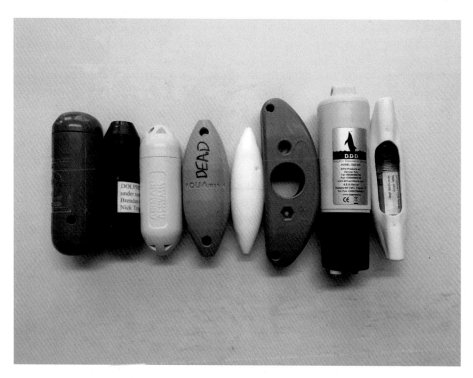

FIGURE 7.2 Various types of pinger have been used to reduce the chance of porpoises becoming entangled in fishing gear. From left to right, these include Dukane Netmark, PICE, AIRMAR, Aquamark 100, Fumunda (now 'Future Oceans'), SAVEWAVE holder (a replaceable pinger unit can be fitted inside), DDD, and banana pinger. *Photo: Nick Tregenza/Chelonia Ltd.*

contact leading to net entanglement. These are usually referred to as acoustic deterrent devices (ADDs), although in some cases they may simply act by alerting the animal to the net. Studies on captive porpoises were undertaken at the Fjord & Bælt Centre in Denmark (Figs. 7.3 and 7.4) to better understand their foraging behaviour, whilst others developed an ADD unit that could be safely and easily attached to the floatline and which would produce high frequency pings well within the hearing range of a porpoise with a varied sequence of pings to reduce the likelihood of animals habituating to it. Studies in the Gulf of Maine and Bay of Fundy had shown that pingers could reduce bycatch rates by up to 92% (Kraus et al., 1997; Trippel, Strong, Terhune, & Conway, 1999).

Regulation of fisheries in the region comes under the European Union (EU) Common Fisheries Policy, and in response to the above concerns, Council Regulation 812/2004 was issued in 2004 (European Union, 2004). It specified that it was mandatory for ADDs (at maximum spacing of 100 or 200 m depending upon the type) to be used on fishing vessels of 12 m length or more deploying bottom-set gillnets or entangling nets; gave technical specifications and conditions of use of the devices; and issued a requirement for at-sea observer schemes on vessels of 15 m length or more, and for smaller vessels, Member States should take the necessary steps to collect scientific data on incidental catches of cetaceans by means of appropriate scientific studies or pilot projects.

The Regulation was well intentioned but unfortunately has not been wholly successful. First, it has not focused upon controls for all high-risk fisheries, which increasingly have included vessels of smaller size than 12 m, and those vessels operating gear other than bottom-set gillnets which nevertheless entrap cetaceans. Second, the deployment of pingers has not been without its problems and should have been made more practicable for fishers. Problems have included operational failure, low durability, high cost, practical issues of deployment, health and safety issues, and in ensuring compliance.

The imposition of a regulation without the products being available to enable fishers to comply is not a very workable arrangement. This includes situations where fishers have fitted pingers with no guarantee of working over a period of time, no simple method for testing pingers onboard, and an inadequate monitoring programme to ensure continuity of pinger coverage on set nets. Fitting a faulty pinger is worse than not fitting pingers at all as fishers quickly lose confidence in their use, and for a community that was not entirely in support of measures that made life more difficult for them, effectiveness of this form of mitigation was paramount. The fishing communities needed to participate fully and from the start in developing solutions. In this context, it would be better not to prescribe which measures to take but instead set targets that need to be met in each region, thus allowing fishers to find the most suitable solutions for their specific situations. However, this requires adequate financial resources for the improvement of devices like pingers, and for research into alternative mitigation measures and gear types, as well as incentives for the development of more environment-friendly, sustainable, fishing methods. Fisheries and marine mammal biologists need to work in close partnership with fishers if mitigation measures are to be successful.

The development of more powerful pingers that can be deployed at wider spacing has been a useful advance and has also had some success in pelagic trawl fisheries that have a bycatch of other species such as common and striped dolphin.

Dolphin bycatch in trawl fisheries continues to be an issue. Excluder devices such as separation grids, rope and tunnel barriers, guiding and escape panels have been trialled but most have been rather ineffective (20% reduction in bycatch at best). Their positioning has proved critical; they have on occasions led to fish catch reductions; and they have posed handling difficulties in big pelagic trawls.

Different gear types and fishing methods call for different solutions. In the case of bycatch of baleen whales, the addition of weights to groundlines attached to pots causing them to sink, can be effective, as are weak links at the floatline and buoy, and the reduction in the number of vertical lines.

The success of a bycatch mitigation measure cannot be assessed without a monitoring scheme that can derive robust estimates of bycatch in relation to population size. Throughout the ASCOBANS Agreement Area, the level of bycatch monitoring has been inadequate to properly assess bycatch rates. This is largely down to the financial resources available to place observers on vessels, but it is also related to the building of trust within the fishing community to accept having observers aboard. In the case of small vessels, it may not be practicable to have an extra person on board. In an attempt to address this, countries such as Denmark and the Netherlands have been trialling the use of remote electronic monitoring (REM) with some notable success (Fig. 7.5). Danish studies have found them to be both cost-effective and more efficient at recording bycatch (including other nontarget species such as seabirds). Cameras need to be positioned carefully to record animals falling out of the net as it is hauled, and to ensure privacy for the fishers. If incentives are offered (such as increased quotas for those that agree to use them), they are more likely to be accepted by fishers. Their deployment can pose challenges on some vessels and lead to them being costly, but they have strong potential for future development.

FIGURE 7.3   The Fjord&Bælt Centre (A), with its seawater pens created in the harbour of Kerteminde, Denmark (B), has been used for a number of studies of harbour porpoise behaviour (C, D). *Photos: (A–C) Peter GH Evans; (D) Fjord & Bælt - www.fjord-baelt.dk.*

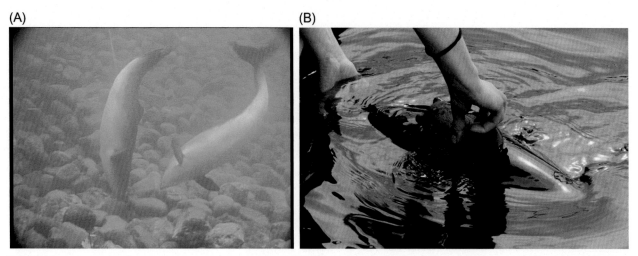

FIGURE 7.4 (A) The manner by which harbour porpoises use echolocation when foraging on bottom-living prey, studied at the Fjord & Bælt Centre, aided the development of pingers to alert them to the presence of a fishing net. (B) Porpoise hearing and vocalisations are currently being studied in captivity using DTag deployments. *Photo: (A) M. Amundin, (B) Peter GH Evans.*

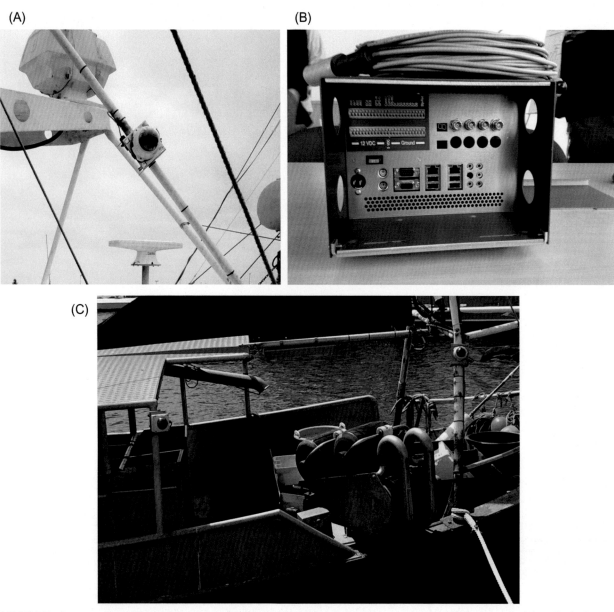

FIGURE 7.5 A remote electronic monitoring camera (A) directly downloading digital video to a computer (B) has the potential to effectively record bycatch of cetaceans and other marine species, particularly on small vessels that do not have the space to take an extra human observer. These can be deployed (C) in ways to avoid encroaching upon the privacy of the crew. *Photos: Meike Scheidat.*

The Common Fisheries Policy and its Data Collection Framework (DCF) have recently been undergoing revisions. There is concern that, as has been the case, they will not gather information at the level of detail needed to properly assess bycatch. Bycatch estimates are currently inaccurate not only due to insufficient sampling effort (i.e. observer coverage, vessel size, gear type), but also the difficulty of extrapolation to fleet level based solely on days at sea. In order to improve bycatch estimation there should be a more comprehensive sampling effort (by dedicated observer programmes, REM or some other means, covering the appropriate types of fishing gear and vessel sizes), and the following parameters should be included in the new DCF: fishing gear/activity at the appropriate level, target species, immersion duration of gear (soak time for set nets), and net dimensions (total length of set nets, aperture of trawl for those fisheries known to cause bycatch).

Monitoring needs to take a regional approach (e.g. inner Baltic Sea; western Baltic, Belt Sea and Kattegat; North Sea; and northeast Atlantic) focusing upon the medium to high-risk fisheries and tracking where these are operating in relation to the changing distributions of those cetacean species at risk, so that mitigation measures can be applied in a timely manner. This would be a more cost-effective way to minimise cetacean bycatch.

Information builds awareness

A Rat asks a Hamster, 'Listen, brother, why, since we are so similar to each other, do we have such different experiences with people – they love you, and they hate me'. And the Hamster replies, 'Good PR, my friend, good PR…!'

When my country ratified ASCOBANS, I discovered a similar backdrop of public awareness to this joke. In a country full of dolphin-lovers, many reacted to the word 'harbour porpoise' with surprise, sometimes with distaste, asking, 'What is this "morświn"? Is it a sea pig?'[1]

And so it was clear that the dolphins no longer needed better PR. They are adored and famous, although they do not have natural habitats in the Baltic Sea. For harbour porpoises, it was, and is, otherwise. The name of these animals has unpleasant connotations for many. Nobody knows of their beauty and intelligence. There is no way even to present them, never mind raise public awareness of them. How can we persuade people that harbour porpoises need protection? How can we explain that they have lived here for thousands of years; that in the past they were numerous; that today they are critically threatened with extinction; and that we are making it impossible for them to rebuild their resources. They are nearly absent from the sea, and nearly absent from public awareness. Before ASCOBANS entered into force, their image in the Baltic Region was barely a memory even for the local fishermen.

Although the Agreement was created to protect all small cetaceans of the Baltic and North Seas, during its initiation the specialists mainly talked of the harbour porpoise. You can understand why. The knowledge deficit, and the assessment of the opportunities for survival – particularly their Baltic resources – contrasted with the possibility of protecting them. It soon became clear that to halt their local extinction, it was necessary to tell the public the story of their life and highlight the problems and value of their existence (Fig. 7.6).

For the biologist and naturalist these issues are clear and basically unambiguous. For others for whom the sea itself is alien, and even more rare and unknown the species in it – it is a difficult and unbelievable story. In the southern Baltic Sea in the second half of the 20th century, the beautiful seals and the charismatic sturgeon had been forgotten, and the small 'Baltic dolphin' survived in the memories of just a few.

To be able to tell someone about something convincingly, to gain an ally for an issue, it is best to show the objective you are working towards. But how can we talk about the harbour porpoise, which so few have seen? After all, the only and most recent live porpoise I have seen was in the early 1970s. In Puck, Kashubia, together with the fishermen, we released it from the fishing net to freedom. Since then, the team at the Hel Marine Station IO UG and I have only seen their remains in the dozens (Fig. 7.7).

I do not think that this Polish story differs greatly from the Baltic Finnish, Russian, Estonian, Latvian or Lithuanian stories. On our sea, there are only three countries whose nationals – Danes, Germans and Swedes – with similar problems on the Baltic coast, can see harbour porpoises around the Danish Isles or in the North Sea. They are, however, individuals from different, much larger populations. Baltic harbour porpoises are not a well-known species today.

Therefore to do anything to increase public awareness and support their protection, we have had to make strenuous efforts, starting from the simplest possible activity – showing them. It was therefore not by accident that the first poster released by the ASCOBANS Secretariat featured the image of this species and was addressed to the inhabitants of the Baltic countries.

Following this in many countries there were also special flyers and brochures. Researchers waited for the effects of this campaign. They were hoping for reports of sightings, of finding harbour porpoises on the shore or while fishing. Slowly they started to receive the relevant information, but not as much as expected. Harbour porpoises continued to dwindle.

(Continued)

---

1. The word 'morświn' (Eng. harbour porpoise, Lat. *Phocoena phocoena*) in Polish was formed from the conjunction of two words morska (sea) + ś winia (pig). Similar names for this species also appear in other languages.

(Continued)

Funding for public information campaigns for environmental protection is poorer than for commercial campaigns. This is a pity, because information and education are key to the effective protection of the environment, and money for this type of project should be much more. Good advertising rapidly shows the existence on the market of a new product pushing it into the public consciousness, and adding the appropriate interest. Much more difficult is the 'advertising' of wildlife topics, for other reasons, too. This kind of information needs to be far more reliable and responsible. We cannot trick the public into appropriate behaviour. Error in the narration, selection of facts or sequence of images may turn the audience against the purpose of protection. The message created should be selected with reference to the basic level of audience perception. They have to know that up-to-date information is only a stage in the process they will witness, and if they trust us they can support it. Conscious understanding of the information content is the result of a prior process of education. And this, too, takes time.

FIGURE 7.6   The first ASCOBANS Outreach and Education Award was made to the Hel Marine Station. *Photo: ASCOBANS Secretariat.*

It is worth remembering that in the day of sensational and spectacular information, ours cannot be. Achieving the objective of species protection is a process requiring many years of consistent and enduring involvement by all parties.

The current unfavourable status of certain species requires sustained public interest in them. But how to maintain media interest in nature, when every day they are full to the brim with news of celebrities, politics, economics or sports? This is a difficult task. It is necessary to seek new and original approaches. Information on nature must periodically find a path to its recipients, and this despite the fact that for the media a message of species extinction is more attractive than the fact that it requires the protection or has been saved.

It is difficult to explain to the average person how valuable the harbour porpoises are. They will spend more time thinking about the cost of a kilogram of their meat or how much it will cost to rebuild its population, and the necessary habitats to provide the porpoises with security.

(Continued)

(Continued)

FIGURE 7.7 Porpoise stranded on the Baltic Sea coast. *Photo: Prof. Krzysztof Skóra Hel Marine Station, Department of Oceanography and Geography, University of Gdańsk.*

FIGURE 7.8 Harbour porpoise monument in Gdynia, Poland. *Photo: Prof. Krzysztof Skóra Hel Marine Station, Department of Oceanography and Geography, University of Gdańsk.*

(Continued)

(Continued)

Gdynia had a different idea. In the place where these animals have been sighted, the city authorities have decided that the harbour porpoises are worth a monument(!). Before they appear again on the Gdynia coast, they can return to human memory this way (Fig. 7.8).

But ASCOBANS is not only to rescue the porpoises. It also protects several species of dolphins, and small whales. The agreement has made many countries which have signed up to it feel responsible for them, regardless of whether a given species lives on their coasts or in another part of the European seas.

Over the 25 years that ASCOBANS has existed, millions of people have become aware of the dangers to the survival of small cetaceans in our seas. I am sure that most have nothing against helping the animals. When we started we had no mobile phones, there were no pingers on fishing nets, and no hydroacoustic detectors in the sea. There were far fewer photographs, movies were rare and the transmission of information was not as widespread as it is today with the development of the internet. Now these modern technologies are a marvellous way to encourage conservation. Modern techniques, however, require new, better transmission of information on wildlife. It must be fast, certified and verifiable.

When further countries joined the ASCOBANS Agreement, its area of action expanded and further recommendations arose. Conservation Plans have been prepared and adopted. We also have hundreds, if not thousands, of scientific publications, reports and databases; there is simply a lot more knowledge than 25 years ago. We know, as professionals, that this is available. It is probably also recognised by those to whom the Agreement's information campaigns are addressed. It is a pity that the species, which ASCOBANS protects, know nothing about this, although I may be wrong…

FIGURE 7.9 A human chain forms the shape of a stranded Harbour Porpoise on the beach at Klaipedia, Lithuania, to raise awareness of how the species was once a common member of the native fauna. *Photo: Prof. Krzysztof Skóra Hel Marine Station, Department of Oceanography and Geography, University of Gdańsk.*

By carrying out this work to disseminate knowledge and raise public awareness, ASCOBANS fulfils the obligations our countries' governments signed up for. It may be possible to reverse the negative trend and learn how to manage the sea so as to not to decimate surfacing dolphins, and make it possible to observe porpoises in the Baltic Sea. After 25 years, we are breathless after our activity. However, there are not too many indications that others understand us and want the ASCOBANS Agreement's regulations to help.

ASCOBANS is the strongest political spokesperson for small cetaceans in Europe. It is loved and admired, sometimes ignored, and also disliked when its objectives concern and affect various human activities that endanger these animals. It is extremely pleasant to participate in ASCOBANS meetings, where competent professionals determine a position, equipping the

(Continued)

(Continued)

Agreement's documents with the best knowledge for the conservation of dolphins and porpoises. The Agreement's first 20 years have been spent on the promotion of its objectives, on broadening the knowledge of the public, who are not familiar with the issue of small cetaceans, and on educational effort addressed to the young. Now the challenge is to develop a sustainable and effective mechanism for dialogue with stakeholders — the group who 10, 15 and 20 years ago were young people in school, and who today are adults — fishermen, administrators and politicians — who practically decide whether the coexistence of small cetaceans and humans in European seas is possible.

There are countries where this dialogue is probably developing relatively well. And there are those where it is missing. The spread of this democratic standard should be an element in raising public awareness. This requires not only the social aspect of the cost of nature conservation. The responsibility for the situation of endangered species and the ineffectiveness of the law in protecting them against the rest of society will mostly be borne by the users of the sea.

Social solidarity in protecting European nature is as necessary as in the pursuit of economic stability and peace (Fig. 7.9).

For humans, dolphins are the most charismatic residents of the seas and oceans. They arouse wide public interest, millions want to protect them, and this is probably in large part the contribution of those who work towards the ASCOBANS Agreement's goals.

Why does the same image not include the Baltic harbour porpoises? Why do some people not like ASCOBANS? Is the PR OK? As advocates for these animals and the Agreement, we have clearly not created good public relations everywhere (Fig. 7.10). In the Baltic Sea porpoises still die as a result of human activity, and such activity does not accept their existence and does not want to protect them. As a feedback this awareness and these objections to the protection of harbour porpoises may force yet greater development of public awareness projects.

Living in ignorance, without the awareness of problems, provides many people with psychological comfort. But for small cetaceans in European seas, human ignorance is the greatest danger.

FIGURE 7.10 Fishermen protest against porpoise protection. *Photo: Prof. Krzysztof Skóra Hel Marine Station, Department of Oceanography and Geography, University of Gdańsk.*

Prof. Krzysztof E. Skóra[†]
Director, Hel Marine Station, University of Gdańsk
Member of Polish delegation
[†]Deceased; 1950–2016

Chemical pollutants

The control of production of hazardous chemicals in Europe rests largely with the EU Integrated Pollution Prevention and Control Directive and the EU Marketing and Use Directive. Over the last 30 years, the Oslo and Paris Convention (OSPAR) has also played an important role in monitoring and regulating releases of a wide range of pollutants: heavy metals including the phase-out of tributyltin (TBT), organohalogens [such as polychlorinated biphenyls (PCBs) and the pesticides dichloro-diphenyl-trichloroethane (DDT) and dieldrin] and polycyclic aromatic hydrocarbons (PAHs). Its most recent Quality Status Report (2010) identified more than 300 substances considered to be of possible concern for the marine environment, of which 26 (substances or groups of substances) were considered to pose a particular risk due to their patterns of use. In 2010 the phase-out of a third of these was well underway in the OSPAR Area, which embraces all of the ASCOBANS Agreement Area with the exception of the inner Baltic Sea (where HELCOM has been active in the control of discharge and emissions - HELCOM, 2010, 2018). The various uses of the six OSPAR priority pesticides (including lindane) have been progressively phased out since 1998, resulting in a general reduction in concentrations in fish and shellfish.

However, the outlook for 2020 was deemed unfavourable for 16 of the priority chemicals. Included in this list of particular relevance to cetaceans were cadmium, lead and organic lead compounds, mercury and organic mercury compounds, some organotin compounds (other than TBT), PCBs, certain brominated flame retardants Polybrominated diphenyl ethers (PBDEs) and Hexabromocyclododecane (HBCD), and PAHs.

As noted in Chapter 5, Conservation threats, problems arise from the continued leakage of some chemicals such as PCBs, as well as wide-scale circulation in the atmosphere and in water. There is also a need for strengthening of implementation of existing measures. For example overall emissions of cadmium and mercury have remained relatively constant, although lead emissions continued to fall. On the other hand, progress on reducing air emissions of cadmium, mercury and lead have varied across OSPAR countries and industries. In 2007 around 900 tonnes of lead and 40 tonnes each of cadmium and mercury were released by OSPAR countries, largely through combustion processes in power plants and industry (Fig. 7.11).

Thus we see that although there have been international efforts in the ASCOBANS Agreement Area to reduce levels of contaminants of concern, there is still much to be done before we can view its marine environment as safe from the impacts of contaminants.

Marine litter

Marine litter has been identified as an environmental concern only relatively recently. The main sources from land include tourism, sewage, fly-tipping, local businesses, and unprotected waste disposal sites. The main sea-based sources are shipping and fishing, including abandoned or lost fishing gear. Microplastics (particles smaller than 5 mm) can come from a variety of sources including industrial abrasives, exfoliants and even the fabrics of clothing, and may also be formed secondarily by the breakdown of larger plastic material.

Monitoring of beach litter in northwest Europe indicates that about 65% of items are plastic. These degrade very slowly over hundred-year timescales and are prone to break up into small particles where they may be taken up by filter-feeding organisms.

International and EU legislation addressing sources of litter include the MARPOL Convention (Annex V) and the EU Port Waste Reception Facilities Directive. In a number of countries within the ASCOBANS Agreement Area, largely at the initiative of individuals or nongovernmental organisations, routine beach cleans (Fig. 7.12) have started to remove litter that could enter the sea, and there are also schemes to collect litter at sea (e.g. OSPAR, in 2007 published Guidelines for the implementation of Fishing for Litter projects in the OSPAR area).

Since 1998 OSPAR has monitored levels of beach litter, through voluntary monitoring programmes, and in their 2010 Quality Status Review, they reported an average of 712 litter items per 100 m of beach with levels remaining relatively constant over the previous decade, but with a slight increase in input from the fishing industry.

Although so far there is no strong evidence for any population level impact on a European cetacean species, it is clear that there is a need for better surveillance of the issue, improved monitoring of litter both on beaches and at sea, and more programmes for the collection and safe disposal of litter not only through beach cleans but active removals at sea and tackling the problem at source — reducing the amount of disposable plastic we use.

FIGURE 7.11   The Mersey Estuary in northwest England is known as one of the most polluted estuaries in Europe. It is a manufacturing centre with a history of growth and development of the British Chemical Industry, surrounded by the major conurbations of Manchester, Warrington/Widnes and Liverpool. *Photo: Peter GH Evans.*

Noise disturbance

One of the most active areas for research and conservation action has been with the aim to limit possible impacts of noise on marine mammals. The oil and gas industry now routinely has marine mammal observers accompanying seismic surveys alerting crews to the presence of animals within a particular zone so they can temporarily cease using airgun arrays. Since 2006 the International Association of Oil & Gas Producers, through its Joint Industry 'Sound & Marine Life' Programme, has contributed around US$50 million to fund research on the acoustic output of industry sources, and the effects of that output on marine animals. They have also funded investigations into the development of alternative technologies for seismic exploration of the seabed, such as marine vibroseis (marine vibrators, or MV). Airguns used in seismic surveys put a large amount of energy into the ground in a short amount of time. The alternative is to place a small amount of energy over a longer period of time. There is no way to adjust the bandwidth of airgun pulses but with vibroseis, there is greater control of the signal amplitude and other characteristics to suit the particular situation. In airgun surveys, 30% of the emitted sound energy (frequencies over 120 Hz) is wasted, that is not used by industry or geophysical researchers (Pramik, 2013). By contrast, with marine vibrators there is almost no energy above 100 Hz which means that there should be minimal impact on toothed whales, dolphins and porpoises that have greatest hearing sensitivities well above that (Diederichs et al., 2014; Weilgart, 2010). However, before these can be introduced, it is important that any potential environmental impacts are fully assessed since there could be masking effects for low-frequency specialists (LGL & MAI, 2011), and studies are currently underway to examine this.

FIGURE 7.12 Local communities increasingly are engaging in regular beach cleans to remove the large quantities of litter dumped in our rivers and oceans. *Photo: Molly Czachur.*

In the meantime several countries within the ASCOBANS Agreement Area have guidelines and regulatory controls. The UK Joint Nature Conservation Committee introduced guidelines (JNCC, 2010a) for mitigating the effects of noise on cetaceans from seismic activities which included ramp up (also termed soft start), whereby the sound source starts at a lower level in order to give animals a chance to move beyond the zone of potential physical injury, and these have been widely adopted around Europe and beyond. They have also issued mitigation guidelines for other activities producing underwater noise such as pile driving (JNCC, 2010b) and explosives (JNCC, 2010c). The Netherlands introduced a ban on pile driving during the first 6 months of the year to protect fish larvae from being killed, and Germany has limited impulsive noise during offshore wind farm construction to 160 dB re 1 μPa^2 s (sound exposure level) and 190 dB re 1 μPa (peak-peak) at 750 m distance from piling in their waters to protect especially harbour porpoises from being injured.

Mitigation measures that have been applied have varied between countries and for different sources of noise (OSPAR, 2014). In order to reduce sound pressure levels, during pile driving activities for wind farm construction, some countries like Germany have employed big bubble curtains (BBCs) — this is a ring of perforated pipes positioned on the sea floor around pile driven foundations such as frame constructions (jackets, tripods) and smaller monopiles, from which bubbles of compressed air rise and reduce the amplitude of the radiated sound wave by means of scattering and absorption effects (Fig. 7.13). Double or triple BBCs offer options for larger monopiles. These have been shown to reduce noise by up to approximately 12−14 dB in the 2 kHz frequency range (Diederichs et al., 2014; Dähne,

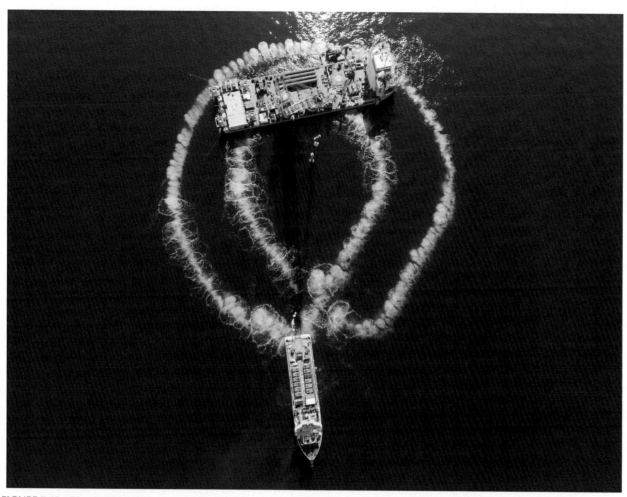

FIGURE 7.13 Big (double) bubble curtain deployed around a piling operation in the southern North Sea. Shown here are the bubble hose, the compressors and the winch. *Photo: Hydrotechnik Lübeck GmbH.*

Tougaard, Carstensen, Rose, & Nabe-Nielsen, 2017; Nehls, Rose, Diederichs, Bellmann, & Pehlke, 2016). However, currently BBCs can only be deployed in waters up to 40 m in depth.

Isolation casings shield the noise radiating pile (Fig. 7.14). The fact that they are reusable makes them attractive economically, although because they are attached directly to the piling frame, this can influence construction times and costs. Although still very much in the development phase, they have potential to achieve considerable noise reduction.

ADDs such as seal scarers have been used in combination with bubble curtains, in order to protect porpoises from hearing loss by scaring them away from the piling activity. Porpoise occurrence, quantified by echolocation signals, decreased when the seal scarer was engaged, during pile driving and up to 5 hours after pile driving stopped. This effect extended out to 12 km (Dähne et al., 2017). On the other hand, the strong reaction to the seal scarer has raised concern that it may exceed the reactions to the pile driving noise itself, when operating with bubble curtains, and so Dähne et al. (2017) propose a reevaluation of the specifications of seal scarer sounds before they are used on a wide scale.

Another noise reduction device is the cofferdam — a large rigid steel tub surrounding the pile from seabed to surface. The water is removed from the space in between so that pile driving takes place in air thus decoupling it from the body of water. A piled steel construction has been proposed that can be safely anchored in the North Sea at water depths of 30 m (OSPAR, 2014).

FIGURE 7.14   IHC noise mitigation casing (A) shows part of the jackup platform with the crane and the IHC-pipe (the noise mitigation pipe) that is being positioned over the monopile, which can be seen sticking out of the water below the pipe. The yellow structure on the platform is the piece that will be installed on top of the pile and will serve as a foundation for a wind measuring station inside the wind farm Nordsee Ost 1. In (B) the hydraulic hammer is hanging from the crane above the jacket foundation, which is installed on the sea floor via the monopiles that are driven into the ground through this jacket structure. *Photos: BioConsult SH GmbH&Co.KG.*

FIGURE 7.14   (Continued).

Hydro Sound Dampers consist of air filled elastic balloons and robust PE-foam elements fixed to nets or frames, which are placed around the pile. The frequencies at which the maximum noise reduction is provided are adjustable by variations in the size of the elements. High-energy absorption is reached by means of material damping. A system using the identical principle of Encapsulated Bubbles is under development in the United States, with the aim of reducing the low-frequency components of pile driving noise where maximum energy is emitted and has achieved a 14 dB reduction in the 100–300 Hz range. These systems are cheaper than bubble curtains because they do not require compressors.

Vibration pile driving (Vibropiling) is a technique used to make the pile oscillate at a low frequency of about 20 Hz. For large piles a number of vibratory hammers can be linked. They can work well in shallow waters such as those prevailing in the southern North Sea where low frequencies propagate to a much less extent, and noise reduction of 15–20 dB has been recorded compared with impact piling.

Several alternative foundation types are available or under development. With these, wind turbines can be founded without impact pile driving, thus generating less underwater noise. Included amongst these are drilled foundations, which work well in bedrock, sandstone or limestone; gravity base foundations, which have been used predominantly in the Baltic Seas at depths of up to 20 m, although there are plans to develop their use at depths up to 45 m; bucket foundations which are founded on the seabed (of sand, silt or clay) by suction pumps and require no pile driving; and floating wind turbines on platform types using different stabilisation mechanisms. A floating wind turbine (HYWIND) was installed in 2009 off the Norwegian coast in a depth of 220 m. In this case noise emissions are limited to transport and the anchoring process.

Other noise mitigation concepts in development include high-frequency–low-energy piling, mandril piles, silt piles and even silent pile driving.

There are also noise-quieting technologies being developed for shipping, and the International Maritime Organisation (IMO) has recently been giving this attention (IMO, 2014). The following is a summary of their recommendations. The greatest opportunities occur during the initial design of the ship, and in particular for the hull and propeller design to be adapted to each other. Propellers should be designed and selected in order to reduce cavitation since this is the dominant radiated noise source and may increase underwater noise significantly. This can be achieved by optimising propeller load, ensuring as uniform water flow as possible into propellers, and careful selection of the propeller characteristics such as diameter, blade number, pitch, skew and sections. The hull form (with its appendages) should be designed to ensure that the wake field is as homogeneous as possible since this will reduce cavitation. Consideration should be given to the selection of onboard machinery along with appropriate vibration control measures, proper location of equipment in the hull, and optimisation of foundation structures that may contribute to reducing underwater radiated and onboard noise. The adoption of a diesel—electric system should be considered as it may facilitate effective vibration isolation of the diesel generators. The use of high-quality electric motors may also help to reduce vibration being induced into the hull.

The most common means of propulsion on board ships is the diesel engine. The large two-stroke engines used for the main propulsion of most ships are not suitable for consideration of resilient mounting. However, for suitable four-stroke engines, flexible couplings and resilient mountings should be considered, and where appropriate, may significantly reduce underwater noise levels. Four-stroke engines are often used in combination with a gearbox and controllable pitch propeller. For effective noise reduction, consideration should be given to mounting engines on resilient mounts, possibly with some form of elastic coupling between the engine and the gearbox. Vibration isolators are more readily used for mounting of diesel generators to foundations.

For existing vessels, there are also ways to make them quieter. Propeller cleaning (to remove marine fouling) helps to reduce propeller cavitation. Maintaining a smooth underwater hull surface and smooth paintwork may also improve a ship's energy efficiency by reducing the ship's resistance and propeller load. This will not only reduce underwater noise but also increase fuel efficiency. For ships equipped with fixed pitch propellers, reducing ship speed can be a very effective operational measure for reducing underwater noise, especially when it becomes lower than the cavitation inception speed. For ships that have controllable pitch propellers, the alternative is to ensure optimum combinations of shaft speed and propeller pitch.

It should be noted that reducing noise pressure levels alone may not necessarily reduce negative impacts on cetaceans. When evaluating the effects of underwater sound sources, any or all of the following may be important: peak pressure, received energy (received sound exposure level), signal rise times, signal duration, spectral type, frequency (range), kurtosis, duty cycle and directionality.

Vessel strikes

Much of the effort to reduce the risk of cetacean mortality from vessel strikes has been in eastern North America, and has centred upon large whales, particularly the North Atlantic right whale where ship strikes have contributed significantly ($>50\%$) to mortality of this endangered population. Ideally one wants to prevent ships and whales being in the same place at the same time and so much effort and resources have focused on vessels being alerted to the presence of an oncoming whale in time to take aversive action. To protect right whales, there is mandatory ship routing and ship reporting (using spotter plans) in a number of areas in eastern United States (Silber et al., 2013), and seasonal management areas have been created where vessel speed restrictions of 10 knots or less are required during times when right whales are likely to be present. Passive acoustic monitoring along with the use of a 'Whale Alert' iPad application has also been introduced to facilitate vessels detecting whales early. On the west coast of the United States, around the Channel Islands in California, a traffic separation scheme operates to reduce collisions with blue whales.

Other methods that have been employed include the use of active acoustics, radar and thermal imaging such as infrared technology to detect whales in the path of the vessel, but these can be very expensive and still rely upon human judgment to interpret the signals obtained, and to do so in time to take aversive action.

In Europe a traffic separation scheme was implemented in the Strait of Gibraltar, and the IMO issued notices to mariners for speed reductions, in order to lessen the chance of ships hitting sperm whales (Silber et al., 2013). Within the Pelagos Sanctuary in the Ligurian Sea, a whale reporting scheme called REPCET has been established, whereby the locations of fin whale sightings are shared between ships.

(A)

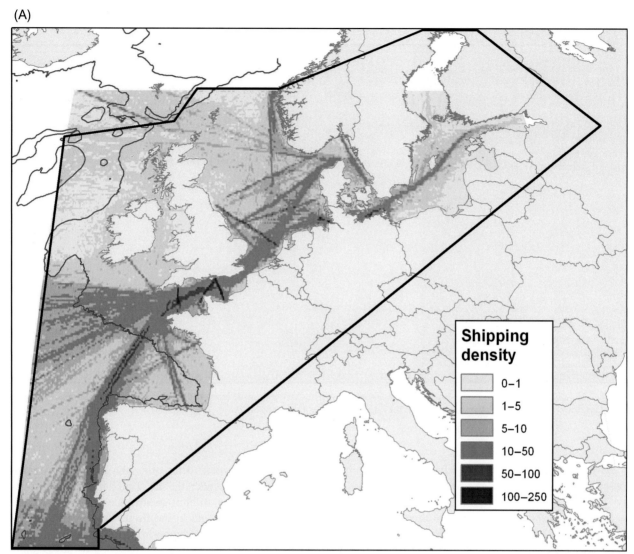

FIGURE 7.15 (A) Density of shipping in northwest Europe *(reproduced from Evans, P. G. H., Baines, M. E., & Anderwald, P. (2011). Risk assessment of potential conflicts between shipping and cetaceans in the ASCOBANS region (32 pp.). ASCOBANS AC18/Doc. 6-04 (S).*

In the ASCOBANS Agreement Area, the number of reported vessel strikes has been too low for conservation action to be taken, with most of the high vessel density areas (e.g. English Channel and southernmost North Sea; see Fig. 7.15) not being areas utilised by large whales, the more vulnerable group. However, an ASCOBANS study (Evans, Baines, & Anderwald, 2011) using predictive modelling, highlighted parts of the Bay of Biscay as areas where strike risk is greatest, and working with World Wide Fund for Nature (WWF), international discussions were held with a major Norwegian shipping company to adopt speed restrictions within those locations. As yet, however, this has not been put into practice.

Recreational activities

Much of the negative impacts of recreational activities on cetaceans can be counteracted by the raising of awareness of how vessels can cause disturbance and educating people on safe ways to approach animals.

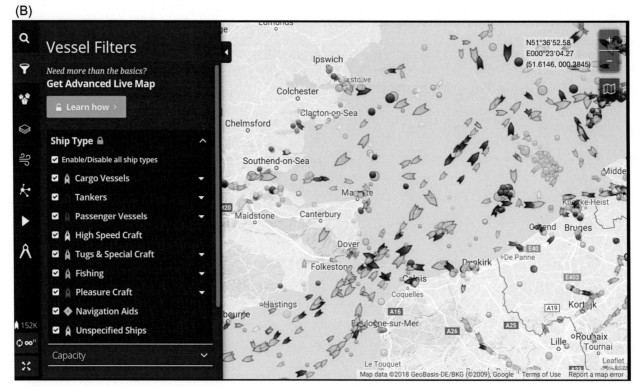

FIGURE 7.15 (B) The Strait of Dover is one of the busiest waterways in the world, as demonstrated by this screen image of vessel locations revealed by AIS. *(reproduced with permission from Marine Traffic website, 29 December 2018).*

Marine protected areas (MPAs) that have been established specifically to protect cetaceans, such as Special Areas of Conservation (SACs) established for bottlenose dolphin or harbour porpoise under the EU Habitats Directive, often have codes of conduct for wildlife watching and for other recreational activities, to limit disturbance. Examples include the SACs established in the Moray Firth and Cardigan Bay (United Kingdom) and in the Ile de Sein marine reserve (France) for the bottlenose dolphin. Some aspects of these codes or guidelines may be enshrined in law making it an offence to deliberately or recklessly disturb protected species like cetaceans, and violations have led to prosecutions (e.g. in the United Kingdom).

Codes of conduct have been developed in many countries, and although they may differ in the detail, they follow broadly similar lines: do not chase animals, do not drive a vessel directly towards them, or encircle them. Wherever possible, let them approach you. If they choose to bow-ride, maintain a steady course and speed (Fig. 7.16). Do not respond to them by changing course or speed in a sudden or erratic manner; slowing down or stopping suddenly can confuse and alarm animals as much as sudden acceleration. Allow groups to remain together. Avoid deliberately driving through, or between, groups of cetaceans. Avoid close approach to individuals with young as you risk disrupting mother−calf bonds and expose inexperienced young to stress and possible boat strikes. Do not throw rubbish or food near or around cetaceans. Always allow cetaceans an escape route. avoid boxing them in between vessels. Ensure that no more than two vessels are within a kilometre of cetaceans at any one time and no more than one boat within close proximity (i.e. radius of 100 m). Refrain from calling other vessels to join you. If other vessels in the vicinity are interested in watching the cetaceans, limit your presence to 20 minutes. Move away slowly if you notice signs of disturbance, such as repeated avoidance behaviour, erratic changes in speed and direction, or lengthy periods underwater (Würsig & Evans, 2002).

Possible sources of noise disturbance can be avoided by ensuring speeds are never greater than 10 knots, and by keeping the engine and propeller well maintained. On the other hand, care should be taken to avoid collisions with cetaceans when using sailing boats or boats with a low engine noise, as the animals are less likely to hear the vessel until it is close to them.

FIGURE 7.16   If some simple rules are followed, there is no reason why recreational vessels cannot coexist with cetaceans. *Photo: Peter GH Evans.*

There is no progress without public awareness

You have to be very lucky to see a harbour porpoise in the wild. These small marine mammals are extremely agile. They zigzag like rabbits. It is impossible to tell where they will pop up again for their next breath. No wonder that some 20 years ago, few people knew about the existence of this tiny, beautiful small cetacean, which is also the only cetacean inhabiting the Baltic Sea (Fig. 7.17). It has little chance of survival as long as every year more porpoises are dying than are born. Reason enough for UNEP and several countries to contemplate how to save the small whale in big trouble.

The list of causes for its demise is long: toxic substances from industry and agriculture, habitat destruction by sand and gravel extraction, noise pollution and the fishing industry are all taking a toll on these sensitive marine mammals. Too many are dying as so-called bycatch in fishing nets laid for other purposes. They are neither able to see nor echolocate such modern synthetic nets. They become entangled and die a slow death from suffocation, even if nowadays no fisherman intends to catch harbour porpoises.

To save the Baltic harbour porpoise, a first discussion on the need to develop a recovery plan took place in 2000 during the 7th meeting of the ASCOBANS Advisory Committee in Bruges, Belgium (2000). A subsequent expert meeting chaired by Randall Reeves (from Canada) took place in Jastarnia, Poland, in 2002. The first version of the plan was finalised in 2002 and 'welcomed' but not yet adopted by the Parties at the 4th Meeting of the Parties (MOP4) in Esbjerg, Denmark, in 2003. Finally, in 2009 the second edition of the 'Jastarnia Plan' was adopted together with the new 'Conservation Plan for the Harbour Porpoises in the North Sea' at the 6th Meeting of ASCOBANS Parties in Bonn, Germany.

One of the recommendations of the 'ASCOBANS/Jastarnia Plan' (2009) was to switch from fishing techniques that endanger porpoises to less dangerous methods. Nonetheless, until now no real action has been taken, although there are at least some studies going on in some Scandinavian countries. All in all, the ASCOBANS member states have not taken sufficient countermeasures to prevent the extinction of the harbour porpoises in the Baltic Sea. This continuously negative development is particularly dramatic in the Baltic Proper (east of the isle of Rügen) where total numbers have diminished to approximately 500 individuals. Therefore the World Conservation Union (IUCN) had to classify the Baltic population as 'critically endangered' in 2008 (Hammond et al., 2008).

FIGURE 7.17   For several countries around the Baltic, most members of the public had never actually seen a harbour porpoise so it was very important to raise awareness about its presence. *Photo: Prof. Krzysztof Skóra Hel Marine Station, Department of Oceanography and Geography, University of Gdańsk.*

(Continued)

(Continued)

Presently, however, alarm bells are also ringing from the 'Belt Sea' (Kattegat, Belt, Sound and western Baltic), where another distinct population still occurs fairly frequently. Appearances can be deceiving, and recent Danish scientific publications suggested a decline in the population (Teilmann, Sveegaard, & Dietz, 2011). Reason enough for ASCOBANS to develop a recovery plan for the 'gap area' or 'Belt Sea' (between the North Sea and the Baltic Proper) as well.

Since it is well-known that people do not see a need to protect or conserve what they do not know, the Jastarnia Group — and ASCOBANS Parties — agreed also on the need for public awareness and education. The small whale in big trouble needs headlines! Thus it was decided to create the 'International Day of the (Baltic) Harbour Porpoise', which has taken place every third Sunday in May since 2003. At least once a year, it provides the occasion for press releases and campaigns to help the porpoises — and their environment.

Along these lines the Society for the Conservation of Marine Mammals (GSM e.V.) developed its project 'Sailors on the Lookout for Harbour Porpoises' in 2002. The project is well respected by authorities and the public. Over the years it has proven valuable as a basis for scientific publications under 'Opportunistic Sightings of Harbour Porpoises in the Baltic Sea' (Cooke, Deimer, & Schütte, 2006; Deimer, Schütte, & Wilhelms, 2003; Loos, 2009). In May 2011 the project was transferred to the 'Deutsches Meeresmuseum' (German Oceanographic Museum) in Stralsund, Germany: www.meeresmuseum.de/de/Wissenschaft/Sichtungen.html.

With the growing public perception that the small whale is in big trouble, members of the public not only record their sightings of healthy porpoises in the sea (Fig. 7.18) but also dead animals found along the beaches (Fig. 7.19). The alarming results: in recent years up to 170 harbour porpoise carcasses were found on German Baltic Sea beaches alone — 138 were found in 2010. No population can cope with such a drain. Some years ago the German Federal Government already presented research results, which substantiated a bycatch quota of at least 47% of all dead harbour porpoises found along the German Baltic coast. This means that 2%–8% of the domestic harbour porpoises were perishing in fishing nets (Herr, Siebert, & Benke, 2009; Koschinski & Pfander, 2009).

Such alarming results underscore not only the urgent need for mitigation but, furthermore, highlight the need to collect not only data of 'bycaught' cetaceans (on fishing vessels) as requested by several international bodies, such as the Scientific Committee of the IWC, but to collect data from strandings as well. Experience shows that strandings may consist largely of undetected bycatches.

The question remains: how long does the Federal Republic of Germany — and other member countries of ASCOBANS — want to continue its passive attitude towards bycatch and to avoid implementing the Jastarnia Plan?

Dr Petra Deimer-Schütte.
President, The Society for the Conservation of Marine Mammals (GSM)

FIGURE 7.18 Anyone who goes to sea can contribute information on our porpoises, dolphins and whales by reporting their sightings to public sightings schemes which now exist in many European countries. *Photo: Peter GH Evans.* (*Continued*)

(Continued)

FIGURE 7.19 Harbour porpoise stranding on a remote beach. The public reporting of strandings aids national monitoring schemes to investigate causes of mortality. *Photo: Sea Watch Foundation Photo Library.*

Climate change

The negative effects of climate change upon marine mammals are not easily addressed directly so the major efforts of national governments have been to try to reduce carbon dioxide emissions and other activities causing global warming. But, of course, we can all do something in our daily lives to help with energy conservation.

Marine protected areas

In the last 25 years there has been increasing focus upon the establishment of MPAs for particular species and habitats (Fig. 7.20). In Europe this has been primarily through the Natura 2000 network under the EU Habitats Directive, with provision made specifically for harbour porpoise and bottlenose dolphin under Annex II. Those two species were selected because of their perceived vulnerability to human activities particularly within 12 nm of the coast. There is a case for having MPAs for additional cetacean species that also might benefit from spatial protective measures, but it is probably best first to ensure the current Natura 2000 system is effective. This is where opinions diverge (Evans, 2018). Some believe that such measures are inappropriate for species that are highly mobile, as applies to most cetaceans, and that it would be better to focus upon issue based mitigation measures that could be applied to specific human activities such as fisheries. Nevertheless, the last two decades have seen sustained efforts globally to establish MPAs, and much existing conservation legislation at both a national and international level encourages their designation. Once designated, they tend to receive greater attention in terms of human and financial resources devoted to status assessment, monitoring and management actions. For this reason, it is important that selection is based upon robust evidence.

FIGURE 7.20   Bottlenose dolphins in the Iroise Sea Marine Park, Brittany. *Photo: Oceanopolis-Brest.*

In the early years of Natura 2000 some sites for marine mammals were proposed before the full range of a species population had been systematically surveyed or analysed so it was not assured that they were indeed persistent hotspots (Evans, 2008), but in the last 10 years, this aspect has been an increasing focus of attention. This should help ensure that the most appropriate sites are selected. If they are not, then it rather defeats the purpose. If at all possible one should try to identify what makes a site important for one or other of the Annex II species. If it is a stable habitat feature such as complex undersea topography, a mix of currents around a group of islands, or an open estuary, then one might be more confident that the site will continue to be important over a long period of time so long as humans maintain the environment in a good state. Analysis of data collected across many years will help to test that the area is indeed important. Nevertheless, populations may shift their range or favoured feeding or breeding areas, and a system of MPAs needs to be responsive to this.

Over the years EU Member States have been a bit slow in establishing Natura 2000 sites for bottlenose dolphin and, particularly, harbour porpoise, and it is clear that one of the reasons has been the conflicting pressures from other stakeholders. SACs are still regarded widely as nature reserves where no other potentially harmful human activity should take place, rather than the consideration of specific management measures that could be applied in those areas to mitigate their effects. For seas to be managed effectively, particularly in the coastal zone where pressures are greatest, the concept of marine spatial planning needs to be better incorporated to take account of environmental concerns. This should apply both within and outside of MPAs (Evans, 2018).

At the time of writing, in the Atlantic and Baltic biogeographic regions that span the ASCOBANS Agreement Area, there were 236 Sites of Community Importance with harbour porpoise, and 84 with bottlenose dolphin, but only 121 and 56, respectively, where populations of those species were classed as significant (i.e. a qualifying feature), and most are rather small (see Fig. 7.21). Furthermore, few of these have yet to possess management plans for the species.

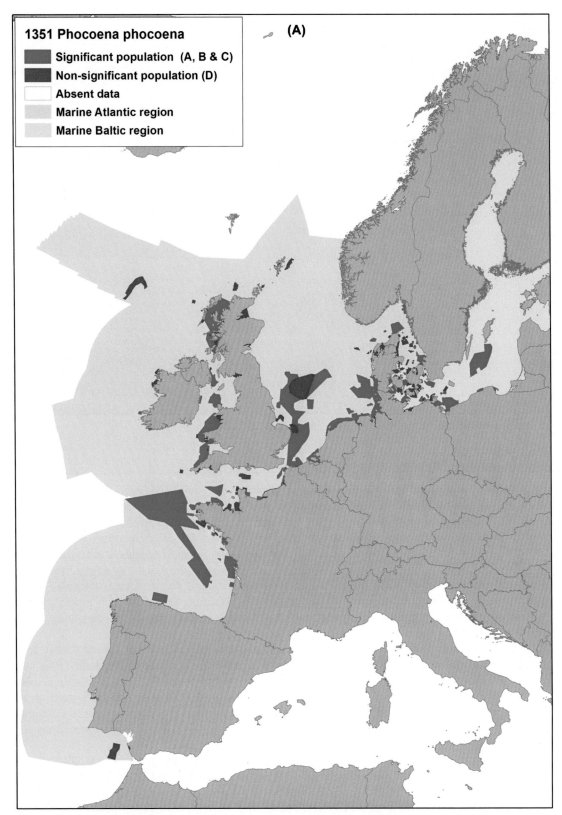

FIGURE 7.21 (A) Natura 200 sites designated by countries for harbour porpoise in fulfilment of the EU Habitats Directive *(from European Topic Centre on Biological Diversity, Paris, courtesy of Laura-Patricia Gavilan Iglesias).*

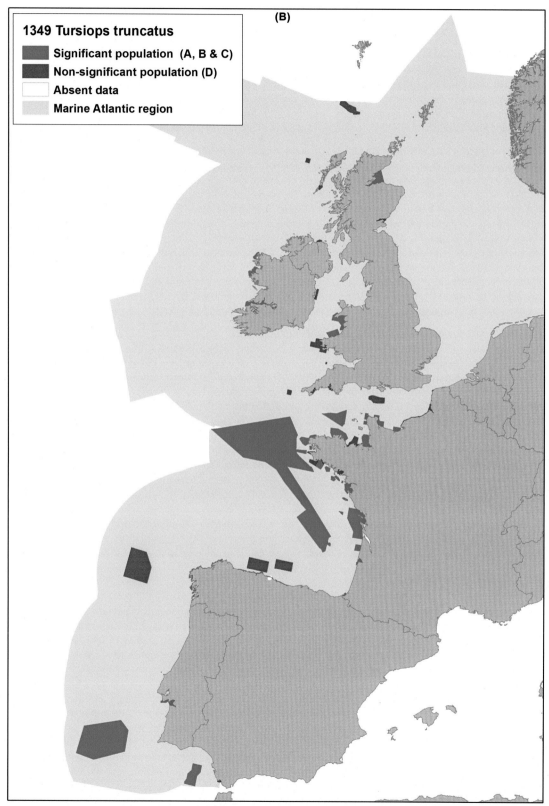

FIGURE 7.21   (B) Natura 2000 sites designated by countries for bottlenose dolphin in fulfilment of the EU Habitats Directive *(from European Topic Centre on Biological Diversity, Paris, courtesy of Laura Patricia Gavilan Iglesias).*

The significance of the harbour porpoise for ASCOBANS and vice versa

The harbour porpoise (*Phocoena phocoena*) is the only cetacean species that reproduces in all ASCOBANS waters, and for some of the 10 Parties to the Agreement, it is the only species they still encounter on a regular basis in their national waters. Therefore it is not surprising to find that this species has been of particular significance to the Agreement, resulting in two existing international plans for the protection of the Baltic Proper population and for the conservation of the North Sea population(s), as well as a third plan to help preserve the Belt Sea population. These plans are among the most prominent outputs of the Agreement and thus provide an excellent means to contemplate the importance of the species for the Agreement as well as the impact of the conservation measures afforded by ASCOBANS using its longest existing plan as an example.

The harbour porpoise in northern Europe and early concerns for its conservation

Of the 16 potential populations of harbour porpoise across the North Atlantic, the ASCOBANS area is home to at least five distinct ones in the North Sea, in the Belt Sea and Kattegat, in the Baltic Proper, around Ireland and the western British Isles, and along the Iberian coast and in the Bay of Biscay. Of these the conservation of the porpoise population in the inner Baltic Sea (i.e. Baltic Proper) has always been of major concern. Already in March of 1996 the Baltic Sea Range States as Contracting Parties to the Helsinki Convention adopted HELCOM Recommendation 17/2 calling for the 'Protection of Harbour Porpoise in the Baltic Sea Area' and 'recognising the relevance of the ASCOBANS' Agreement. In particular, the recommendation called for (a) the avoidance of bycatch; (b) data collection on distribution, abundance, stock identity, 'and threats such as pollutant levels, bycatch mortality, disturbance by shipping (e.g. underwater noise)'; (c) the establishment of MPAs; and (d) triennial progress reports on the implementation of the above recommendations. Consequently, the Baltic Sea Action Plan adopted by HELCOM in 2007, aimed for an improved conservation status of the Baltic harbour porpoise by 2015. In particular, its goal was a significant reduction of harbour porpoise bycatch rates to close to zero within a period of 3 years.

How did the Baltic harbour porpoise galvanise ASCOBANS into action?

One year after the adoption of HELCOM Recommendation 17/2, at the second Meeting of the ASCOBANS Parties (MOP2 in Bonn, Germany, 1997), the idea of a recovery plan for the Baltic harbour porpoise was raised. During the 7th meeting of the ASCOBANS Advisory Committee in Bruges, Belgium (2000), a discussion ensued over a 'process to develop a recovery plan for harbour porpoises in the Baltic', as annex 10 of the final report outlines. The Secretariat strongly promoted this endeavour, and a decisive expert meeting was convened in Jastarnia, Poland, in early 2002. Following this meeting, Randy R. Reeves (Canada), who had also acted as facilitator in Jastarnia, finalised the Recovery Plan for the Baltic Sea Harbour Porpoise, nowadays also known as the Jastarnia Plan. Although the plan was widely praised (e.g. by HELCOM, IWC, and ICES), the subsequent Meeting of ASCOBANS Parties (MOP4 in Esbjerg/Denmark, 2003) did not adopt it due to the intervention of the European Commission, which contested the competence of the Agreement to adopt any plan promoting measures affecting fisheries. This was only rectified 6 years later, when the 6th Meeting of ASCOBANS Parties (MOP6 in Bonn/Germany, 2009) unanimously adopted the second edition of the Jastarnia Plan together with the new 'Conservation Plan for Harbour Porpoises in the North Sea' that had been under preparation since MOP4.

Although Parties had felt unable to adopt the Jastarnia Plan in 2003, they still went ahead and installed the working group that had been called for in the plan, to help implement the measures proposed. The so-called Jastarnia Group (Fig. 7.22) met for the first time in Bonn in March 2005 and has continued to meet annually since then. Every year this group has recorded the progress of the implementation of the measures that the Parties agreed upon — or lack thereof — and made recommendations for the further implementation of the plan.

Thanks to scientists such as Per Berggren (Sweden) and Andrew Read (United States) the precarious status of the porpoise population in the Baltic Proper has long been well-known, and in 2008 this population was eventually officially recognised by IUCN as 'critically endangered'. Previous aerial surveys of the most densely populated part of the Baltic Proper resulted in abundance estimates of somewhere between 200 and 3300 animals in July 1995, and 10—460 animals in July 2002. These extrapolations served as a basis for a population model that showed that any anthropogenic take (removal) of more than one individual per year would make a recovery of this population highly unlikely. These results were adopted by the Jastarnia Plan and hence serve as the basis for the mitigation measures agreed upon by the six governments.

The three impacts already listed by HELCOM Recommendation 17/2 (i.e. pollutants, bycatch and disturbance) had been widely recognised as the main threats for a large number of cetacean populations worldwide. These three factors include a variety of man-made impacts from persistent organic pollutants to heavy metals, from bottom-set gillnets to other types of fishing gear in other seas and from noise pollution to collision with fast-moving vessels. The Jastarnia Plan always concentrated on mitigating bycatch in set nets, as it was perceived to be by far the most important anthropogenic mortality factor for porpoises in the Baltic Sea. In addition to bycatch reduction, the Jastarnia Plan calls for research and monitoring, the creation of MPAs, the raising of public awareness and the cooperation between the Agreement and other bodies.

Did the Jastarnia Plan motivate conservation activities of other organisations?

(Continued)

(Continued)

FIGURE 7.22   Jastarnia Group meeting in Gothenburg, Sweden, April 2013. *Photo: ASCOBANS Secretariat.*

Shortly after preventing the adoption of the Jastarnia Plan in August of 2003, the European Commission was the third international body — after HELCOM and ASCOBANS — to become involved in the international protection of the harbour porpoise in the Baltic Sea. On paper this species had been protected in the European Union since 1992, when the Habitats Directive (92/43/EEC) placed the harbour porpoise on its Annex II (together with the bottlenose dolphin) and on Annex IV, together with all other cetacean species. However, the creation of MPAs for porpoises (Article 3 of the Habitats Directive) has been extremely slow and in the Baltic Sea was largely based on areas previously nominated as Baltic Sea Protected Areas by HELCOM. More importantly none of these so-called Natura 2000 sites have yet been equipped with a management plan that actually mitigates anthropogenic impacts on porpoises. General conservation measures (required for very mobile species according to Article 12 of the Habitats Directive) in the entire distribution area are also still lacking.

Another more recent EU Directive, the Marine Strategy Framework Directive (2008/56/EC), was created in 2008 to establish 'a framework for community action in the field of marine environmental policy'. Harbour porpoise populations could serve easily as 'qualitative descriptors for determining good environmental status', especially for the maintenance of biological diversity (descriptor 1; e.g. bycatch), for the normal abundance and diversity of marine food webs (descriptor 4; e.g. prey depletion due to overfishing), for contaminant concentrations (descriptor 8), for marine litter (descriptor 10; e.g. derelict fishing gear) and for underwater noise (descriptor 11). Harbour porpoises would be excellent indicators and would certainly profit themselves from any improvement in the environmental status of these descriptors.

In 2002 the European Commission revised its Common Fisheries Policy thus calling also for bycatch mitigation. Subsequently this process led to the creation of Regulation EC 812/2004 'laying down measures concerning incidental catches of cetaceans in fisheries' in April 2004. The two main tools of this regulation to protect porpoises from being bycaught were the

(Continued)

(Continued)

use of acoustic deterrents (frequently also called 'pingers') and of dedicated onboard observers to quantify bycatch and ultimately improve the efficiency of acoustic deterrents in mitigating bycatch. So far, however, acoustic deterrents only need to be used by large vessels, whereas approximately 95% of set-netters in the Baltic Sea use vessels shorter than 12 m. Even when disregarding the many other problems with the large-scale use of deterrents in the attempt to protect a critically endangered population, the low number of remaining Baltic Proper porpoises were deemed worthy of protection only in the extreme southwestern part of the Baltic Proper (i.e. ICES 'subdivision 24') thus leaving the vast majority of the Baltic Proper and all its connected waters unprotected by Regulation 812/2004 (i.e. no mitigation measures required there) (Fig. 7.23). With the next revision of the Common Fisheries Policy formulated in the last few years, it remains to be seen how the European Commission will improve its bycatch mitigation efforts to make them more effective for the harbour porpoise and other (small) cetaceans — and not only in the Baltic Sea.

FIGURE 7.23 The island of Rügen in the German Baltic forms part of a transition zone between the porpoises of the Belt Seas/western Baltic that seasonally migrate into the Baltic Proper where the small endangered Baltic population resides. *Photo: Helena Herr.*

How significant is the work of ASCOBANS for harbour porpoise conservation?

In summary, it appears that in the first 25 years of its existence, ASCOBANS has worked hard to protect harbour porpoises in the Baltic Sea — in particular through the dedication of the Jastarnia Group members — and also by influencing the policies of other organisations. The same is likely to happen elsewhere, especially with the establishment of the North Sea Porpoise Conservation Plan and the Conservation Plan for the Belt Sea Porpoise being developed. However, additional efforts are still needed by the respective governments to implement the bycatch mitigation measures of the Jastarnia Plan. Although success is usually difficult

(Continued)

(Continued)

to quantify in nature conservation attempts, the work of ASCOBANS appears to have been instrumental at least in raising general awareness about the dire needs of porpoises. Hence, porpoises and other (small) cetaceans are now also considered when new impacts arise, such as the emission of impulsive (and potentially lethal) sounds from pile driving, seismic exploration or underwater explosions.

Dr Stefan Bräger
Chair of the Advisory Committee, 2007–10
Chair of the Jastarnia Group, 2006–07

Conservation plans

Parties to ASCOBANS have had a special interest in the harbour porpoise. It is the only small cetacean species recorded in all the Range States; its predominant shelf sea distribution makes it particularly vulnerable to human activities; and the agreement was founded on concerns for its status. It is therefore appropriate that conservation plans for the species should have been developed in different parts of the Agreement Area. In January 2002 a meeting in the Polish town of Jastarnia launched a recovery plan for the species within the inner Baltic Sea (ASCOBANS, 2002). This became known

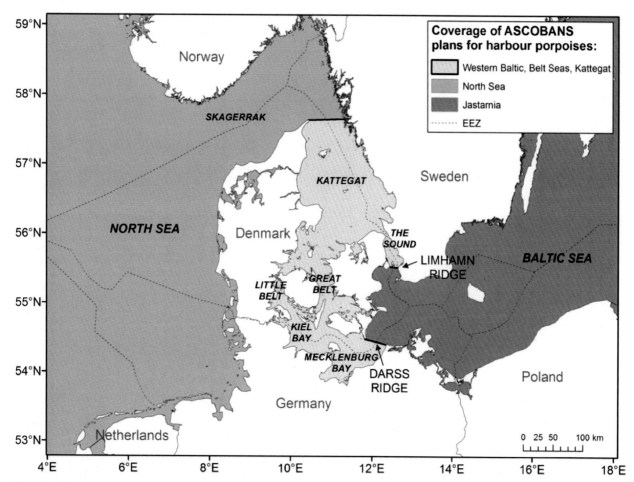

FIGURE 7.24   Map of the North Sea and the Baltic indicating where the geographical area covered by the Plan for the population in the western Baltic, the Belt Sea and the Kattegat adjoins that of the ASCOBANS North Sea Plan and the ASCOBANS Jastarnia Plan. The dashed line indicates the national borders of the Exclusive Economic Zone (EEZ). *From ASCOBANS 2012, courtesy of ASCOBANS Secretariat.*

FIGURE 7.25   Joint meeting of the North Sea and Jastarnia Groups, Wilhelmshaven, Germany, June 2017. *Photo: ASCOBANS Secretariat.*

FIGURE 7.26   Robin Hoods Bay, North Yorkshire. Harbour porpoises in the central North Sea have experienced significant mortality from bottom-set gillnetting. *Photos: Peter GH Evans.*

FIGURE 7.27   White-beaked dolphins are confined to the North Atlantic, with important populations in the North Sea and around Scotland. *Photo: Steve CV Geelhoed.*

FIGURE 7.28   Risso's dolphins occur in small localised populations along the Atlantic coasts of Europe and in the Mediterranean. *Photo: Peter GH Evans.*

as the Jastarnia Plan, and a steering group (called the Jastarnia Group) was formed to develop and help implement conservation actions. Since 2005 the group has met annually, and with new information the plan was revised in 2009 (ASCOBANS, 2009a) and again in 2016 (ASCOBANS, 2016). Around the same time (March 2002), at the 5th International Conference for the Protection of the North Sea in Bergen (Norway), a conservation plan was proposed for the harbour porpoise in the North Sea. This evolved from a Recovery Plan (ASCOBANS, 2006) into a Conservation Plan (ASCOBANS, 2009b) and was adopted in September 2009. A Steering Group was established with annual meetings initiated. The activities of the group received significant impetus with the appointment of a part-time coordinator from 2011 to 2015.

Whereas the two plans cover much of the original ASCOBANS Agreement area, the sea areas from the western Baltic through the Belt Sea to the Kattegat were not addressed by either. It was therefore proposed in February 2011 that another conservation plan should be developed for this so-called Gap Area (Fig. 7.24), and has now been implemented by the Jastarnia Group (ASCOBANS, 2012).

Although the conservation plans for harbour porpoise are well-established at least for the North Sea and Baltic (Fig. 7.25), they do not guarantee that the Range States take the actions that their respective groups recommend. Nevertheless, they have focused attention on regional needs in terms of filling information gaps (e.g. the SAMBAH project for Baltic harbour porpoise) and canvassing for specific actions (e.g. improved bycatch monitoring using various approaches including remote electronic monitoring methods); Fig. 7.26. There are also periodic reviews to establish progress against conservation targets (ASCOBANS, 2018a, b, c). The time has now come to consider other small cetacean species, and a conservation plan is currently being developed for the common dolphin (ASCOBANS, 2018d), with consideration also for white-beaked dolphin (Fig. 7.27), Atlantic white-sided dolphin, bottlenose dolphin and Risso's dolphin (Fig. 7.28) in the future.

The Dutch Conservation Plan for the Harbour Porpoise

The Conservation Plan for the Harbour Porpoise in the Netherlands is the result of the concerted effort of its authors, the Advisory Committee, NGOs and interested stakeholders. The commitment of the stakeholders with the recommendations, which were to come out of the process, was a precondition in the assignment given to the authors, Marije Siemensma and Kees Camphuysen.

There were several reasons why this plan was initiated. The large increase in the number of harbour porpoises in Dutch waters, the large numbers of stranded animals, the perceived lack of good data, the obligations from international agreements (including ASCOBANS and the European Marine Strategy Framework Directive), the expected growth in economic activities in the North Sea and the prospective negative impacts – these all led to the wish to come up with an all-embracing plan to protect these beautiful animals.

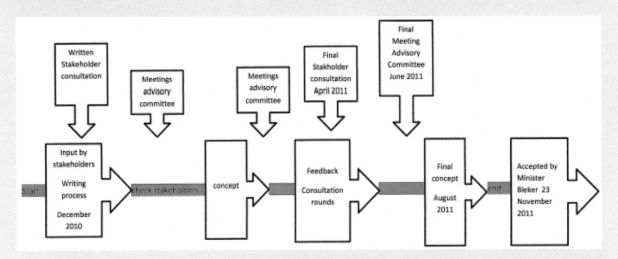

Therefore a process was started to consult stakeholders. This process is broadly outlined in the graphic above. Diverse fishery sectors, the offshore industry, NGOs such as WWF and the North Sea Foundation were contacted and given the opportunity to help shape the plan. In the Advisory Committee, representatives from stakeholders, concerned Ministries and Departments (Nature, Environment, Defence and Fishery), NGOs and from the scientific community were invited to participate (Fig. 7.29). Also, international experts were consulted on parts of the plan during the process.

(Continued)

(Continued)

FIGURE 7.29 A session investigating the diet of the harbour porpoise in the Netherlands. *Photo: Marije Siemensma.*

The plan resulted in a prioritised list of recommendations for research, policy and mitigation measures. The main recommendations in the Harbour Porpoise Conservation Plan to ensure the favourable preservation status are:

- Installation of an independent national scientific research steering group.
- Continue monitoring of the distribution of the population.
- Develop and prioritise relevant research questions on ecology, physiology, pathology and behaviour.
- Continuing and strengthening of a coordinated stranding network.
- Monitoring of bycatch aboard of all fleets with passive gear.
- Controlled use of acoustic devices (not random usage; season, time of day and location).
- Policy measures for acoustic disturbance; loud explosive sounds including seismic surveys.
- Development of general regulation rules; no specific closed areas for this highly mobile species.
- The international fine-tuning of the recommendations.

This plan, with the support of the stakeholders, was presented to the Minister of EL&I on 23 November 2011. The plan met his approval and he emphasised the need for good research in order to base policy measures on sound scientific evidence.

Folchert van Dijken
Member, later Head of Dutch delegation, Advisory Committee, 2009–13

The Conservation Plan for Harbour Porpoises in the North Sea

If there is a small cetacean that is a familiar figure in the North Sea, it is definitely the little harbour porpoise. Not so easy to observe though. One or two glimpses of a small triangular dark shiny dorsal fin glinted at the surface and often no more. Unfortunately strandings on North Sea beaches are rather commonplace and allow a longer sight of the small animal.

(Continued)

(Continued)

The harbour porpoise occurs throughout the North Sea, although its distribution is not static in space or time. Observations and strandings decreased in the southern North Sea and the Channel in the 1980s and 1990s, although the trend seems to have reversed now. The North Sea is a very busy piece of water, and porpoises face a variety of anthropogenic threats due to the heavy maritime traffic, fishing and industrial activities, and the resulting habitat degradation, including overfishing, pollution and noise.

The extent and the effects of these threats vary among areas and are not always well documented. Entanglement in fishing gears is however considered to be by far the most severe, with annual bycatch, at least in some years, likely to have exceeded 10,000 porpoises for the North Sea alone. Harbour porpoises are particularly vulnerable to bycatch in bottom-set gillnet fisheries.

In spite of relatively high population estimates, with over 300,000 porpoises in the North Sea, there have been notable declines in some areas in certain periods (e.g. southern North Sea and English Channel). The high bycatch rates were deemed unsustainable, and strong concerns were raised over the combined effects of high levels of pollution and elevated underwater noise.

A Conservation Plan (ASCOBANS, 2002) was developed as the result of a call by the 5th International Conference for the Protection of the North Sea (Bergen, March 2002). It aims to restore and/or maintain North Sea harbour porpoises at a favourable conservation status, whereby

- population dynamics data suggest that harbour porpoises are maintaining themselves at a level enabling their long-term survival as a viable component of the marine ecosystem;
- the range of harbour porpoises is neither reduced, nor is it likely to be reduced in the foreseeable future;
- habitat of favourable quality is, and will be, available to maintain harbour porpoises on a long-term basis; and
- the distribution and abundance of harbour porpoises in the North Sea are returned to historic coverage and levels wherever biologically feasible.

These objectives incorporate the long-term goal of ASCOBANS to restoring and/or maintaining populations at 80% or more of the carrying capacity (ASCOBANS, 1997, 2000b).

In the shorter term a pragmatic minimum objective is to at least maintain the present situation and, if possible, improve it.

The actions recommended by the Conservation Plan are listed below, together with their prioritisation.

The Conservation Plan is articulated around 12 specific prioritised actions (see Table 7.1), built upon three general considerations:

- Major information gaps need to be filled for fully assessing the situation and being able to recommend effective and adequate conservation measures, both with respect to the harbour porpoise itself (feeding ecology, behaviour around nets) and the human activities it is subject to and their actual/potential impact.
- Monitoring is essential for informing trends in the conservation status (i.e. in the species, the threats, the implementation and efficiency of the mitigation measures) and informing the effectiveness of the management actions, and if necessary to adjust them, to achieve the established conservation aims.
- The Plan needs to be reviewed periodically to adjust the actions based on the information provided by the monitoring.

The successful implementation of this plan can only be achieved through the active involvement and political will of a number of stakeholders (intergovernmental and national authorities, scientists, representatives from the fishing industry, local communities, NGOs, etc.), some of whose livelihoods may be affected (e.g. fishermen). A variety of challenges lie ahead, especially with regards to obtaining a good estimate of the trend in abundance and distribution of harbour porpoises in the North Sea and a proper estimate of bycatch numbers. Although progress has been made in some areas, the information on bycatch delivered to ICES from member countries is often incomplete in terms of areas and gears, and sometimes delivered in a format or with a lack of detail preventing extrapolation to the overall fleet. The situation of the harbour porpoise in the North Sea remains very uncertain and the need for a new international SCANS survey was high. Parallel to SCANS-III, which focused on gathering data on abundance and distribution, it is essential to increase the effort allocated to obtaining reliable and comparable bycatch data for the entire North Sea (Fig. 7.30). Additionally, the understanding of the effects of inshore fisheries on harbour porpoises is still lacking as these are not covered by EU regulations and currently are at best sparsely monitored, although it is required under the EU Habitats Directive and monitoring should have been developed under this framework. Optimising the synergy between monitoring programmes (e.g. between EU directives, the Marine Strategy Framework Directive and Habitats Directive) would result in more comprehensive information on the habitats and habitat needs of harbour porpoises.

A part-time coordinator was appointed in July 2011, working under the guidance of a Steering Group, composed of representatives of the countries around the North Sea and NGOs, and reporting to the ASCOBANS Advisory Committee. One of its main tasks has been to monitor the implementation of the Plan in the different countries and to produce status updates. This appointment had to cease in 2015[2] because resources from Parties were not available for its continuation, and, instead, the meetings have relied upon the ASCOBANS Secretariat and the Chair of the Steering Group (Peter Evans) to develop agendas and review progress.

(Continued)

---

2. It was reinstated in 2018, with a part-time coordinator responsible for all three Plans.

(Continued)

TABLE 7.1 The North Sea harbour porpoise Conservation Plan.

Management actions	1	Implementation of the Conservation Plan: coordinator and Steering Committee	High
	2	Implementation of existing regulations on bycatch of cetaceans (e.g. Habitats Directive, EU Regulation 812/2004)	High
	3	Establishment of bycatch observation programmes on small vessel (<15 m) and recreational fisheries	High
	4	Regular evaluation of all relevant fisheries with respect to extent of porpoise bycatch	High
Research action	5	Review of current pingers, development of alternative pingers and gear modifications	High
Research and management action	6	Finalise a management procedure approach for determining maximum allowable anthropogenic removals in the region	High
Research actions	7	Monitoring trends in distribution and abundance of harbour porpoises in the region	High
	8	Review of the stock structure of harbour porpoises in the region	High
	9	Collection of incidental catch data through stranding networks in the region	Medium
	10	Investigation of the health, nutritional status and diet of harbour porpoises in the region	Medium
	11	Investigation of the effects of anthropogenic sounds on harbour porpoises	Medium
	12	Collection and archiving of data on anthropogenic activities and development of a North Sea-wide GIS based database	Medium

FIGURE 7.30 Alongside the Steering Group meetings for the regional Porpoise Conservation Plans, the ASCOBANS Bycatch Working Group has periodically held Expert Workshops to address the issue of "Unacceptable Interactions" and bycatch, as took place in Bonn, Germany, February 2017. *Photo: A Bahramlouian/ASCOBANS.*

(Continued)

This appointment, demanded by the Conservation Plan, represented a positive step in the implementation of the Plan, as it provided an overview of the situation in the different countries involved and overall for the North Sea. It certainly revealed that although progress had been accomplished in different areas, much remained to be done, and the information remained very patchy in most areas. The implementation of the Plan was far from being fulfilled.

The will and further cooperation of all stakeholders would enhance the implementation of the Conservation Plan and ensure the conservation of the harbour porpoise in the North Sea. A wider implementation of the plan seems particularly needed at a time when harbour porpoises face the cumulative impact of a diverse array of direct and indirect anthropogenic threats upon a background of climate and environmental change. Indeed, reproductive impairments in harbour porpoises have been directly related to the high level of PCBs contaminants in European waters.

Dr Geneviève Desportes
Coordinator, North Sea Harbour Porpoise Conservation Plan, 2011–15
General Secretary, North Atlantic Marine Mammal Commission, 2015 to present

References

ASCOBANS (1997). Cetacean by-catch issues in the ASCOBANS area, unpublished report of the ASCOBANS advisory committee working group on by-catch. Report No. ASCOBANS/MOP/2/DOC.1 rev. 25 pp.

ASCOBANS (2000a). *Report of the 7th meeting of the Advisory Committee.* Bruges, Belgium, 13-16 March 2000. <https://www.ascobans.org/en/document/report-7th-meeting-advisory-committee>.

ASCOBANS (2000b). Resolution No. 3. Incidental Take of Small Cetaceans. Pp. 93-96. In: Proceedings of the Third Meeting of Parties to ASCOBANS. Bristol, UK, 26–28. July 2000, 108 pp. <https://www.ascobans.org/sites/default/files/document/MOP3_2000-3_IncidentalTake_1.pd>.

ASCOBANS (2002). Recovery plan for Baltic harbour porpoises (Jastarnia Plan) (26 pp.). In *9th Advisory Committee meeting, Hindås, Sweden, 10-12 June 2002* (39 pp.). Doc. AC9/Doc. 7(S). <https://www.ascobans.org/sites/default/files/document/AC9_7_DraftRecoveryPlanAnnotated_1.pdf>.

ASCOBANS (2006). Recovery plan for harbour porpoises in the North Sea (NSRP). In *ASCOBANS 13th Advisory Committee meeting, Tampere, Finland, 25-27 April 2006* (93 pp.). Doc. AC13/Doc. 18(S). <https://www.ascobans.org/sites/default/files/document/AC13_18_NSRP_1.pdf>.

ASCOBANS (2009a). Recovery plan for Baltic harbour porpoises. Jastarnia Plan (Revision). In *6th Meeting of the Parties, UN Campus, Bonn, Germany, 16-18 September 2009* (47 pp.). MOP6/Doc. 7-01 (AC). <https://www.ascobans.org/sites/default/files/document/MOP6_7-01_RevisionJastarniaPlan_1.pdf>.

ASCOBANS (2009b). Conservation plan for harbour porpoises in the North Sea. In *6th Meeting of the Parties, UN Campus, Bonn, Germany, 16-18 September 2009* (31 pp.). MOP6/Doc. 7-02 (AC). <https://www.ascobans.org/sites/default/files/document/MOP6_7-02_NorthSeaConservationPlan_1.pdf>.

ASCOBANS (2012). Conservation plan for the harbour porpoise population in the western Baltic, the Belt Sea and the Kattegat. In *7th Meeting of the Parties to ASCOBANS, Brighton, UK, 22-24 October 2012* (46 pp.). MOP7/Doc7-01 (AC). <https://www.ascobans.org/sites/default/files/document/MOP7_7-01_DraftResolution_HarbourPorpoiseConservation_1.pdf>.

ASCOBANS (2016). Recovery plan for Baltic harbour porpoises. Jastarnia Plan (Revision). In *ASCOBANS Draft Resolution 3 (with annex), 8th Meeting of the Parties to ASCOBANS, Helsinki, Finland, 30 August-1 September 2016* (94 pp.). <https://www.ascobans.org/sites/default/files/document/MOP8_In-Session_DraftResolution3_JastarniaPlan_inclAnnex_0.pdf>.

ASCOBANS (2018a). Progress report on the recovery plan for Baltic Harbour porpoises (Jastarnia Plan). In *24th Meeting of the Advisory Committee, Vilnius, Lithuania, 25-27 September 2018* (49 pp.). Doc. AC24/Doc. 3.1.b. <https://www.ascobans.org/en/document/progress-report-recovery-plan-baltic-harbour-porpoises-jastarnia-plan>.

ASCOBANS (2018b). Progress report on the conservation plan for harbour porpoises in the North Sea. In *24th Meeting of the Advisory Committee, Vilnius, Lithuania, 25-27 September 2018* (40 pp.). Doc. AC24/Doc. 3.2.b. <https://www.ascobans.org/en/document/progress-report-conservation-plan-harbour-porpoises-north-sea>.

ASCOBANS (2018c). Progress report on the conservation plan for the Harbour Porpoise population in the western Baltic, the Belt Sea and Kattegat (WBBK). In *24th Meeting of the Advisory Committee, Vilnius, Lithuania, 25-27 September 2018* (38 pp.). Doc. AC24/Doc. 3.3. <https://www.ascobans.org/en/document/progress-report-conservation-plan-harbour-porpoise-population-western-baltic-belt-sea-and>.

ASCOBANS (2018d). Draft species action plan for the North-East Atlantic Common Dolphin. In *24th Meeting of the Advisory Committee, Vilnius, Lithuania, 25-27 September 2018* (72 pp.). Doc. AC24/Doc. 3.4. <https://www.ascobans.org/en/document/draft-species-action-plan-north-east-atlantic-common-dolphin>.

Cooke, J. G., Deimer, P., & Schütte, H. J. (2006). *Opportunistic sightings of harbour porpoises (Phocoena phocoena) in the Baltic Sea: 3rd and 4th Seasons (2004-5).* AC 13/Doc. 23(P). Bonn: ASCOBANS.

Dähne, M., Tougaard, J., Carstensen, J., Rose, A., & Nabe-Nielsen, J. (2017). Bubble curtains attenuate noise from offshore wind farm construction and reduce temporary habitat loss for harbour porpoises. *Marine Ecology Progress Series*, 580, 221–237.

Deimer, P., Schütte, H. J., &Wilhelms, S. (2003). *Opportunistic sightings of harbour porpoises (Phocoena phocoena) in the Baltic Sea.* AC10/Doc.8 (P). Bonn: ASCOBANS.

Diederichs, A., Pehlke, H., Nehls, G., Bellmann, M., Gerke, P., Oldeland, J., ... Rose, A. (2014). Entwicklung und Erprobung des Großen Blasenschleiers zur Minderung der Hydroschallemissionen bei Offshore-Rammarbeiten. Bioconsult-SH, Husum. <http://bioconsult-sh.de/site/assets/files/1312/1312.pdf>.

European Union (2004). Council Regulation (EC) No. 812/2004 of 26 April 2004 laying down measures concerning incidental catches of cetaceans in fisheries and amending regulation (EC) No. 88/98. *Official Journal of the European Union*, Series L 150, 30.4.2004, 12−31. <http://eur-lex.europa.eu/legal-content/EN/TXT/PDF/?uri = CELEX:32004R0812&from = EN>. Corrigendum <http://eur-lex.europa.eu/legal-content/EN/TXT/PDF/?uri = CELEX:02004R0812-20070719&from = EN>. Summary <http://eur-lex.europa.eu/legal-content/EN/TXT/HTML/?uri = LEGISSUM: l66024&from = EN>.

Evans, P.G.H. (Editor) (2008). *Selection criteria for marine protected areas for cetaceans.* Proceedings of the ECS/ASCOBANS/ACCOBAMS Workshop held in San Sebastían, Spain, 22 April 2007, *European Cetacean Society Special Publication Series*, 48: 1-104.

Evans, P. G. H. (2018). Marine protected areas and marine spatial planning for the benefit of marine mammals. *Journal of the Marine Biological Association of the United Kingdom*, 98(5), 973−976. Available from https://doi.org/10.1017/S0025315418000334.

Evans, P. G. H., Baines, M. E., & Anderwald, P. (2011). R*isk assessment of potential conflicts between shipping and cetaceans in* the ASCOBANS *region* (32 pp.). ASCOBANS AC18/Doc. 6-04 (S).

Hammond, P. S., Bearzi, G., Bjørge, A., Forney, K., Karczmarski, L., Kasuya, T., ... Wilson, B. (2008). *Phocoena phocoena (Baltic Sea subpopulation)*. IUCN Red List of Threatened Species. <www.iucnredlist.org>.

HELCOM (2010). Hazardous substances in the Baltic Sea - an integrated thematic assessment of hazardous substances in the Baltic Sea. Executive Summary, Baltic Sea Environment Proceedings No. 120B (116 pp.)

HELCOM (2018). State of the Baltic Sea Second HELCOM holistic assessment 2011-2016. In: Baltic Sea Environment Proceedings 155 (155 pp.)

Herr, H., Siebert, U., & Benke, H. (2009). *Stranding numbers and bycatch implications of harbour porpoises along the German Baltic Sea Coast.* AC16/Doc.62(P). Bonn: ASCOBANS.

International Maritime Organisation (IMO) (2014). *Guidelines for the reduction of underwater noise from commercial ahipping to address adverse effects on marine life* (8 pp.). London: International Maritime Organisation. MEPC. 1/Circ.833. 7 April 2014.

Joint Nature Conservation Committee (JNCC) (2010a). *JNCC guidelines for minimising the risk of injury and disturbance to marine mammals from seismic surveys* (16 pp.). Peterborough: JNCC.

Joint Nature Conservation Committee (JNCC) (2010b). *Statutory nature conservation agency protocol for minimising the risk of injury to marine mammals from piling noise* (13 pp.). Peterborough: JNCC.

Joint Nature Conservation Committee (JNCC) (2010c). *JNCC guidelines for minimising the risk of injury to marine mammals from using explosives* (10 pp.). Peterborough: JNCC.

Koschinski, S., & Pfander, A. F. (2009). *Bycatch of harbour porpoises (Phocoena phocoena) in the Baltic coastal waters of Angeln and Schwansen (Schleswig-Holstein, Germany)*. AC16/Doc.60(P). Bonn: ASCOBANS.

Kraus, S. D., Read, A. J., Solow, A., Baldwin, K., Spradlin, T., Anderson, E., & Williamson, J. (1997). Acoustic alarms reduce porpoise mortality. *Nature (London)*, *388*, 525.

LGL, & MAI. (2011). *Environmental assessment of marine vibroseis* (207 pp.). LGL Rep. TA4604-1; JIP contract 22 07-12. Report from LGL Ltd., Environ. Res. Assoc., King City, Ont., Canada, and Marine Acoustics Inc., Arlington, VA, U.S.A., for Joint Industry Programme, E&P Sound and Marine Life, Intern. Assoc of Oil & Gas Producers, London, UK.

Loos, P. (2009). *Opportunistic sightings of harbour porpoises in the Baltic at large - Kattegat, Belt Sea, Sound, Western Baltic and Baltic Proper* (Bachelor thesis). Hamburg: University of Hamburg. Downloadable via <www.gsm-ev.de> in the English section under "Publications".

Nehls, G., Rose, A., Diederichs, A., Bellmann, M., & Pehlke, H. (2016). Noise mitigation during pile driving efficiently reduces disturbance of marine mammals. *Advances in Experimental Medicine & Biology*, *875*, 755−762.

OSPAR Commission. (2014). *OSPAR inventory of measures to mitigate the emission and environmental impact of underwater noise*. London: OSPAR Commission, 40 pp.

Pramik, B. (2013). Marine Vibroseis: Shaking up the industry. *First Break*, *31*, 67−72.

Silber, G. K., Vanderlaan, A. S. M., Tejedor Arceredillo, A., Johnson, L., Taggart, C. T., Brown, M. W., ... Sagarminaga, R. (2013). The role of the International Maritime Organization in reducing vessel threat to whales: Process, options, action and effectiveness. *Marine Policy*, *36*, 1221−1233.

Teilmann, J., Sveegaard, S., & Dietz, R. (2011). Status of a harbour porpoise population - Evidence of population separation and declining abundance. In S. Sveegard 2010. *Spatial and temporal distribution of harbour porpoises in relation to their prey* (Ph.D. thesis). Aarhus: University of Aarhus.

Trippel, E. A., Strong, M. B., Terhune, J. M., & Conway, J. D. (1999). Mitigation of harbour porpoise (*Phocoena phocoena*) by-catch in the gillnet fishery in the lower Bay of Fundy. *Canadian Journal of Fisheries and Aquatic Science*, *56*, 113−123.

Weilgart, L.S. (Editor) [2010]. *Report of the Workshop on Alternative Technologies to Seismic Airgun Surveys for Oil and Gas Exploration and their Potential for Reducing Impacts on Marine Mammals*. Monterey, California, USA, August 31- September 1, 2009. Okeanos-Foundation for the Sea. Retrieved from www.sound-in-the-sea.org/download/AirgunAlt2010_en.pdf.

Würsig, B., & Evans, P. G. H. (2002). Cetaceans and humans: influences of noise. In P. G. H. Evans, & J. A. Raga (Eds.), *Marine Mammals: Biology and Conservation* (pp. 555−576). London: Kluwer Academic/Plenum Press, 630 pp.

Chapter 8

Focus on the future

A quarter of a century has passed since the ASCOBANS Agreement came into force. A similar time has elapsed since the Rio Earth Summit [Convention on Biological Diversity (CBD)], the European Union (EU) Habitats Directive, and OSPAR (The Convention for the Protection of the Marine Environment of the North-East Atlantic), all reflecting the increasing environmental awareness and general concern for the need for conservation action. It is therefore perhaps a good time to take stock of how much progress has been made and look into the future. The next few pages are very much a personal view, and since different people will have different perspectives, I have made a point of inviting others from a variety of backgrounds to contribute their own views alongside mine.

As you may have guessed from my first chapter, when I was growing up, and most impressionable, two books influenced me more than any other: 'Silent Spring' by Rachel Carson drawing attention to the dangers of pesticides and other forms of pollution, and a lesser known paperback book called 'A Blueprint for Survival', a radical manifesto for change that had filled an issue of the newly formed periodical, 'The Ecologist'. Reading these it seemed so obvious to me what needed to be done. This applied to reversing human population growth, moving away from fossil fuels as a source of energy, stopping the global destruction of rainforests, and ceasing the commercial exploitation of whales. Many years later I have reconciled myself to the stark fact that change takes a long time (Fig. 8.1), that not everyone has the same opinions about the environment or the same priorities (Fig. 8.2). To me, this is the fundamental reason why a conservation agreement may not progress as far as one might as an idealist hope for, or as fast.

To assess whether any agreement aimed at conservation of wild/life is proving effective, the ultimate question should always be 'will the species at which it is aimed actually experience a difference?' That is difficult to judge as often we are fighting a rearguard action as pressures upon the environment mount. Maintaining the status quo is difficult enough.

FIGURE 8.1 Bottlenose dolphins in Pen Llŷn a'r Sarnau Special Area of Conservation (SAC), west Wales. The Habitats Directive came into force in 1992 but it was not until 2004 that the SAC was formally designated to protect this Annex II species. *Photo: Pia Anderwald.*

European Whales, Dolphins, and Porpoises. DOI: https://doi.org/10.1016/B978-0-12-819053-1.00008-9

FIGURE 8.2 At present, species such as the common dolphin are in practice not given full protection. This species continues to suffer significant bycatch from trawl fisheries. In 2019, ASCOBANS launched a conservation plan for the species. *Photo: Peter GH Evans.*

FIGURE 8.3 The minke whale is the main species of baleen whale occurring in the ASCOBANS Agreement Area and yet, for political reasons, has been excluded from consideration. *Photo: Peter GH Evans.*

For me, the great value of ASCOBANS and its regular meetings is that it creates a forum for a mix of specialists on marine mammals and individuals representing national governments within the region to meet, exchange views and reach consensus on proposed actions. Each meeting reviews progress and identifies where targets have not been met. No other forum provides quite the same opportunity for a focus upon the small cetacean species inhabiting northwest Europe. I wish it would consider all cetaceans (Fig. 8.3) rather than just the smaller ones. The arguments against this are purely political. There is no reason why its sister agreement, ACCOBAMS, can address all cetacean species and it cannot.

25 years of ASCOBANS — How would our seas look without it?

I started working as World Wide Fund for Nature (WWF) marine conservation officer in the late 1980s, addressing the North Sea Ministerial Conferences in the first instance but expanding the scope of my work and projects to the northeast Atlantic Oslo & Paris Convention (OSPAR) and European marine environmental issues, policies and frameworks in the following years since then. Literally from my first day in office, human pressures on small cetaceans became part of my working agenda and I learned that our harbour porpoise populations were in bad shape and dire straits. Discussions to bring a regional agreement on the conservation of small cetaceans into being had just emerged at various levels, hence the beginning of my career got inextricably linked with the harbour porpoise and ASCOBANS.

In the early 1990s WWF actively and strongly supported the idea of such a new instrument, advocating certain governments as well as the North Sea Conference as a whole, which adopted a Memorandum of Understanding on the issue and drew attention to excessive harbour porpoise bycatch rates in the Quality Status Reports. We also initiated or funded tentative reporting schemes for harbour porpoise strandings, sightings and bycatch in cooperation with research institutions, with a view to underpinning the need for urgent conservation measures. We also produced the first information tool kits about the harbour porpoise for wider audiences and communicated it as a flagship species for the protection of the North Sea and campaigns for the reform of the Common Fisheries Policy (CFP). Finally, as a matter of course, we engaged with ASCOBANS as a nongovernmental observer once the agreement had entered force and its working structure was operational.

To make a long success story short, in the early and mid-1990s:

- there was no legally enforceable instrument to reduce small cetacean bycatch in our waters;
- recovery plans for North and Baltic Sea harbour porpoise populations did not exist;
- only a few coastal and hardly any offshore marine protected areas had been designated for the specific purpose of protecting the harbour porpoise or other small cetaceans;
- decision makers' awareness of the threat to cetaceans by toxic and persistent chemicals including endocrine disruptors was still limited; and
- the significant impact of underwater noise was scientifically obvious but had not entered the political agenda yet.

A quarter of a century later, the political situation has changed considerably — while recovery at population and ecosystem level is still sadly lagging behind:

- Fisheries regulations to mitigate incidental catches of small cetaceans are in place *but* poorly implemented.
- Threshold levels of unacceptable anthropogenic interaction or disturbance have been defined for bycatch as well as underwater noise and introduced into other intergovernmental frameworks and Directives (OSPAR and HELCOM Ecological Quality Objectives; EC Marine Strategy Framework Directive, MSFD). However, bycatch levels in certain regions, in particular the western Baltic and Baltic Proper, remain unacceptably high. Precautionary noise levels continue to be exceeded by pile driving for offshore installations and seismic exploration activities at sea, including in or near protected areas.
- A number of EU Member States have nominated or designated offshore sites with cetacean breeding and feeding grounds as marine Natura 2000 sites at an early stage while others kept failing to do so, according to evaluations and judgements by the European Commission in 2009/10, and again in 2016. This process fortunately resulted in new designations of large areas for the conservation of the harbour porpoise quite recently.
- However, in protected areas already designated, fisheries management such as the required ban on gill and entangling nets and other conservation measures are not in place yet. The mechanism to introduce fisheries measures in marine Natura 2000 sites based on new provisions of the reformed CFP turns out to be very time-consuming and is leading to the lowest common denominator amongst Member States concerned.

One could refer to many underlying circumstances and parameters which fostered and promoted the progress and positive developments described above, such as the availability of European Union funds for the comprehensive SCANS and/or SAMBAH surveys, the application of the EU Habitats Directive to the offshore limit of national jurisdiction and/or Exclusive Economic Zone, and the extension of ASCOBANS' geographic scope itself. Most notably, however, it is the specific merit of ASCOBANS to bring all these approaches and strands together, to come up with scientifically corroborated proposals for necessary actions and convey them to other fora and competent authorities, and to serve as a think-tank and a pressure group for cetacean conservation measures.

Therefore WWF has the pleasure to congratulate ASCOBANS and all players involved on a quarter of a century of hard work to safeguard small whales, dolphins and porpoises in our oceans. They have paved the way and set the scene for improved protection of these amazing creatures at policy and technical level. Despite these achievements, however, progress is often

(Continued)

(Continued)

hampered by competency struggles as well as lack of political willingness, follow-up and compliance. There is still a long way to go and plenty of perseverance needed to make the real-world change, too. Best of luck for the next 25 years!

Stephan Lutter
Senior Policy Advisor Marine Protected Areas & Cetaceans, WWF Germany

Signing up to a conservation agreement

Not all Range States are Parties to the Agreement and that is also a shame. The obvious absentees are Norway, Ireland, Spain and Portugal. Norway once attended the meetings even though it had not signed up to the Agreement, and its input was greatly valued (Fig. 8.4). The sticking point is that Norway still wishes to hunt whales which is not necessarily compatible with a conservation agreement, albeit on small cetaceans. Ireland has always been something of an anomaly. In June 1991 the then Taoiseach of Ireland, Charles Haughey, declared Irish territorial waters a whale and dolphin sanctuary. In so doing, the announcement was that 'it was a clear indication of Ireland's commitment to contribute to the preservation and protection of these magnificent creatures in their natural environment, and to do everything possible to ensure that they should not be put in danger of extinction but should be preserved for future generations'. Under this legislation, the hunting of all whale species, including dolphins and porpoises, was totally banned within the exclusive fishery limits of the State, that is out to 200 miles from the coast. However, by that time there was no hunting of cetaceans in Irish waters anyway so this gesture, though sounding splendid, was not affecting the status quo. It is ironic that more than a quarter of a century later, Ireland will still not sign up to this important international conservation agreement, nor even send a government representative to participate in meetings.

In 2008 an Atlantic extension of the Agreement area came into force westwards and southwards to the southwest tip of Portugal. Both Spain and Portugal were approached to sign up as Parties to the Agreement. Spain was already a Party to ACCOBAMS (from 2001) because of its Mediterranean territory, and in 2005 Portugal decided to go the same way. Thus there is an area of sea in the Atlantic that is shared by both Agreements. One of the arguments put forward was that it made better sense for the waters surrounding the Iberian Peninsula to be considered in the same forum as the Mediterranean. This is clearly not true since the more mobile populations of most of the Atlantic species range up and down that ocean and do not appear to enter the semienclosed Mediterranean Sea, and the human pressures such as

FIGURE 8.4 Many ASCOBANS faces at its Advisory Committee meeting in Brest, France, April 2005. *Photo: ASCOBANS Secretariat.*

fisheries do likewise. For Spain, which has territorial seas in the Bay of Biscay, to participate only in an agreement about the Mediterranean and Black Sea really does not make sense. Fortunately the two sister Agreements increasingly are working together, particularly on specific issues. A joint Noise Working Group has been established, and a similar one on bycatch is currently in the process of being set up. They have also shared support for a number of workshops at annual European Cetacean Society conferences whose venues alternate between southern and northern Europe.

ASCOBANS — a personal perspective

ASCOBANS has been close to my heart since before the Agreement was signed — I helped in the negotiation of the Agreement and was then present at every Advisory Committee and every session of the Meeting of Parties (MOP) until 2011. I was chair or vice-chair of the Advisory Committee from the first to the 13th meeting. I chaired the 7th session of the Meeting of the Parties in 2012.

ASCOBANS was founded in an era when pollution was regarded as the main risk to small cetaceans, but it is noticeable that even at the time of drafting the Agreement, other risks had been identified including bycatch and disturbance. To me, it was very obvious, however, that bycatch was THE issue. Pollution was being (and still is being) tackled through mechanisms such as the Oslo and the Paris Agreements (later brought together as OSPAR), and there seemed little (other than keeping a good watch on progress) that the new Agreement could do towards that. Bycatch on the other hand was killing small cetaceans directly — and judging by some of the scientific papers published around the time of the formation of the Agreement, in some numbers. A big problem was (and still is) the lack of ability to get adequate statistics on numbers killed and ability to put those numbers into a population perspective. This latter issue is important as I have long felt that an objective to reduce bycatch to zero is unattainable and tends to inhibit the negotiation needed to reduce bycatch. In the early years of ASCOBANS it was also plain that some Parties did not wish bycatch to be considered as they felt that it might harm the economics of their fishing fleets. There was also very obvious antipathy from the relevant parts of the European Commission — who needed to be persuaded if there are to be any fishing regulations brought in for any seas outside those near the coasts in ASCOBANS waters (Fig. 8.5).

FIGURE 8.5   The European Commission in Brussels has played an important role in shaping conservation legislation around Europe, but there remains much to be done. The Directorate General mainly responsible for environmental issues is DG Environment (b) although DG MARE, responsible for fisheries, also has a major role to play. *Photos: Peter GH Evans.*

It took until 2004 before a vital European regulation was brought in, and key in that work was the advice provided to the European Commission by the International Council for the Exploration of the Sea (ICES). ICES provides impartial scientific advice on many marine matters to the European Commission and to many others. Advice was also received by the Commission from its own scientific advisors. Without the advice of both of these bodies on the apparent unsustainability of the harbour porpoise

(Continued)

(Continued)

bycatch in certain fisheries, no regulation would have been drafted and agreed. It is essential that Parties to ASCOBANS keep collecting information on bycatch and keep taking actions that derive from the implications of that information. There is no doubt that the 2004 regulation is flawed, but the only way that it will be improved is through argument based on facts. I believe that ASCOBANS could be doing a lot more to encourage its Parties to continue to invest in research to determine bycatch rates in their fisheries, and that it then needs to ensure that those bodies (e.g. ICES) that can impartially assess that information be requested to do so.

Another reflection is on the vital importance of individuals. Things have not happened in cetacean conservation solely due to the presence of the Agreement or some (often long-argued) Resolution. Progress has been made when an individual (or sometimes small group of individuals) makes up their mind to move an issue on. That individual (or individuals) needs to figure out the human dimension of any necessary action, and work with those other individuals. Thus fisheries measures ideally need the understanding (if not cooperation) of the individuals that might be affected from the fishing industry. The Regulators are also individuals that have many pressures upon them, so that working to reduce (or at a minimum without increasing) those pressures will be more likely to be successful. Further individual understanding and cooperation are needed to negotiate and implement, and actions are needed to meet any regulatory requirement. There are still too many in the conservation community who believe that just passing a law is sufficient to bring success – it is not, and probably never will be. A lesson from this is that Parties need to find and nurture those individuals willing to go that bit further to work with and through the many interactions with others to achieve progress. It is also essential that ASCOBANS be 'owned' by its Parties. It is NOT an agreement of Convention on Migratory Species (CMS) or the Secretariat – it is an Agreement of the Parties. The implication of this is that the Secretariat should not control the Agreement or its actions, however much Parties may wish to offload work onto them – burdens and responsibilities should fall upon Parties.

There are many things that I would like to see in the small cetacean world. The most obvious would be the recovery of those populations that have either declined in number or lost parts of their range. Harbour porpoise should be much commoner in the Baltic Sea and off northwest Iberia; bottlenose dolphin ought to be present off many more coastlines than it is currently. In the case of the Baltic Sea, the Jastarnia Plan (and any revisions) should be driving this recovery – those responsible for this plan need to consider whether the present strategy will lead to recovery – and if not, why not? We do not know why bottlenose dolphins are absent in many areas in which they were recorded historically – studies to examine these reasons seem interminably held up by the inability of many groups that presently look after 'their' local bottlenose dolphin populations to come together and make a coherent plan, although there are recent encouraging signs that this might be changing.

Memories (for me) tend towards the funnier end of the scale. One that sticks in my mind was the impossibility of holding meetings of the Advisory Committee while the teams of certain Parties were playing their matches in the 2002 World Cup – essentially the delegations of those Parties refused to be present in the meeting. There was the occasion at the 7th AC in Bruges where our Belgian hosts kindly took us to the meeting dinner in a brewery – where the amount of beer that could be drunk was free and limitless. I am sure that many delegates and the vice-chair (who was acting chair) of that meeting were running on autopilot the next day. On a much happier note, I remember observing to Peter Reijnders (the chair of the AC that day) that two delegates seemed to be missing from the meeting and being amused when they came in late, but not quite synchronously – it is great that they are now married and have two children – ASCOBANS has done a lot more than conserve cetaceans! Long may it continue to do so.

I am now less involved in the affairs of ASCOBANS (having retired from my United Kingdom job with the Joint Nature Conservation Committee), but I am very fond of the Agreement, and miss working with many friends to try to enhance cetacean conservation through the Agreement. My best wishes for the next 25 years.

Mark Tasker
Vice-Chair of the Advisory Committee, 1995–2001,
Chair of the Advisory Committee, 2001–07,
Chair of the Meeting of the Parties, 2012–16

ASCOBANS – An NGO's personal perspective

My beard has gone increasingly grey over the years that I have been involved with ASCOBANS. (These things are perhaps not unrelated.) I helped with the early drafting of the agreement. At that time I worked for Greenpeace and they, WWF and Sea Watch were all influential in pushing for a new agreement for cetaceans in the region and advising on what it should contain. Sweden was the principal country involved and I recall vividly my grave disappointment when I viewed the revised agreement text on its return from a closed consultation between potential Party nations. The main conservation actions were relegated to an annex and significantly 'softened'. The Agreement's mandate was now limited to 'small cetaceans'. The rumour at the time

(Continued)

(Continued)

was that this was intended to facilitate Norway joining, which it never has. I suspect that the general softening of the final agreement may have suited many Parties.

In an article in *The Pilot* (the Newsletter of UNEP's Marine Mammal Action Plan), I raised the spectre of ASCOBANS being only a 'paper-tiger', noting 'only time will reveal the full (if any) potential of a new treaty, but an early indicator must be whether or not its text contains suitable reference to measures which should clearly be present [and that] the disproportionate emphasis on research measures does not bode well' (Simmonds, 1993). Published alongside this was a robust response to my 'pessimistic and provocative' comments authored by the first ASCOBANS Secretary and the then executive secretary of CMS. They commented 'what is needed for the future implementation of ASCOBANS [and similar future treaties] are more individuals who are involved in research and monitoring the species concerned and are thus in a position to make qualified proposals for conservation...' (Lockyer & Müller-Helmbrecht, 1993). I would not have disagreed and am certainly not opposed to science (far from it), but my point about the draft agreement was that this should not be the primary focus of the treaty. Lockyer and Muller-Helmbrecht also noted some early signs of the success of ASCOBANS, including the intent of the United Kingdom to apply it to all its waters (which happened) and that the Republic of Ireland wished to negotiate membership (which has never happened, still leaving a big hole in the middle of the revised treaty area).

FIGURE 8.6 Mark Simmonds attending the ASCOBANS Meeting of the Parties in Helsinki, Finland, September 2017. *Photo: Mark Simmonds.*

I have attended every ASCOBANS MOP (Fig. 8.6) and Advisory Committee, bar one. ASCOBANS meetings are typically good humoured, with a good sense of partnership, and nongovernmental contributions have usually been welcomed. Many participants have worked with the Agreement for many years and the annual meetings are, in some respects, like an annual reunion of old friends. Being an NGO observer is sometimes a difficult role. Generally, we can only encourage and offer advice (and undertake tasks when requested) and we appreciate that the Parties will only go as far as their mandates — dictated from somewhere else — allow (although we often sense that there is more than a little personal discretion in this). Likewise, the Parties mainly understand that the NGOs' role is to do the best for the animals. NGOs do not always get our input or tone right, and there have certainly been times when I have put a big foot in my mouth and wished the ground would swallow me up.

Despite ASCOBANS (and the European law that has followed), the conservation of small cetaceans in the Agreement area has been described as only limited and patchy (Parsons, Clark, & Simmonds, 2010; Parsons, Clark, Warham, & Simmonds, 2010). For instance, although major abundance surveys (SCANS 1–3) have taken place, there has been only limited government-funded research on population size, structure, distribution and trends. Moreover, ASCOBANS has yet to address some issues, most notably prey availability, and this should now of course be considered in the light of climate change.

(Continued)

(Continued)

Arguably, European cetaceans survive not because of anything that we have done to help them but just because of their inherent robustness. ASCOBANS has certainly identified and explored threats. It has had some success in getting the key players together (although the fishing community remains somewhat shy), and its influence on matters such as military noise may have been greater than we shall ever know with such a sensitive topic. Importantly ASCOBANS was the main forum to address fisheries bycatch issues[1] and it helped underpin the development and eventual agreement of European Union Regulation EC 812/2004. It established a precautionary approach to bycatch, and here the NGO community certainly played a role; I would like to give due credit to my then Whale and Dolphin Conservation (WDCS)-colleague Ali Ross and to Helen McLachlan [at that time with the Royal Society for the Prevention of Cruelty to Animals (RSPCA)] for their efforts in contributing to this.

Overall, however, in many ways ASCOBANS has failed to live up to our original hopes. Part of the reason for this could be the aforementioned loose nature of the treaty text, including how its subsidiary bodies are constituted, but also the complex political world of Europe now dominated by EU legislation, in which ASCOBANS has yet to find a clear role. However, the primary reason is probably that many of those with interests in activities that threaten cetaceans (see Parsons, Clark, & Simmonds, 2010; Parsons, Clark, Warham, et al., 2010) are very powerful, including for example the fossil fuel and closely related marine renewables industries and fisheries sector. Whales and dolphins are also not part of any electoral caucus in Europe, so why should politicians bother with them?

The NGO community could have done so much more. In particular it has become clear that it is not enough to help set up an agreement and then play very little further part in how this agreement works. This issue is not unique to ASCOBANS, and a colleague and myself recently noted how well a coalition of organisations had once helped to address the threats posed by industrial development at sea for a certain period, and then how the issue drifted into trouble again when the groups concerned moved their attention elsewhere (Green & Simmonds, 2008); continuity in such matters is key.

The nongovernmental community could also have injected more effort, including funds, into ASCOBANS, and developed more projects in partnership. Funding, of course, is always tight on our side; it cannot compare with what governments can provide and there are many competing issues we might want to spend it on, including within the developing world. Nonetheless, and with hindsight, some small strategic grants from us could have really helped move things along; and this is still the case. There has also been a change in the guiding philosophy of at least some NGO players. This is difficult to substantiate but for some at least the species-focus (and ASCOBANS is seen as a species-focused treaty) is now viewed as old-fashioned. Instead, the 'ecosystems' approach and the philosophy of sustainable use have come to dominate the thinking of some. Personally I believe that the need to conserve intact fully functioning ecosystems can be complementary to a focus on particular species; by contrast the whole philosophy of sustainable use leaves me cold. This is something of a confidence trick. At best this line of reasoning leads to a big experiment without a control. How can we say in our unstable and rapidly changing world that any use of wildlife may prove to be sustainable at some unknown point in the future? The risks are too great. We need to return to a more precautionary approach and act in favour of the animals where there is doubt. Sustainable-use dogma also pushes towards the only value of wildlife being in their economic value. Is this really how we should view our world and the creatures we share it with? Cetaceans are extraordinary animals; they deserve and need our protection.

ASCOBANS is also important in the fact that it is the first of the CMS-derived treaties for cetaceans. ACCOBAMS followed hard on its heels and then the West African and South Pacific cetacean Memoranda of Understanding (MOUs). If ASCOBANS had failed miserably, it is arguable that none of the others would have followed. In the 21st century, it faces three big challenges to make it fit for purpose. It should now reengage with its original intended mandate and extend to all cetacean species. The threats are largely common to all, and the bigger species would certainly benefit. I know that there are concerns that this would detract from work on the small species and perhaps particularly the endangered Baltic porpoise. I do not believe it would do this. I think it would make the agreement more complete and more functional. It also needs to find a proper role in the new Europe. The Habitats Directive requires actions for cetaceans, and ASCOBANS can help with this. Finally, I think it would be helpful if the Scientific and Administrative parts of intersessional work are separated (but still accessible to observers from civil society). ASCOBANS' slightly younger sister – ACCOBAMS – works better in this regard, and a scientific body that is populated exclusively by scientists would serve ASCOBANS better.

Good luck ASCOBANS and congratulations on your first 25 years. Words alone are totally inadequate; words must lead to actions. So please help to set comprehensive standards for cetacean conservation that should set the standards and inspire similar efforts elsewhere in the world – which was always our dream for you! I also encourage those with new ideas – and preferably energy and resources that they can put into underpinning this important little treaty – to come forward and help!

Mark Simmonds
International Director of Science, Whale & Dolphin Conservation, 1993–2012,
Senior Marine Scientist, Humane Society International, 2012 to present

---

1. ICES now take a major role in bycatch issues as well.

The mechanics of a conservation agreement

When I first attended ASCOBANS Advisory Group meetings, I was struck by the amount of time spent on getting wording of resolutions acceptable to all Parties. My first reaction was that more effort was being devoted to words than to executing actions. Over the years I have come to understand why this has to be but, nevertheless, there is a danger that because it is easier to play with words than to carry out the actions expressed in those words, humans sometimes lose their way a bit on how best to progress.

Most of the problems that we face in the world are created by humans, and it is the management of ourselves which is the greatest challenge. That happens at all levels. At an ASCOBANS meeting, the input and decisions arising are very much determined by the composition of the group, and factors such as their personalities, backgrounds, knowledge, and even personal views can come into play (Fig. 8.7). The group is speaking on behalf of countries, national and international nongovernmental organisations, and the scientific community, but in reality it is just a group of people. This is not unique to an agreement such as ASCOBANS, it applies to OSPAR, HELCOM, ICES, International Whaling Commission (IWC) or any other body. Put a different group of people together, even if drawn from the same communities, and you will likely get some different emphases, and proposals may be shaped differently. Often it is largely the same groups of individuals participating in the different fora if they are on related issues but the composition may change somewhat and so the outcomes may do so also. That said, there is nevertheless usually broad agreement on what are key issues and how to address them. One might argue that if the various fora include a lot of the same people and are addressing similar subjects, is there a need for such duplication? After all resources in the form of time and money are limited. There is a case for ensuring greater synergy between the different advisory bodies, directives and agreements.

Participation in ASCOBANS Advisory Committee meetings has fluctuated somewhat over the years. In my view the most effective composition is an even mix of marine mammal specialists, representatives of the NGO community and advisors to government drawn from both environment and fisheries departments (since they are often separated). That has not always been the case, and across the years the contribution from marine mammal scientists has lessened somewhat, which I think is counterproductive. Decisions must be evidence based and it is therefore important that there is as much scientific input as possible to make the best judgement over recommendations.

FIGURE 8.7 The 7th Meeting of the Parties to ASCOBANS in Brighton, United Kingdom in October 2012. *Photo: ASCOBANS Secretariat.*

FIGURE 8.8 Throughout the history of the ASCOBANS Agreement, the harbour porpoise has been the focus of attention, occurring in the waters of all Member States, and well known to experience significant mortality from fisheries bycatch. *Photo: Wouter Jan Strietman.*

FIGURE 8.9 The killer whale or orca carries some of the highest levels of contaminant burdens of any marine mammal. *Photo: Christopher Swann.*

Focus on species and issues

Within the Agreement, the harbour porpoise (Fig. 8.8) has received by far the most attention compared with other small cetacean species, not surprisingly given that it is the one species that has occurred in all the Range States. There has also been the strongest evidence for the effects of human pressures upon the species through high bycatch and pollutant levels. However, that in parts reflects our greater knowledge due to its relative abundance and wide distribution. Several of the other small cetacean species are very poorly known, and pressures upon them consequently may be less obvious.

The Agreement is starting to focus attention upon another abundant and widely distributed species, the common dolphin, that also shows evidence of negative effects of human activities. Whilst conservation action for this species is desirable, one should not forget other less common species such as white-beaked and Risso's dolphin, the coastal populations of bottlenose dolphin that are so vulnerable to the effects of pollution and human disturbance, the small populations of killer whale in the region (Fig. 8.9), and the beaked whales that inhabit the deep canyons of our offshore seas.

Impressions on ASCOBANS

When I started to work on harbour porpoises in the beginning of the 1990s, only a fraction of today's knowledge about these small cetaceans was available. Especially little was known about their migration. On the other hand, shocking news emerged of enormous numbers of porpoises bycaught in the Danish driftnet fishery. Also, there were scarcely any sightings in the Baltic where porpoises used to be abundant. Clearly, the harbour porpoise is a classic example of a migrating animal, and the importance of international cooperation both in research and also in conservation measures is obvious. ASCOBANS was, and still is, instrumental in achieving this by bringing scientists together with NGOs and governmental agencies.

The results I was getting with the porpoise team of the University of Kiel through our survey flights, and the analysis of the stranding data increasingly showed that a high number of calves were in the waters around the islands of Sylt (Fig. 8.10) and Amrum. I came to the conclusion that we had found a calving ground or nursery area for porpoises (Sonntag, Benke, Hiby, Lick, & Adelung, 1999). This was an important reason, according to ASCOBANS and the EU Habitats Directive, to give the area a high protection status, and I started a public debate at the time.

FIGURE 8.10 The waters around the Island of Sylt were identified in the 1990s as important breeding grounds for harbour porpoise. *Photo: Helena Herr.*

In the very controversial discussions between environmental NGOs, scientists and people who used the area commercially, Germany's membership of ASCOBANS became very significant. The conclusion of the political discussion was in favour of a cetacean sanctuary and resulted in the inclusion of the porpoise breeding grounds off Sylt and Amrum into the Wadden Sea National Park. In the end, ASCOBANS was instrumental both in identifying the area and in protecting this important habitat for harbour porpoises.

On a more general subject: at the time there was also a recognised need for more scientific data including studies of population abundance and behaviour in addition to technological solutions to reduce entanglement risk. Visual surveys for harbour porpoises were proving difficult because of their inconspicuous behaviour, but it was known that they also produced high-frequency echolocation clicks. International Fund for Animal Welfare (IFAW) scientists had previously developed acoustic survey methods for sperm whales and began an intensive research effort to develop acoustic techniques to monitor harbour porpoises. This work was intended to support conservation initiatives through ASCOBANS and other bodies. These acoustic survey methods were used also during the SCANS survey in 1994. This was the first comprehensive survey of small cetacean abundance in northern Europe and provided estimates that could be compared to the ASCOBANS conservation criteria for bycatch. Acoustic surveys have subsequently been used in surveys conducted by the research vessel Song of the Whale, the SCANS-II survey in 2005, and by many other groups in Europe and worldwide.

Working in coordination with ASCOBANS and local research groups, IFAW conducted dedicated acoustic and visual surveys for harbour porpoises with Song of the Whale in the Baltic Sea proper in 2001 and 2002, providing the only abundance

(Continued)

(Continued)

data available at the time for Polish waters. Results were made widely available through ASCOBANS and fed into the Jastarnia workshop for the recovery of the Baltic porpoise. More recently, in 2011 surveys for porpoises were conducted in other areas for which relatively little dedicated survey data existed, including the English Channel, and, in winter, the Dogger Bank in the southern North Sea, a candidate marine protected area at the time.

I have also closely followed ASCOBANS initiatives on the issue of underwater noise and its impacts on small cetaceans. Acoustic monitoring is now an important tool in environmental impact assessments and for mitigation of activities that can cause disturbance, including studies of porpoise distribution and diving behaviour.

In my opinion there is an ongoing competition between the effects of climate change and the increasing utilisation of the oceans on the one hand, and the need for conservation of the marine environment on the other — basically the necessity for people to have healthy oceans. Although it is desirable for everybody to speak with one voice and to agree conservation priorities, the ambition in ASCOBANS to always have consensus can be used to veto important decisions and thus may favour economic interests. In an agreement to protect small cetaceans a stronger mandate on fishery issues would also be helpful.

The CMS instruments such as the different MOUs and the ASCOBANS Agreement are very important in helping to bundle and coordinate protection measures in the oceans and other ecosystems. This could be even more influential if the member states provided more legal and political power as well as funding, in order to increase the chances of nature winning this race against time.

Dr Ralf Sonntag
Senior consultant, Marine Conservation,
Director, IFAW Germany, 2003—15

ASCOBANS — Personal perspective from a Government Representative

As the Danish representative for the Ministry of Environment I joined the ASCOBANS family at the 10th meeting of the Advisory Committee in Bonn in April 2003. Denmark would host the 4th Meeting of Parties later that year so it was important for me to learn as much as possible about ASCOBANS very fast (Fig. 8.11). My first impression of ASCOBANS was an agreement

FIGURE 8.11 The harbour of Kerteminde, Denmark. Danes have long played an important role in conservation research and action towards the harbour porpoise. *Photo: Peter GH Evans.*

(Continued)

(Continued)

consisting of a group of dedicated participants with a huge knowledge about small cetaceans, especially harbour porpoises. It was also an agreement with a very active and engaged participation of especially environmental NGOs.

The 4th Meeting of Parties went well and the major achievements reached at the MOP were to extend the Agreement area to include the northeastern Atlantic and the adoption of a North Sea Harbour Porpoise Action Plan. Being responsible for the organisation of the meeting I got a lot of support from a dedicated Secretariat of the Agreement.

Quickly I got familiar with the ASCOBANS work and enjoyed the meetings. Then at the 13th meeting of the Advisory Meeting, the situation changed suddenly. The Agreement was apparently in a critical economic situation and, to address this, it was proposed to integrate the Secretariat into CMS. ASCOBANS then became a piece in a bigger political picture and for a period of time the atmosphere changed and became less friendly. The outcome was a farewell to a well-functioning ASCOBANS Secretariat and its staff members, and a hello to a merge with the CMS Secretariat. The dust settled after some years and I feel that ASCOBANS is now back on track, and the good atmosphere has been reestablished.

I find that the most important outcome from ASCOBANS is the two plans for harbour porpoises in the Baltic and the North Sea, respectively. I also find that ASCOBANS is facing a new future with all the Member States also being members of the European Union. The years to come will show if and how ASCOBANS is able to make a difference for the small cetaceans in the Agreement area. In this process it is vital to attract all players to address bycatch, pollution, noise and other disturbance.

Maj Munk
Head of Danish delegation, 2003—15

ASCOBANS — Twenty-five years! — Personal perspective from the present Co-Chair of the Advisory Committee

Twenty-five years old, this is a very nice age for the ASCOBANS Agreement!

Personally, I have been involved in this agreement for 18 years. I started first as an observer for the French Ministry of the Environment. If I remember correctly my first meeting was in Aberdeen. Indeed from 1999 to 2005 France was not a Party. Nevertheless, I have tried to contribute to the Agreement, collecting the relevant data and participating in the discussions and the working groups. And step by step, after regular exchanges with our authorities and NGOs, the ratification was decided. To introduce this the first Advisory Committee organised in France was held in Brest in 2005. It was for me a very exciting experience and a complete satisfaction to prepare this meeting — a fitting location given its proximity to an important area for bottlenose dolphins which we had long studied at Oceanopolis (Fig. 8.12). During these 13 years significant changes have occurred. One of them was the merging in 2007 of the Secretariat with the UNEP/CMS Secretariat, and the southwestward extension of the Agreement area, which entered into force in 2008.

FIGURE 8.12 Bottlenose dolphins within a marine reserve off the coast of Brittany, France. *Photo: Oceanopolis-Brest, courtesy of Sami Hassani.*

(Continued)

(Continued)

Among the most striking subjects during the last decade, one has been the problem of accidental capture of cetaceans in fisheries, generating long exchanges on bycatch. Hard and successful work was the development of two plans to the harbour porpoise: the Baltic Recovery Plan (Jastarnia Plan) and the North Sea Conservation Plan. Recently a Conservation Plan for the Harbour Porpoise Population in the western Baltic, the Belt Sea and the Kattegat was implemented and more recently a conservation plan for the common dolphin initiated.

If I were to say something on the atmosphere of the work at the ASCOBANS Advisory Committees, it is the respect between the delegates, the NGOs and all the participants, that creates a fruitful and friendly ambience. I would like also to underline the active role, the investment, and the contributions of the NGOs to improve our knowledge, and to go further in the Agreement and for conservation, in concert with the Parties' engagements.

For the future ASCOBANS will certainly continue to have a prominent place in initiating actions in favour of the conservation of cetaceans by maintaining dialogue with other bodies and agreements. Accordingly I imagine a very close effective cooperation with ACCOBAMS, the sister agreement. Even if the concerned areas are different, there are common species and common problems, so common solutions can be found. I can also imagine, transcending the difficulties and any expenditure involved, that large cetaceans will be taken into consideration.

I hope that, in the near future, the Jastarnia Plan can be moved into a Conservation Plan, and that those conservation plans will have had a positive impact, that the threats such as bycatch, pollutants, ship strikes and noise disturbance will have been reduced considerably, or solved.

And I like to imagine an agreement that will maintain a close watch and a public awareness effort so as to avoid the mistakes of the past, and to identify the emergence of any other threats to cetaceans.

Dr Sami Hassani
Co-Chair of the Advisory Committee, 2010 to present

ASCOBANS – Personal perspective from the present Co-Chair of the Advisory Committee

When I started my studies in biology at Turku University, I had a dream to be a whale biologist, but as frequently happens, your dreams do not always come true. For me, they have been partly fulfilled, because I am now working at the Finnish Ministry of the Environment and to some extent on whale-related issues.

When Finland joined the ASCOBANS Agreement in 1999, I represented the country at the AC meeting in Bruges, Belgium. It is now over 18 years and a lot of things have happened since then. There were some doubts when Finland became a Party to the Agreement, especially because we did not even know if harbour porpoises were to be found anymore in our waters (Fig. 8.13). This was one reason why the Ministry of the Environment, together with some other institutions such as the Särkänniemi Dolphinarium, the Finnish Museum of Natural History and WWF Finland, started a harbour porpoise campaign in 2001. Our first aim was to raise awareness among people that one small whale species, namely the harbour porpoise, has lived in our waters. Secondly, we wanted people to notify us of harbour porpoise sightings. Concerning the first aim we were hopeful, but for the second one we were not so sure of the outcome. But it was a success in both cases – people are now aware that the Baltic Sea has its own small whale species and every summer people send in a form, specially made for harbour porpoise sightings, to inform us that they have seen or think they have seen a harbour porpoise swimming in our waters. We have continued the campaign every year. At the beginning of each summer, we put out a press release reminding people to look out for harbour porpoises and to inform us if they see any.

The year 2017 was important not only because it is the 25th anniversary of the ASCOBANS Agreement, but also because it was the 15th anniversary of the Jastarnia Plan. In 2002 about 40 scientists, fishermen, civil servants and other interested parties met in Poland in a little village called Jastarnia, on the coast of the Baltic Sea (Fig. 8.14). The meeting resulted in the Recovery Plan for the Baltic Sea Harbour Porpoise (Jastarnia Plan). One of the most challenging recommendations in the Plan is bycatch reduction. We cannot solve this problem ourselves; rather, we need the fishery sector to be involved in the process. Together we can do this, but it needs a lot of work.

One important project for the entire Baltic Sea was the EU-funded LIFE+ SAMBAH Project. This was a good example of what can happen when enthusiastic people meet and share a common goal. The aim of the project was to estimate population densities and total abundance, and to produce distribution maps. Instead of doing surveys either from boats or planes, the project used static acoustic monitoring devices called Cetacean – Porpoise Detectors (PODs) to survey the Baltic Sea harbour porpoise. The project was a success! We have now a more solidly based population estimate of approximately 500 animals and we have maps showing the distribution of porpoises in time and space, based on modelling. One of the best outcomes in the project was the finding of the major breeding area around the Midsjö offshore banks southeast of Öland, where porpoise presence was previously virtually unknown.

(Continued)

(Continued)

FIGURE 8.13 The coast of Finland at Kopparnäs. The waters around Finland once hosted a large population of harbour porpoise. *Photo: Penina Blankett.*

FIGURE 8.14 Puck Bay and the town of Jastarnia, where the Recovery Plan for the Baltic Harbour Porpoise was established. *Photo: Prof. Krzysztof Skóra Hel Marine Station, Department of Oceanography and Geography, University of Gdańsk.*

(Continued)

(Continued)

Based on these facts the Swedish Government has designated a new Natura 2000 site in the central Baltic Proper for protecting the harbour porpoise. This is a good example of how to use the outcome from the SAMBAH project. It is great, but we need also changes in fishing practices to avoid bycatch.

There have been for a long time discussions and different opinions as to whether the harbour porpoise population in the Baltic Sea is a genetically separate species or not. Some research results indicate that there is relatively little genetic difference between harbour porpoises in the Baltic Sea and the North Sea. But then some genetic and morphological studies indicate that there are indeed differences between Baltic Proper harbour porpoises and harbour porpoises in Danish waters, as well as harbour porpoises in the North Sea. Now the SAMBAH project has shown a clear distinction between the population inhabiting the Baltic Proper, and the more abundant population in the western Baltic, Belt Seas and Kattegat area during the months important for reproduction.

ASCOBANS has done a lot to raise public awareness; for example, we have a leaflet in 15 different languages, the International Day of the Baltic Harbour Porpoise and the ASCOBANS exhibition. But some other countries have also done a great job; I especially admire what my Polish colleagues have done during these years to raise public awareness. At the same time, I think Finland has also done much work in public outreach, and especially the Särkänniemi Dolphinarium did a great job, before they closed the dolphinarium in 2016.

For Finland, and I suppose for other Baltic Sea countries as well, the work that HELCOM is doing to protect the marine environment of the Baltic Sea from all sources of pollution through intergovernmental cooperation, is valuable. One of HELCOM's visions is that of a Baltic Sea with diverse biological components functioning in balance. Therefore it is important to have good cooperation between ASCOBANS and HELCOM. Since 1996 HELCOM has had a harbour porpoise recommendation, and its aims are similar to those of the Jastarnia Plan. In 2007 the HELCOM Ministerial meeting agreed on an ambitious programme, called the Baltic Sea Action Plan, where the aim is to restore the good ecological status of the Baltic marine environment by 2021. Harbour porpoises and the bycatch issue are also mentioned in the plan, as is the cooperation with ASCOBANS on development of the harbour porpoise database and reporting issues.

The harbour porpoise database is part of the HELCOM map and data service. The database contains a wide range of data on activities and pollution loads affecting the Baltic Sea, and also information on biological features.

I hope that the cooperation between ASCOBANS and HELCOM will continue to be as good in the future as it always has been.

At the moment, I think we have an agreement with competent people working together towards the same goals. This is good, because we need these people and their skills to successfully confront the major ongoing challenges such as bycatch, and also new threats like underwater noise, marine litter, and the continuously growing multiple use of marine areas and resources.

The work that HELCOM and ASCOBANS are doing for a better Baltic Sea is of utmost value in our efforts to achieve a good ecological status in our seas. In particular the EU's Marine Strategy Framework Directive gives us tools to reach this goal.

I have a new dream now – that before I retire, I will see a clean or cleaner Baltic Sea, and I will see the achievement of the goals set by ASCOBANS, the Jastarnia Group, and HELCOM. We will no longer need recovery plans, because the harbour porpoise population in the Baltic Sea will be thriving and this special species will be a common sight all around the Baltic Sea, and I especially hope to see them in our waters here in Finland. Let's hope that this dream will come true.

Penina Blankett
Co-Chair of the Advisory Committee, 2010 to present

Converting conservation legislation into action

The greatest stumbling block to effecting conservation action comes at the stage between making recommendations and resolutions, and countries taking those away and actually executing them. One reason for this is that we are largely dealing with voluntary actions; much of what is in environmental agreements and directives may be considered soft law. The United Nations Convention on the Law of the Sea is binding law, but it has no management associated with it. The International Maritime Organisation on the other hand can apply binding management. UNEP with its CMS and CBD applies management but is nonbinding in the legal sense. Unfortunately much of conservation falls into the nonbinding category. That is one of the virtues of the EU Habitats Directive and Marine Strategy Framework Directive. Unlike some of my countrymen, I view the little island off the coast of continental Europe as very much part of Europe, and having binding laws that go beyond national boundaries can be an important asset.

The other obvious block to translating conservation recommendations into actions is that whereas the focus of an ASCOBANS meeting will be small cetacean conservation, national governments have a much wider set of considerations. So by the time that its recommendations reach a government minister, their importance will have been considerably lessened as they juggle the needs and demands of health, education and economic development for other sectoral interests like fisheries and the energy industries. The environment has never been viewed as a particularly strong vote catcher. And even when as serious an issue as climate change has been demonstrated to the world, there remain powerful sceptics who view this as a barrier to economic progress. This is where public opinion can count for a lot. Raising public awareness and informing people about biodiversity and the conservation pressures that our wildlife and their habitats face is extremely important (Fig. 8.15). ASCOBANS through its Secretariat and many of its members is playing an increasingly strong role in this respect. Events like the International Day of the Baltic Harbour Porpoise bring these animals to the attention of the wider public. Those types of activities need further support and strengthening. Whales, dolphins and porpoises are iconic animals in the eyes of the public, and by promoting their conservation and the protection of their habitats, there comes protection for the wider ecosystem. In this way they serve as flagship species for a better marine environment.

FIGURE 8.15 Raising public awareness of the existence of whales, dolphins and porpoises in European seas and the need for their conservation is a fundamental first step to achieving action. *Photo: Kathy James/Sea Watch Foundation.*

FIGURE 8.16 Our seas around Europe would be a poorer place without the three dozen species of cetaceans that inhabit them. *Photo: Peter GH Evans.*

Success or failure?

Looking back over the last quarter of a century, there are a number of successes for which ASCOBANS has played its part: the pressure to address bycatch issues resulting in Regulation 812/2004 within the EU CFP; the evidence gathering exercises in the form of international sightings surveys such as SCANS, SCANS-II, and III, CODA, the DEPONS Project, and the ambitious acoustic survey project, SAMBAH, to better determine the distribution and abundance of the endangered Baltic harbour porpoise; and the establishment of conservation plans for porpoises in the North Sea and Baltic. These are just some of the actions that should contribute to a safer place for our small cetaceans. Of course there have been challenges throughout and not everything has been achieved that one would like. But I believe the seas around northwest Europe have indeed benefited from ASCOBANS, and with concerted effort continuing, it should benefit further. I dedicate this book to all those who worked tirelessly to achieve those aims over the last quarter of a century, and to the animals that if they were not here, would leave us the poorer (Fig. 8.16).

References

Boxed contributions

Green, M., & Simmonds, M. P. (2008). Riding the Waves — Lessons from Campaigning on Oil and Gas. *ECOS*, *29*(3/4), 72−79.

Lockyer, C., & Müller-Helmbrecht, A. (1993). ASCOBANS—Conservation Progress. The Pilot, July 1993, 14−16.

Parsons, E. C. M., Clark, J., & Simmonds, M. P. (2010). The Conservation of British Cetaceans: A review of the threats and protection afforded to whales, dolphins, and porpoises in UK Waters, Part 2. *Journal of International Wildlife Law & Policy*, *13*(2), 99−175.

Parsons, E. C. M., Clark, J., Warham, J., & Simmonds, M. P. (2010). The Conservation of British Cetaceans: A review of the threats and protection afforded to whales, dolphins, and porpoises in UK Waters, Part 1. *Journal of International Wildlife Law & Policy*, *13*(1), 1−62.

Simmonds, M. (1993). New International agreements or empty pledges. The Pilot, July 1993: 13−14.

Sonntag, R. P., Benke, H., Hiby, A. R., Lick, R., & Adelung, D. (1999). Identification of the first harbour porpoise (*Phocoena phocoena*) calving ground in the North Sea. *Journal of Sea Research*, *41*(3), 225−232.

Appendix

Agreement On The Conservation Of Small Cetaceans Of The Baltic, North East Atlantic, Irish And North Seas

ASCOBANS

Disclaimer: This text combines the original treaty text of 1992 with the amendment agreed in 2003. This version is to be used for information purposes only and cannot serve as a legal document. Official documents are held by the Secretary General of the United Nations, who acts as Depositary for this Agreement.

The Parties

Recalling the general principles of conservation and sustainable use of natural resources, as reflected in the World Conservation Strategy of the International Union for the Conservation of Nature and Natural Resources, the United Nations Environment Programme, and the WorldWide Fund for Nature, and in the report of the World Commission on Environment and Development,

Recognizing that small cetaceans are and should remain an integral part of marine ecosystems,

Aware that the population of harbour porpoises of the Baltic Sea has drastically decreased,

Concerned about the status of small cetaceans in the Baltic and North Seas,

Recognizing that by-catches, habitat deterioration and disturbance may adversely affect these populations,

Convinced that their vulnerable and largely unclear status merits immediate attention in order to improve it and to gather information as a basis for sound decisions on management and conservation,

Confident that activities for that purpose are best coordinated between the States concerned in order to increase efficiency and avoid duplicate work,

Aware of the importance of maintaining maritime activities such as fishing,

Recalling that under the Convention on the Conservation of Migratory Species of Wild Animals (Bonn 1979), Parties are encouraged to conclude agreements on wild animals which periodically cross national jurisdictional boundaries,

Recalling also that under the provisions of the Convention on the Conservation of European wildlife and Natural Habitats (Berne 1979), all small cetaceans regularly present in the Baltic and North Seas are listed in its Appendix II as strictly protected species, and

295

Referring to the Memorandum of Understanding on Small Cetaceans in the North Sea signed by the Ministers present at the Third International Conference on the Protection of the North Sea, have agreed as follows:

1. **Scope and interpretation**
   1.1. This agreement shall apply to all small cetaceans found within the area of the agreement.
   1.2. For the purpose of this agreement:
       (a) "Small cetaceans" means any species, subspecies or population of toothed whales *Odontoceti*, except the sperm whale *Physeter macrocephalus*;
       (b) "Area of the Agreement" means the marine environment of the Baltic and North Seas and contiguous area of the North East Atlantic, as delimited by the shores of the Gulfs of Bothnia and Finland; to the south-east by latitude 36°N, where this line of latitude meets the line joining the lighthouses of Cape St. Vincent (Portugal) and Casablanca (Morocco); to the south-west by latitude 36°N and longitude 15°W; to the north-west by longitude 15°W and a line drawn through the following points: latitude 59°N/longitude 15°W, latitude 60°N/longitude 05°W, latitude 61°N/longitude 4°W, latitude 62°N/longitude 3°W; to the north by latitude 62°N; and including the Kattegat and the Sound and Belt passages;
       (c) "Bonn Convention" means the Convention on the Conservation of Migratory Species of Wild Animals (Bonn 1979);
       (d) "Regional Economic Integration Organization" means an organization constituted by sovereign States which has competence in respect of the negotiation, conclusion and application of international agreements in matters covered by this agreement;
       (e) "Party" means a range State or any Regional Economic Integration Organization for which this agreement is in force;
       (f) "Range State" means any State, whether or not a Party to the agreement, that exercises jurisdiction over any part of the range of a species covered by this agreement, or a State whose flag vessels, outside national jurisdictional limits but within the area of the agreement, are engaged in operations adversely affecting small cetaceans;
       (g) "Secretariat" means, unless the context otherwise indicates, the Secretariat to this agreement.

2. **Purpose and basic arrangements**
   2.1. The Parties undertake to cooperate closely in order to achieve and maintain a favourable conservation status for small cetaceans.
   2.2. In particular, each Party shall apply within the limits of its jurisdiction and in accordance with its international obligations, the conservation, research and management measures prescribed in the Annex.
   2.3. Each Party shall designate a Coordinating Authority for activities under this agreement.
   2.4. The Parties shall establish a Secretariat and an Advisory Committee not later than at their first Meeting.
   2.5. A brief report shall be submitted by each Party to the Secretariat not later than 31 March each year, commencing with the first complete year after the entry into force of the agreement for that Party. The report shall cover progress made and difficulties experienced during the past calendar year in implementing the agreement.
   2.6. The provisions of this agreement shall not affect the rights of a Party to take stricter measures for the conservation of small cetaceans.

3. **The Coordinating Authority**
   3.1. The activities of each Party shall be coordinated and monitored through its Coordinating Authority which shall serve as the contact point for the Secretariat and the Advisory Committee in their work.

4. **The Secretariat**
   4.1. The Secretariat shall, following instructions provided by the meetings of the Parties, promote and coordinate the activities undertaken in accordance with Article 6.1 of this agreement and shall, in close consultation with the Advisory Committee, provide advice and support to the Parties and their Coordinating Authorities.
   4.2. In particular, the Secretariat shall: facilitate the exchange of information and assist with the coordination of monitoring and research among Parties and between the Parties and international organizations engaged in similar activities; organize meetings and notify Parties, the observers mentioned in Article 6.2.1 and the Advisory Committee; coordinate and circulate proposals for amendments to the agreement and its Annex; and present to the Coordinating Authorities, each year no later than 30 June, a summary of the Party reports submitted in accordance with Article 2.5, and a brief account of its own activities during the past calendar year, including a financial report.

4.3. The Secretariat shall present to each Meeting of the Parties a summary of, <u>inter alia</u>, progress made and difficulties encountered since the last Meeting of the Parties. A copy of this report shall be submitted to the Secretariat of the Bonn Convention for information to the Parties of that Convention.

4.4. The Secretariat shall be attached to a public institution of a Party or to an international body, and that institution or body shall be the employer of its staff.

5. The Advisory Committee

5.1. The Meeting of the Parties shall establish an Advisory Committee to provide expert advice and information to the Secretariat and the Parties on the conservation and management of small cetaceans and on other matters in relation to the running of the agreement, having regard to the need not to duplicate the work of other international bodies and the desirability of drawing on their expertise.

5.2. Each Party shall be entitled to appoint one member of the Advisory Committee.

5.3. The Advisory Committee shall elect a chairman and establish its own rules of procedure.

5.4. Each Committee member may be accompanied by advisers, and the Committee may invite other experts to attend its meetings. The Committee may establish working groups.

6. The Meeting of the Parties

6.1. The Parties shall meet, at the invitation of the Bonn Convention Secretariat on behalf of any Party, within one year of the entry into force of this agreement, and thereafter, at the notification of the Secretariat, not less than once every three years to review the progress made and difficulties encountered in the implementation and operation of the agreement since the last Meeting, and to consider and decide upon:

(a) The latest Secretariat report;

(b) Matters relating to the Secretariat and the Advisory Committee;

(c) The establishment and review of financial arrangements and the adoption of a budget for the forthcoming three years;

(d) Any other item relevant to this agreement circulated among the Parties by a Party or by the Secretariat not later than 90 days before the Meeting, including proposals to amend the agreement and its Annex; and

(e) The time and venue of the next Meeting.

6.2.1. The following shall be entitled to send observers to the Meeting: the Depositary of this agreement, the secretariats of the Bonn Convention, the Convention on International Trade in Endangered Species of Wild Fauna and Flora, the Convention on the Conservation of European Wildlife and Natural Habitats, the Convention for the Prevention of Marine Pollution by Dumping from Ships and Aircraft, the Convention for the Prevention of Marine Pollution from Landbased Sources, the Common Secretariat for the Cooperation on the Protection of the Wadden Sea, the International Whaling Commission, the North-East Atlantic Fisheries Commission, the International Baltic Sea Fisheries Commission, the Baltic Marine Environment Protection Commission, the International Council for the Exploration of the Sea, the International Union for the Conservation of Nature and Natural Resources, and all non-Party Range States and Regional Economic Integration Organizations bordering on the waters concerned.

6.2.2. Any other body qualified in cetacean conservation and management may apply to the Secretariat not less than 90 days in advance of the Meeting to be allowed to be represented by observers. The Secretariat shall communicate such applications to the Parties at least 60 days before the Meeting, and observers shall be entitled to be present unless that is opposed not less than 30 days before the Meeting by at least one third of the Parties.

6.3. Decisions at Meetings shall be taken by a simple majority among Parties present and voting, except that financial decisions and amendments to the agreement and its Annex shall require a three-quarters majority among those present and voting. Each Party shall have one vote. However, in matters within their competence, the European Economic Community shall exercise their voting rights with a number of votes equal to the number of their member States which are Parties to the agreement.

6.4. The Secretariat shall prepare and circulate a report of the Meeting to all Parties and observers within 90 days of the closure of the Meeting.

6.5. This agreement and its Annex may be amended at any Meeting of the Parties.

    6.5.1. Proposals for amendments may be made by any Party.

    6.5.2. The text of any proposed amendment and the reasons for it shall be communicated to the Secretariat at least 90 days before the opening of the Meeting. The Secretariat shall transmit copies forthwith to the Parties.

6.5.3. Amendments shall enter into force for those Parties which have accepted them 90 days after the deposit of the fifth instrument of acceptance of the amendment with the Depositary. Thereafter they shall enter into force for a Party 30 days after the date of deposit of its instrument of acceptance of the amendment with the Depositary.

6.5.4. Any State that becomes a Party to the Agreement after the entry into force of an Amendment shall, failing an expression of a different intention by that State:

(a) be considered as a Party to the Agreement as amended; and

(b) be considered as a Party to the unamended Agreement in relation to any Party not bound by the Amendment.

7. Financing

7.1. The Parties agree to share the cost of the budget, with Regional Economic Integration Organizations contributing 2.5 per cent of the administrative costs and other Parties sharing the balance in accordance with the United Nations scale, but with a maximum of 25 per cent per Party.

7.2. The share of each Party in the cost of the Secretariat and any additional sum agreed for covering other common expenses shall be paid to the Government or international organization hosting the Secretariat, as soon as practicable after the end of March and in no case later than before the end of June each year.

7.3. The Secretariat shall prepare and keep financial accounts by calendar years.

8. Legal matters and formalities

8.1. This is an agreement within the meaning of the Bonn Convention, Article IV (4).

8.2. The provisions of this agreement shall in no way affect the rights and obligations of a Party deriving from any other existing treaty, convention, or agreement.

8.3. The Secretary-General of the United Nations shall assume the functions of Depositary of this agreement.

8.3.1. The Depositary shall notify all Signatories, all Regional Economic Integration Organizations and the Bonn Convention Secretariat of any signatures, deposit of instruments of ratification, acceptance, approval or accession, entry into force of the agreement, amendments, reservations and denunciations.

8.3.2. The Depositary shall send certified true copies of the agreement to all signatories, all non-signatory Range States, all Regional Economic Integration Organizations and the Bonn Convention Secretariat.

8.4. The agreement shall be open for signature at the United Nations Headquarters by 31 March 1992 and thereafter remain open for signature at the United Nations Headquarters by all Range States and Regional Economic Integration Organizations, until the date of entry into force of the agreement. They may express their consent to be bound by the agreement (a) by signature, not subject to ratification, acceptance or approval, or (b) if the agreement has been signed subject to ratification, acceptance or approval, by the deposit of an instrument of ratification, acceptance or approval. After the date of its entry into force, the agreement shall be open for accession by Range States and Regional Economic Integration Organizations.

8.5. The agreement shall enter into force 90 days after six Range States have expressed their consent to be bound by it in accordance with Article 8.4. Thereafter, it shall enter into force for a State and Regional Economic Integration Organization on the 30th day after the date of signature, not subject to ratification, acceptance or approval, or of the deposit of an instrument of ratification, acceptance, approval or accession with the Depositary.

8.6. The agreement and its Annex shall not be subject to general reservations. However, a Range State or Regional Economic Integration Organization may, on becoming a Party in accordance with Article 8.4 and 8.5, enter a specific reservation with regard to any particular species, subspecies or population of small cetaceans. Such reservations shall be communicated to the Depositary on signing or at the deposit of an instrument of ratification, acceptance, approval or accession.

8.7. A Party may at any time denounce this agreement. Such denunciation shall be notified in writing to the Depositary and take effect one year after the receipt thereof.

In witness whereof the undersigned, being duly authorized thereto, have affixed their signatures to this agreement.

Done at New York on _____, the English, French, German and Russian texts of the agreement being equally authentic.

Annex

Conservation and management plan

The following conservation, research, and management measures shall be applied, in conjunction with other competent international bodies, to the populations defined in Article 1.1:

1. **Habitat conservation and management**

   Work towards (a) the prevention of the release of substances which are a potential threat to the health of the animals, (b) the development, in the light of available data indicating unacceptable interaction, of modifications of fishing gear and fishing practices in order to reduce by-catches and to prevent fishing gear from getting adrift or being discarded at sea, (c) the effective regulation, to reduce the impact on the animals, of activities which seriously affect their food resources, and (d) the prevention of other significant disturbance, especially of an acoustic nature.

2. **Surveys and research**

   Investigations, to be coordinated and shared in an efficient manner between the Parties and competent international organizations, shall be conducted in order to (a) assess the status and seasonal movements of the populations and stocks concerned, (b) locate areas of special importance to their survival, and (c) identify present and potential threats to the different species.

   Studies under (a) should particularly include improvement of existing and development of new methods to establish stock identity and to estimate abundance, trends, population structure and dynamics, and migrations. Studies under (b) should focus on locating areas of special importance to breeding and feeding. Studies under (c) should include research on habitat requirements, feeding ecology, trophic relationships, dispersal, and sensory biology with special regard to effects of pollution, disturbance and interactions with fisheries, including work on methods to reduce such interactions. The studies should exclude the killing of animals and include the release in good health of animals captured for research.

3. **Use of by-catches and strandings**

   Each Party shall endeavour to establish an efficient system for reporting and retrieving by-catches and stranded specimens and to carry out, in the framework of the studies mentioned above, full autopsies in order to collect tissues for further studies and to reveal possible causes of death and to document food composition. The information collected shall be made available in an international database.

4. **Legislation**

   Without prejudice to the provisions of paragraph 2 above, the Parties shall endeavour to establish (a) the prohibition under national law, of the intentional taking and killing of small cetaceans where such regulations are not already in force, and (b) the obligation to release immediately any animals caught alive and in good health. Measures to enforce these regulations shall be worked out at the national level.

5. **Information and education**

   Information shall be provided to the general public in order to ensure support for the aims of the agreement in general and to facilitate the reporting of sightings and strandings in particular; and to fishermen in order to facilitate and promote the reporting of by-catches and the delivery of dead specimens to the extent required for research under the agreement.

Index

Note: Page numbers followed by "*f*," "*t*," and "*b*" refer to figures, tables, and boxes, respectively.

A

ABDG. *See* ASCOBANS Baltic Discussion Group (ABDG)
Abundance estimates, 73
AC. *See* Advisory Committee (AC)
ACCOBAMS, 27, 29, 279—280
   relationship between ASCOBANS and sister Agreement ACCOBAMS, 29*b*
Acoustic deterrent devices (ADDs), 239—240, 251
Acoustic monitoring, 221
Acoustics, 220—227
Act Take Reduction Plans, 36
Active acoustics, 254
Active sonar, 184—186
ADDs. *See* Acoustic deterrent devices (ADDs)
Advisory Committee (AC), 2—4
Aerial surveys, 208—209, 210*f*
Agreement for Conservation of Cetaceans of Black Sea, Mediterranean Sea and Contiguous Atlantic Area (ACCOBAMS). *See* ACCOBAMS
Agreement for Conservation of Small Cetaceans of Baltic, Northeast Atlantic, Irish and North Seas (ASCOBANS). *See* ASCOBANS
Airguns, 249
AIS. *See* Automatic Identification System (AIS)
Alarm bells, 259
Ambitious acoustic survey project, 294
AMO. *See* Atlantic Multidecadal Oscillation (AMO)
Antarctic minke whale, 73
Antarctic minke whale. *See Balaenoptera bonaerensis* (Antarctic minke whale)
Arctic cetaceans, 192
"Area of the Agreement", 296
ASCOBANS, 1, 25, 27, 73, 96, 168*b*, 169, 184, 208, 237, 239—240, 246—247, 277, 279—280, 279*b*, 295—300
   from Aberdeen to San Sebastián, 12*b*
   AC chairs and vice-chairs, 6*t*
   Advisory Committee meeting, 10*f*
   Agreement on move, 15*b*
   Area, 45, 46*t*, 73, 87, 91
   ASCOBANS and IWC, 35*b*
   Bruges, 11*f*
   conservation legislation into action, 292—293
   early years of, 7*b*

establishment, 2*f*
executive secretary to, 8*b*
future, 19*b*
harbour porpoises suffer mortality, 22*f*
impressions on, 287*b*
International Conservation Agreement function, 9—23
list of scientific workshops supported by, 11*t*
locations of meetings of parties, 4*t*
and other international fora, 31*b*
Outreach and Education Award, 244*f*
personal perspective, 281*b*
   from government representative, 288*b*
   NGO, 282*b*
   from present co-chair of Advisory Committee, 289*b*, 290*b*
Reijnders, Peter, 20*f*
relationship between ACCOBAMS and, 29*b*
report writer's view, 14*b*
running 'new' secretariat, 18*b*
secretariat staff, 7*t*
secretariat venues, 6*f*
success or failure, 294
25th anniversary of, 21*b*
UN Campus in Bonn, 19*f*
ASCOBANS Baltic Discussion Group (ABDG), 15—16
Atlantic Multidecadal Oscillation (AMO), 191
Atlantic Scotland and Ireland, 60—63
Atlantic spotted dolphin. *See Stenella frontalis* (Atlantic spotted dolphin)
Atlantic white-sided dolphin. *See Lagenorhynchus acutus* (Atlantic white-sided dolphin)
Australia, conservation legislation in, 39—42
Australian Whale Sanctuary, 40
Automatic Identification System (AIS), 188
Availability bias, 208
Axe heads, 159

B

Balaena mysticetus (Bowhead whale), 138—139, 138*f*, 192
Balaenidae
   Balaena mysticetus, 138—139
   Eubalaena glacialis, 136—138
Balaenoptera acutorostrata (Minke whale), 54—55, 143—147, 143*f*, 144*f*, 145*f*, 159—162, 162*f*, 167—168

Balaenoptera bonaerensis (Antarctic minke whale), 145
Balaenoptera borealis (Sei whale), 147—150, 147*f*, 148*f*, 149*f*
Balaenoptera brydei (Bryde's whale), 150—151
Balaenoptera musculus (Blue whale), 155—157, 155*f*, 156*f*
Balaenoptera physalus (Fin whale), 40, 151—155, 151*f*, 152*f*, 153*f*
Balaenopteridae, 139—157
   Balaenoptera acutorostrata, 143—147
   Balaenoptera borealis, 147—150
   Balaenoptera brydei, 150—151
   Balaenoptera musculus, 155—157
   Balaenoptera physalus, 151—155
   Megaptera novaeangliae, 139—143
Baleen whales, 136, 167—168
Baltic Harbour Porpoise Recovery Plan. *See* Jastarnia Plan
Baltic Marine Environment Protection Commission. *See* Helsinki Convention (HELCOM)
Baltic Recovery Plan. *See* Jastarnia Plan
Baltic Sea
   Agreement on conservation of small cetaceans, 295—300
   cetacean species, 45—47
   other species, 47
Baltic Sea Recovery Plan, 14
Barcelona Convention, 30
Bay of Biscay, 64—65, 91—92
BBCs. *See* Big bubble curtains (BBCs)
Beaked whales, 184—186
Beaufort scale, 208
Belt Sea, 47—49, 259
Beluga. *See Delphinapterus leucas* (Beluga)
Beluga whale, 192
Bern Convention (1979), 25
Big bubble curtains (BBCs), 250—251
Binding law, 292
Blainville's beaked whale. *See Mesoplodon densirostris* (Blainville's beaked whale)
Blue whale. *See Balaenoptera musculus* (Blue whale)
A Blueprint for Survival (The Ecologist), 1, 277
BOEM. *See* US Bureau of Ocean Energy Management (BOEM)
Bonn Convention, 296
Bottlenose dolphin, 54—55, 208
   coastal populations, 206*f*

Bottlenose dolphin (*Continued*)
 feeding amongst algal forests, 64*f*
 individuals, 219, 220*f*
 Management Units, 205*f*
 within marine reserve, 289*f*
 North Sea population, 54*f*
 sound spectrogram of bottlenose dolphin whistle, 223*f*
Bottom-set gillnets, 165
Bowhead whale. *See Balaena mysticetus* (Bowhead whale)
Britten Norman BN 2 Islander, 211
Bryde's whale. *See Balaenoptera brydei* (Bryde's whale)
Bycatch, 31, 163−169, 168*b*, 239−247, 243*b*, 299
 of Hector's and Māui dolphins, 42
 of porpoises, 165−166
Bycatch Regulation, 31

C

C-POD, 224*f*, 225, 226*t*
 deploying C-POD from ice in Baltic Sea, 227*f*
 in SAMBAH Project, 226*f*
C-POD-F, 225
Calanus finmarchicus, 191−192
Calanus helgolandicus, 191−192
Canada, conservation legislation in, 37−39
Canada National Marine Conservation Areas Act, 39
Canada National Parks Act, 39
Capbreton Canyon, 64
CAPTURE (software program), 220
Carbon isotope ratios ($\delta^{13}C$ ratios), 230
Cardigan Bay, 57−58, 59*f*
Catch Limit Algorithm (CLA), 33
CBD. *See* Convention on Biological Diversity (CBD)
CCAMLR. *See* Commission for Conservation of Antarctic Marine Living Resources (CCAMLR)
Celtic Sea front, 57−58
Cetacean Offshore Distribution and Abundance in European Atlantic waters (CODA), 211−219, 294
Cetacean Strandings Investigation Programme (CSIP), 187−188, 228
Cetaceans, 159
 European international agreements applicable to, 25−32
 international agreements applicable to, 32−35
 in Northwest Europe, 45
  Atlantic Scotland and Ireland, 60−63
  Baltic, 45−47
  Bay of Biscay, 64−65
  Belt Seas, 47−49
  English Channel, 55
  Iberian Peninsula, 65−68
  Irish Sea, 55−59
  North Sea, 50−55
  Norwegian Atlantic, 50
  Skagerrak and Kattegat, 50

CFP. *See* Common Fisheries Policy (CFP)
Chemical pollutants, 170−172, 248
Circle-back method, 208
CITES. *See* Convention on International Trade in Endangered Species of Wild Fauna and Flora (CITES)
CLA. *See* Catch Limit Algorithm (CLA)
Climate change, 191−192, 260
CMS. *See* Convention on Migratory Species (CMS)
Coastal slow-moving whale species, 159
Coastal surveys, 208
Cod stocks, 163
CODA. *See* Cetacean Offshore Distribution and Abundance in European Atlantic waters (CODA)
Codes of conduct, 256
Cofferdam, 251
Cold-water shelf species, 192
Commercial whaling, 159
Commission for Conservation of Antarctic Marine Living Resources (CCAMLR), 41−42
Common bottlenose dolphin. *See Tursiops truncatus* (Common bottlenose dolphin)
Common Fisheries Policy (CFP), 243, 279
Comprehensive assessment, 33, 237
Computer simulations, 33
Conference of Parties (COP), 19−20
Conservation Act (1987), 41
Conservation actions, 237
 chemical pollutants, 248
 climate change, 260
 conservation plans, 267−274
 hunting, 237
 incidental capture in fishing gear, 239−247
 information builds awareness, 243*b*
 marine litter, 248
 MPAs, 260−267
 noise disturbance, 249−254
 prey reduction from fishing, 237−238
 recreational activities, 255−260
 vessel strikes, 254−255
Conservation agreements, 280−284
 European international agreements applicable to cetaceans, 25−32
 international agreements applicable to cetaceans, 32−35
 MMPA, 32−35
 need for evidence-based consensus at broadest level, 42−43
 overview, 43
Conservation and management plan, 299
Conservation legislation
 into action, 292−293
 in Australia and New Zealand, 39−42
 in Canada, 37−39
Conservation research, 203
 acoustics, 220−227
 counting animals, 208−219
 photo-ID, 219−220
 population structure, 203−208
 postmortem examinations, 228−231
Conservation threats

active sonar, 184−186
chemical pollutants and hazardous substances, 170−172
climate change, 191−192
explosions, dredging and drilling, 180−184
fishing activities, 168*f*
hunting, 159−162
incidental capture in fishing gear, 163−169
lacerations, 165*f*
marine litter, 172−174
noise disturbance, 174−187
prey reduction from fishing, 162−163
recreational disturbance, 189−191
risk assessment, 186−187
seismic surveys, 177−179
shipping, 176−177
TTS/PTS thresholds, 187*t*
vessel strikes, 187−189
Convention for Protection of the Marine Environment of North-East Atlantic, 277
Convention on Biological Diversity (CBD), 277
Convention on International Trade in Endangered Species of Wild Fauna and Flora (CITES), 35−36
Convention on Migratory Species (CMS), 4−9, 282, 288
Coordinated Environmental Monitoring Programme, 171
COP. *See* Conference of Parties (COP)
Copepods, 191−192
Counting animals, 208−219
Critically Endangered Baltic Proper harbour porpoise population (CR Baltic Proper harbour porpoise population), 225
CSIP. *See* Cetacean Strandings Investigation Programme (CSIP)
Cuvier's beaked whale. *See Ziphius cavirostris* (Cuvier's beaked whale)

D

Danish Belt Seas, 47, 50*f*
Data Collection Framework (DCF), 243
DDT, 106−107, 170−171, 248
Delphinapterus leucas (Beluga), 113−114, 114*f*
Delphinidae
 Delphinus delphis, 89−93
 Feresa attenuata, 104
 Globicephala macrorhynchus, 111−112
 Globicephala melas, 108−111
 Grampus griseus, 100−103
 Lagenodelphis hosei, 93−94
 Lagenorhynchus acutus, 97−100
 Lagenorhynchus albirostris, 94−97
 Peponocephala electra, 103
 Pseudorca crassidens, 104−105
 Stenella coeruleoalba, 86−89
 Stenella frontalis, 85−86
 Steno bredanensis, 78−79
 Tursiops truncatus, 79−85
Delphinus delphis (Common dolphin), 89−93, 90*f*, 278*f*

Department of Conservation (DOC), 40–41
DEPONS Project, 294
DFO. *See* Fisheries and Oceans Canada (DFO)
Dieldrin, 170–171, 248
Diesel engine, 254
Directed permits, 35–36
DOC. *See* Department of Conservation (DOC)
Dolphin bycatch in trawl fisheries, 240
Double BBCs, 250–251
'Double team' methods, 212
Dredging, 180–184
Driftnets, 165–167
Drilled foundations, 253
Drilling, 180–184
Dutch Conservation Plan for Harbour Porpoise, 270b
Dwarf sperm whale. *See Kogia simus* (Dwarf sperm whale)

E

Earth's climate, 191
Ecological markers, 205–208
Ecological Quality Objectives (EcoQOs), 27
'Ecologist, The', magazine, 1, 277
EcoQOs. *See* Ecological Quality Objectives (EcoQOs)
ECS. *See* European Cetacean Society (ECS)
EEZs. *See* Exclusive Economic Zones (EEZs)
EIA. *See* Environmental Impact Assessment (EIA)
Encapsulated Bubbles, 253
Endangered blue whale, 40
Endangered Species Act, 36
English Channel, 55
Environment and Climate Change Canada, 39
Environmental Impact Assessment (EIA), 186
Environmental Protection and Biodiversity Act (EPSC), 40
EU. *See* European Union (EU)
EU Habitats Directive, 25, 31, 43
EU Port Waste Reception Facilities Directive, 248
EU-funded LIFE+ SAMBAH Project, 290–291
Eubalaena glacialis (North Atlantic right whale), 36–37, 136–138, 136f, 159, 160f
Eubalaena japonica, 136–137
European Atlantic waters, estimating abundance of cetaceans in, 211b
European cetacean families
  Balaenopteridae, 139–157
  Delphinidae, 78–112
  Kogiidae, 128–131
  Monodontidae, 112–114
  Mysticeti, 136
  Phocoenidae, 73–78
  Physeteridae, 131–136
  Ziphiidae, 115–128
European Cetacean Society (ECS), 1
European international agreements applicable to cetaceans, 25–32
  ASCOBANS and other international fora, 31b

map of Region covered by HELCOM, 28f
relationship between ASCOBANS and sister Agreement ACCOBAMS, 29b
territorial conventions, 27
European Union (EU), 25, 240
  Habitats Directive, 277
Evidence-based consensus, need for, 42–43
Exclusive Economic Zones (EEZs), 40, 237
Explosions, 180–184

F

False killer whale. *See Pseudorca crassidens* (False killer whale)
Farne Islands, 189–190
Fatty acid profile or signature, 229–230, 231f
Feresa attenuata (Pygmy killer whale), 104
Fin whale. *See Balaenoptera physalus* (Fin whale)
Financing, 298
Fish stocks, 162–163
Fisheries Act (1996), 41
Fisheries and Oceans Canada (DFO), 38
Fisheries bycatch, 43
Fishing
  fishing gear, incidental capture in, 163–169
  nets and lines, 168–170
  prey reduction from, 162–163, 237–238
  vessels, 162–163, 163f
Fjord&Bælt Centre, 239–240, 241f
Flamborough Head, 52, 53f
Four-stroke engines, 254
Fraser's dolphin. *See Lagenodelphis hosei* (Fraser's dolphin)
French tuna driftnet fisheries, 166–167

G

Genetics, 203–208
Geometric morphometrics, 228–229
German Wadden Sea, 50–52, 52f
Gervais' beaked whale. *See Mesoplodon europaeus* (Gervais' beaked whale)
Ghost netting, 168–170
Gillnets, 165
Globicephala macrorhynchus (Short-finned pilot whale), 111–112, 112f
Globicephala melas (Long-finned pilot whale), 54–55, 108–111, 109f, 110f
Gonatus fabricii, 135
Grampus griseus (Risso's dolphin), 66, 100–103, 101f
Gravity base foundations, 253
Gray's beaked whale. *See Mesoplodon grayi* (Gray's beaked whale)
Gray whale, 73, 159, 192f

H

Habitat conservation and management, 299
Habitats Directive, 25, 31, 43, 277
Habitats Regulations Appraisal or Assessment, 186
Harbour porpoise. *See Phocoena phocoena* (Harbour porpoise)

Harbour Porpoise Conservation Plan for North Sea, 16
Hazardous substances, 170–172
Hebrides, 62
Hector's dolphin, 42
HELCOM. *See* Helsinki Convention (HELCOM)
Helgoland, 53f, 54
Hells Mouth, beach at, 172, 176f
Helsinki Convention (HELCOM), 27, 292
  HELCOM Ecological Quality Objectives, 279
  HELCOM Recommendation 17/2, 264
Herring stocks, 163, 164f
Hull form, 254
Humpback whale. *See Megaptera novaeangliae* (Humpback whale)
Hunting, 159–162, 237
Hydro Sound Dampers, 253
Hydrogen ratios (δ^2H ratios), 230
Hyperoodon ampullatus (Northern bottlenose whale), 118–121, 118f, 119f, 120f, 159
HYWIND (floating wind turbine), 253

I

Iberian Peninsula, 65–68
Iberian Pole Current (IPC), 65–66
ICES. *See* International Council for Exploration of Sea (ICES)
IDBHP. *See* International Day of Baltic Harbour Porpoise (IDBHP)
IFAW. *See* International Fund for Animal Welfare (IFAW)
IHC noise mitigation casing, 251, 253f
IMO. *See* International Maritime Organisation (IMO)
Incidental capture in fishing gear, 163–169, 239–247
Incidental catches. *See* Bycatch
Incidental permits, 35–36
Independent observers, 208
Initial Management Stocks, 32–33
Inner Hebrides, 62, 63f
Integrated Maritime Policy, 25
International agreements applicable to cetaceans, 32–35
International Conservation Agreement function, 9–23
International Convention for Safety of Life at Sea, 188–189
International Council for Exploration of Sea (ICES), 27, 162–163, 212, 237–238, 281–282
International Day of Baltic Harbour Porpoise (IDBHP), 17
International Fund for Animal Welfare (IFAW), 287
International Maritime Organisation (IMO), 254, 292
International Whaling Commission (IWC), 30, 32–35, 212, 237
  ASCOBANS and, 35b
IPC. *See* Iberian Pole Current (IPC)

Irish Sea, 55—59
  Front, 57—58
  surface tidal currents, 57
Irish Shelf Front, 60—62
Irish tuna driftnet fishery, 166—167
Isle of Lewis, 63*f*
Isolation casings, 251
IWC. *See* International Whaling Commission
  (IWC)

J

Jastarnia Group, 267—270
Jastarnia Plan, 9, 16, 258, 282, 290

K

Killer whale. *See Orcinus orca* (Killer whale)
Kogia breviceps (Pygmy sperm whale),
  128—130
Kogia simus (Dwarf sperm whale), 128, 129*f*,
  130—131, 130*f*
Kogiidae, 128—131
  Kogia breviceps, 128—130
  Kogia simus, 130—131

L

Lagenodelphis hosei (Fraser's dolphin), 93—94,
  94*f*
Lagenorhynchus acutus (Atlantic white-sided
  dolphin), 54—55, 62, 97—100, 98*f*,
  101*f*, 169*f*, 192
Lagenorhynchus albirostris (White-beaked
  dolphin), 94—97, 95*f*, 96*f*, 192
Legislation, 43
Line-transects, 208—219, 211*b*
Lofoten Islands, 189—190
Long-finned pilot whale. *See Globicephala
  melas* (Long-finned pilot whale)
Loud sounds, 174—175
Low-frequency cetaceans, 187

M

Mackerel stocks, 163, 164*f*
Management Units (MU), 75, 203—205
Marine
  construction, 180—184
  ecosystems, 191
  environment, 25
  litter, 172—174, 248
  mammals, 1
  spatial planning, 261
Marine Mammal Protection Act (MMPA) of
  United States. *See* US Marine Mammal
  Protection Act (MMPA)
Marine Mammal Protection Act of New
  Zealand, 39—40
Marine Mammal Regulations, 37—38
Marine protected areas (MPAs), 256, 260—267
Marine Strategy Framework Directive (MSFD),
  25
Marine Strategy Framework Directive, 265
Marine vibrators (MV), 249

Mark—recapture technique, 219
MARPOL Convention, 248
MARPRO Project, 151—152
Masking, 174—175
Maximum sustainable yield (MSY), 32—33
Meeting of the Parties (MOPs), 2—4, 7—8, 15,
  258, 281, 283
Megaptera novaeangliae (Humpback whale),
  40, 139—143, 140*f*, 141*f*, 142*f*,
  167—168, 220, 221*f*
Melon-headed whale. *See Peponocephala
  electra* (Melon-headed whale)
Mersey Estuary in northwest England, 248,
  249*f*
Mesoplodon bidens (Sowerby's beaked whale),
  125—126, 125*f*, 126*f*
Mesoplodon densirostris (Blainville's beaked
  whale), 127—128, 127*f*
Mesoplodon europaeus (Gervais' beaked
  whale), 123—124, 124*f*
Mesoplodon grayi (Gray's beaked whale), 126
Mesoplodon mirus (True's beaked whale),
  122—123, 122*f*, 123*f*
Microplastics, 248
  ingestion in cetaceans, 173
Milford Haven Oil Refinery, 170*f*
Minke whale. *See Balaenoptera acutorostrata*
  (Minke whale)
Mitigation measures, 250—251
Mitochondrial DNA, 80
  control region, 203
MMPA. *See* US Marine Mammal Protection
  Act (MMPA)
Modern whaling, 1, 159—162
Monodon monoceros (Narwhal), 112—113,
  113*f*, 192
Monodontidae, 112—114
Monofilament nets, 1
MOPs. *See* Meeting of the Parties (MOPs)
Morfa Harlech in West Wales, 57—58, 59*f*
MPAs. *See* Marine protected areas (MPAs)
MSFD. *See* Marine Strategy Framework
  Directive (MSFD)
MSY. *See* Maximum sustainable yield (MSY)
MU. *See* Management Units (MU)
MV. *See* Marine vibrators (MV)
Mysticeti, 136

N

NAMMCO. *See* North Atlantic Marine
  Mammal Commission (NAMMCO)
NAO. *See* North Atlantic Oscillation (NAO)
Narwhal. *See Monodon monoceros* (Narwhal)
NASS. *See* North Atlantic sightings survey
  (NASS)
National Marine Fisheries Service (NMFS),
  35—36
National Oceanic and Atmospheric
  Administration (NOAA), 35—36
Natura 2000 system, 260—267, 263*f*
Near continuous active sonar, 186
New Zealand, conservation legislation in,
  39—42

New Zealand Biodiversity Strategy, 41
New Zealand Marine Mammal Action
  Plan, 41
Nitrogen isotope ratios (δ ^{15}N ratios), 230
NMFS. *See* National Marine Fisheries Service
  (NMFS)
NOAA. *See* National Oceanic and Atmospheric
  Administration (NOAA)
Noise disturbance, 174—187, 249—254
  active sonar, 184—186
  large vessels, 177*f*
  marine construction, 180—184
  risk assessment, 186—187
  seismic surveys, 177—179
  shipping, 176—177
  shot point densities, 179*f*
Noise-quieting technologies, 254
Nongovernmental community, 284
Nonpulse sounds, 175
North Atlantic Current, 50
North Atlantic Marine Mammal Commission
  (NAMMCO), 34
North Atlantic Oscillation (NAO), 191
North Atlantic right whale. *See Eubalaena
  glacialis* (North Atlantic right whale)
North Atlantic sightings survey (NASS), 92,
  211—219
North East Atlantic Sea, Agreement on
  conservation of small cetaceans,
  295—300
North Sea, 50—55
  Agreement on conservation of small
    cetaceans, 295—300
  Conservation Plan for Harbour Porpoises in,
    271*b*
  harbour porpoise Conservation Plan, 273*t*
  population of bottlenose dolphins, 54*f*
  seasonal changes in surface temperature, 54
  tidal currents, 52
North Sea Conference, 279
North Sea Conservation Plan, 169, 290
Northern bottlenose whale. *See Hyperoodon
  ampullatus* (Northern bottlenose whale)
Northern Europe, 45
Norwegian Atlantic, 50
Nuclear microsatellite loci, 203

O

ObSERVE surveys of Irish EEZ, 76, 92,
  95—96, 115—116, 217
Oceans Act (1996), 38
Odontoceti (Toothed whales), 178, 296
Offshore wind farm, 180—181, 181*f*
Optimal sustainable population (OSP), 36
Orca. *See Orcinus orca* (Killer whale)
Orcinus orca (Killer whale), 54—55, 105—108,
  105*f*, 106*f*, 107*f*
Organochlorine pesticides, 170
Oslo and Paris Convention (OSPAR), 27, 31,
  162—163, 171, 248, 279, 281
  status of chemical contamination in, 171*f*
OSP. *See* Optimal sustainable population
  (OSP)

OSPAR. *See* Oslo and Paris Convention (OSPAR)
Otoliths of fish species, 228–229, 230*f*
Outer Hebrides, 62, 63*f*
Oxygen isotope ratios (δ^{18}O ratios), 230

P

PAHs. *See* Polycyclic aromatic hydrocarbons (PAHs)
Pan-tropical spotted dolphin, 83
Parks Canada, 39
Partenavia PN68 aircraft, 211
Particle motion component, 175
PBRs. *See* Potential Biological Removals (PBRs)
PCBs. *See* Polychlorinated biphenyls (PCBs)
Peak sound pressure, 175
Peak-to-peak sound pressure, 175
Peponocephala electra (Melon-headed whale), 103
Perception bias, 208
Permanent threshold shifts (PTS), 174–175
Perpendicular distance, 208
Phocoena phocoena (Harbour porpoise), 1, 2*f*, 8, 54–55, 58, 73–76, 74*f*, 166–167, 167*f*, 183*f*, 192, 264
    for ASCOBANS and vice versa, 264*b*
    diet in Netherlands, 271*f*
    distribution, 215*f*
    estimated annual rates of change for species/regions, 218*t*
    fishermen protest against porpoise protection, 247*f*
    human chain, 246*f*
    locations of tagged and hotspots, 207*f*
    Management Units, 204*f*
    monument, 245*f*
    North Atlantic distribution, 74*f*
    North Sea harbour porpoise Conservation Plan, 273*t*
    with satellite radio tag, 203–205, 206*f*
    stranding, 245*f*, 260*f*
Phocoenidae, 73–78
Photoidentification (Photo-ID), 219–220
    of recognisable individuals, 205–208
Physeter macrocephalus (Sperm whale), 131–136, 132*f*, 133*f*, 134*f*, 296
Physeteridae, 131–136
    P. macrocephalus, 131–136
Pile driving, 180–182, 182*f*
Pilot, The, 283
Pingers, 239, 239*f*, 265–266
    deployment, 240
    development, 240
Plastics, 172
    ingestion, 43
Polychlorinated biphenyls (PCBs), 170–171, 248
Polycyclic aromatic hydrocarbons (PAHs), 248
Pop-Ups, 221
Population structure, 203–208
Porpoises, 45, 50, 73–78
    in Kattegat, 51*f*

Portuguese current, 65–66
Postmortem examinations, 228–231, 229*f*
Potential Biological Removals (PBRs), 36
Precautionary approach, 75
Pressure component, 175
Prey depletion, 43
Prey reduction from fishing, 162–163, 237–238
Primary observers, 208
Propellers, 254
    cleaning, 254
Protection Stocks, 32–33
Pseudorca crassidens (False killer whale), 104–105
PTS. *See* Permanent threshold shifts (PTS)
Public awareness, 32, 243*b*, 258*b*
Pulse sounds, 175
Purse seine nets, 167
Pygmy killer whale. *See Feresa attenuata* (Pygmy killer whale)
Pygmy sperm whale. *See Kogia breviceps* (Pygmy sperm whale)

R

Radar imaging, 254
Radio telemetry, 205–208
Ramping up sound, 178
Range State, 296, 298
Recreational activities, 255–260
Recreational disturbance, 189–191
Regional Economic Integration Organization. *See* Range State
REM. *See* Remote electronic monitoring (REM)
Remote electronic monitoring (REM), 240, 242*f*
REPCET (whale reporting scheme), 254
Reproductive system, 228–229
Réseau National d'Echouages (RNE), 228
Revised Management Procedure (RMP), 33–34, 237
Rio Earth Summit, 277
Risso's dolphin. *See Grampus griseus* (Risso's dolphin)
RMP. *See* Revised Management Procedure (RMP)
RMS. *See* Root-mean-square (RMS)
RNE. *See* Réseau National d'Echouages (RNE)
Rock carvings, 159, 160*f*
Root-mean-square (RMS), 175
Rough-toothed dolphin. *See Steno bredanensis* (Rough-toothed dolphin)
Royal Society for the Prevention of Cruelty to Animals (RSPCA), 284

S

SACs. *See* Special Areas of Conservation (SACs)
Saguenay–St. Lawrence Marine Park, 39
Salinity, 45
SAM. *See* Static acoustic monitoring (SAM)
SAMBAH project. *See* Static Acoustic Monitoring of Baltic Sea Harbour Porpoise project (SAMBAH project)

Sandeel stocks, 163
SARA. *See* Species at Risk Act (SARA)
Särkänniemi Dolphinarium, 290
'Save the Whale' campaigns, 237
SCANS survey. *See* Small Cetacean Abundance in North Sea survey (SCANS survey)
SEA. *See* Strategic Environmental Assessment (SEA)
Seal scarers, 251
Secondary microplastics, 172
"Secretariat", 296
Sei whale. *See Balaenoptera borealis* (Sei whale)
Seismic surveys, 177–179, 178*f*
SEL. *See* Sound energy levels (SEL)
Shelf edge species, 192
Ship Strike Rule, 36–37
Ship strikes, 43
Shipping, 176–177
    movements, 188
Short-beaked common dolphin. *See Delphinus delphis* (Common dolphin)
Short-finned pilot whale. *See Globicephala macrorhynchus* (Short-finned pilot whale)
Silent Spring (Carson), 1, 277
Single nucleotide polymorphisms, 203
Skagerrak and Kattegat, 50
SL. *See* Source level (SL)
SLA. *See* Strike Limit Algorithm (SLA)
Small Cetacean Abundance in North Sea survey (SCANS survey), 76, 95–96, 211–219, 279
    revised estimates of abundance from SCANS and SCANS-II, 217*t*
    SCANS-II survey, 76, 214, 287, 294
    of northwest European shelf, 82
    SCANS-III survey, 76, 214–219, 294
    and adjacent surveys, 216*f*
    survey transects, 213*f*
    vessel used for, 212*f*
Small Isles, 63*f*
Snapshot wide-scale surveys, 211–219
Social solidarity, 247
Sound, 175
    traps, 221
'Sound & Marine Life' Programme, 249
Sound energy levels (SEL), 186
Sound pressure levels (SPLs), 175
Source level (SL), 175
Southern right whale, 40
Sowerby's beaked whale. *See Mesoplodon bidens* (Sowerby's beaked whale)
SPA Protocol. *See* Specially Protected Areas Protocol (SPA Protocol)
SPAMI. *See* Specially Protected Areas of Mediterranean Importance (SPAMI)
Special Areas of Conservation (SACs), 256, 260–267, 263*f*, 277*f*
Specially Protected Areas of Mediterranean Importance (SPAMI), 30–32
Specially Protected Areas Protocol (SPA Protocol), 30–31
Species at Risk Act (SARA), 38–39

Sperm whale. *See Physeter macrocephalus* (Sperm whale)
Sperm whales, 54–55, 178
SPLs. *See* Sound pressure levels (SPLs)
Sprat stocks, 163
Stable isotopes, 230
Static acoustic monitoring (SAM), 220–221
Static Acoustic Monitoring of Baltic Sea Harbour Porpoise project (SAMBAH project), 17, 221, 225b, 227f, 279, 294
Stenella coeruleoalba (Striped dolphin), 86–89, 86f, 88f
Stenella frontalis (Atlantic spotted dolphin), 85–86
  adult and young, 85f
Steno, 78
Steno bredanensis (Rough-toothed dolphin), 78–79
Strandings, 203, 228, 299
Strategic Environmental Assessment (SEA), 186
Strike Limit Algorithm (SLA), 34
Striped dolphin. *See Stenella coeruleoalba* (Striped dolphin)
Subbasin, 45
Sulphur isotope ratios (δ^{34}S ratios), 230
Survey designs, 208
Survey method, 208
Surveys and research, 208–219, 211b, 299
Sustained Management Stocks, 32–33

T

T-POD. *See* Timed porpoise detector (T-POD)
Tangle nets, 165
Taxonomy, 73
TBT. *See* Tributyltin (TBT)
Temporary threshold shifts (TTS), 174–175
TFEU. *See* Treaty on Functioning of European Union (TFEU)

Thermal imaging, 254
Timed porpoise detector (T-POD), 222–223, 223f, 225
Tiumpan Head, 63f
TNASS survey, 132–133
Toothed whales. *See Odontoceti* (Toothed whales)
Towed hydrophones, 220–221
Trace metals, 172
Traffic separation scheme, 254
Trammel net, 165
Transects, 208
Transport Canada, 39
Trawls, 165
Treaty on Functioning of European Union (TFEU), 17
Tributyltin (TBT), 248
Triennium Work Plan, 9
Triple BBCs, 250–251
True's beaked whale. *See Mesoplodon mirus* (True's beaked whale)
TTS. *See* Temporary threshold shifts (TTS)
Tursiops truncatus (Common bottlenose dolphin), 79–85, 80f
  North Atlantic distribution of, 81f

U

UK Joint Nature Conservation Committee, 250
UN Convention on Law of Sea (UNCLOS), 43
Underwater explosions, 180
Underwater noise, 31–32
United Nations Convention on Law of Sea, 292
Unstable weather patterns, 191
US Bureau of Ocean Energy Management (BOEM), 36
US Fish and Wildlife Service, 36
US Marine Mammal Protection Act (MMPA), 32–35

V

Vessel strikes, 187–189, 254–255
Vibration isolators, 254
Vibration pile driving, 253
Vibropiling, 253
Vibroseis, 249

W

Whale Alert' iPad application, 254
Whale and Dolphin Conservation (WDC), 284
Whale Protection Act of Australia, 39
White whale. *See Delphinapterus leucas* (Beluga)
White-beaked dolphin. *See Lagenorhynchus albirostris* (White-beaked dolphin)
Wind farms, 183b, 183f, 184
Wind turbines, 253
Working group on bycatch of protected species (WGBYC), 27
Working group on marine mammal ecology (WGMME), 27
World Wide Fund for Nature (WWF), 255, 279–280

Z

Ziphiidae, 115–128
  Hyperoodon ampullatus, 118–121
  Mesoplodon bidens, 125–126
  Mesoplodon densirostris, 127–128
  Mesoplodon europaeus, 123–124
  Mesoplodon grayi, 126
  Mesoplodon mirus, 122–123
  Ziphius cavirostris, 115–118
Ziphius cavirostris (Cuvier's beaked whale), 115–118, 115f, 116f, 117f

Printed in the United States
By Bookmasters